Essentials of Apoptosis

Essentials of Apoptosis

A Guide for Basic and Clinical Research

Edited by

Xiao-Ming Yin, MD, PhD

University of Pittsburgh School of Medicine, Pittsburgh, PA

and

Zheng Dong, PhD

Medical College of Georgia, Augusta, GA

Humana Press ✳ Totowa, New Jersey

© 2003 Humana Press Inc.
999 Riverview Drive, Suite 208
Totowa, New Jersey 07512

www.humanapress.com

This publication is printed on acid-free paper. ∞
ANSI Z39.48-1984 (American Standards Institute) Permanence of Paper for Printed Library Materials.

Production Editor: Robin B. Weisberg.

Cover illustrations: The phase micrograph shows dying hepatocytes shrinking and rounding up with intensive membrane blebbing and boiling. One fluorescence micrograph shows one healthy hepatocyte (lower right corner) with mitochondrial cytochrome c (red) and one apoptotic cell (upper left corner) with cytochrome c released into the cytosol (red) and with caspase activation (green). The other photo shows dying cells losing mitochondria transmembrane potentials (red) and their nuclei in fragments (blue), as compared to the healthy cells (lower left corner). Cover artwork courtesy of Xiao-Ming Yin and Wen-Xing Ding, Department of Pathology, University of Pittsburgh School of Medicine.

Cover design by Patricia F. Cleary.

For additional copies, pricing for bulk purchases, and/or information about other Humana titles, contact Humana at the above address or at any of the following numbers: Tel.: 973-256-1699; Fax: 973-256-8341; E-mail: humana@humanapr.com or visit our Website: humanapress.com

Printed in the United States of America. 10 9 8 7 6 5 4 3 2 1

Library of Congress Cataloging in Publication Data

Main entry under title:

Essentials of apoptosis : a guide for basic and clinical research / edited by Xiao-Ming Yin and Zheng Dong.
    p. cm.
  Includes bibliographical references and index.
  ISBN 1-58829-146-4 (alk. paper); 1-59259-361-5 (e-book)
   1. Apoptosis. I. Yin, Xiao-Ming. II. Dong, Zheng , PhD.

QH671 .E85 2003
571.9'36--dc21   2002027356

Preface

Life and death are topics that no one takes lightly. In the cell, death by apoptosis is just as fundamental as proliferation for the maintenance of normal tissue homeostasis. Too much or too little apoptosis can lead to developmental abnormality, degenerative diseases, or cancers. Although apoptosis, or programmed cell death (PCD), has been recognized for more than 100 years, its significance and its molecular mechanisms were not revealed until recently.

We have witnessed rapid progress in apoptosis research in the last decade. Apoptosis can now be defined not only by morphology, but also by molecular and biochemical mechanisms. As a result, there has been an information explosion in the field. On one hand, this has dramatically expanded our understanding of the role of apoptosis in both biology and medicine; on the other hand, it has made the study of apoptosis quite complicated, and sometimes confusing. One often wonders whether findings from other laboratories can be generalized or whether the methods used can be made applicable to other systems.

Studies of apoptosis are unusual in that the common focus on a basic process that is driven by specific sets of biochemical machinery is studied in an array of very diverse research areas. Investigators from different fields have documented their views of apoptosis in numerous review articles. These reviews, published in various scientific journals, are aimed at either summarizing the latest findings or providing brief introductions to apoptosis. However, essential information about apoptosis, such as its mechanisms and pathophysiological roles, has yet to be presented in a systematic and concise way. This has posed a great hurdle to many investigators who want to enter this field or to apply the knowledge to their own research, and are not sure where and how to begin.

Essentials of Apoptosis: A Guide for Basic and Clinical Research serves as a starting point for those investigators who are relatively new to apoptosis research. Therefore, instead of describing detailed findings in one specific field, we present the concepts, the molecular architecture (the molecules and the pathways), and the pathophysiological significance of apoptosis. Controversial results are presented only if they are related to the essential process. In addition, standard biochemical and cellular approaches to apoptosis research are described as a guideline for bench work. *Essentials of Apoptosis: A Guide for Basic and Clinical Research* is intended to provide readers with the basics of apoptosis in order to stimulate their interests and to prepare them for the commencement of apoptosis-related research in their chosen areas. We hope that *Essentials of Apoptosis: A Guide for Basic and Clinical Research* will prove useful reading for all those interested in apoptosis research.

Xiao-Ming Yin, MD, PhD
Zheng Dong, PhD

Contents

Contributors

BRUNO ANTONSSON • *Protein Biochemistry, Serono Pharmaceutical Research Institute, Geneva, Switzerland*

LI BAI • *Department of Pathology, University of Pittsburgh School of Medicine, Pittsburgh, PA*

SUJIN BAO • *Department of Molecular Biology and Pharmacology, Washington University School of Medicine, St. Louis, MO*

FERMIN BRIONES • *Department of Molecular Pathology, The University of Texas M.D. Anderson Cancer Center, Houston, TX*

ROSS L. CAGAN • *Department of Molecular Biology and Pharmacology, Washington University School of Medicine, St. Louis, MO*

NIKHIL S. CHARI • *Department of Molecular Pathology, The University of Texas M.D. Anderson Cancer Center, Houston, TX*

JUN CHEN • *Department of Neurology, University of Pittsburgh School of Medicine, Pittsburgh, PA*

SONG H. CHO • *Department of Molecular Pathology, The University of Texas M.D. Anderson Cancer Center, Houston, TX*

MILE CIKARA • *Department of Immunology, University of Colorado Health Sciences Center, Denver, CO*

J. JOHN COHEN • *Department of Immunology, University of Colorado Health Sciences Center, Denver, CO*

MARIA D. DEVORE • *Department of Immunology, University of Colorado Health Sciences Center, Denver, CO*

WEN-XING DING • *Department of Pathology, University of Pittsburgh School of Medicine, Pittsburgh, PA*

ZHENG DONG • *Department of Cellular Biology and Anatomy, Medical College of Georgia, Augusta, GA*

ELIZABETH A. DOWLING • *Department of Immunology, University of Colorado Health Sciences Center, Denver, CO*

WAI GIN FONG • *Molecular Genetics, Children's Hospital of Eastern Ontario Research Institute, Ottawa, Ontario, Canada*

GREGORY J. GORES • *Division of Gastroenterology and Hepatology, Mayo Medical School, Clinic, and Foundation, Rochester, MN*

MARIA EUGENIA GUICCIARDI • *Division of Gastroenterology and Hepatology, Mayo Medical School, Clinic, and Foundation, Rochester, MN*

REBECCA L. HAMM • *Department of Molecular Pathology, The University of Texas M.D. Anderson Cancer Center, Houston, TX*

MARJA JÄÄTTELÄ • *Apoptosis Laboratory, Institute for Cancer Biology, Danish Cancer Society, Copenhagen, Denmark*

YOSHIHIKO KADOWAKI • *Department of Molecular Pathology, The University of Texas M.D. Anderson Cancer Center, Houston, TX*

ROBERT G. KORNELUK • *Molecular Genetics, Children's Hospital of Eastern Ontario Research Institute, Ottawa, Ontario, Canada*

CHIA-YI KUAN • *Division of Developmental Biology, Children's Hospital Medical Center, Cincinnati, OH*

KEISUKE KUIDA • *Vertex Pharmaceuticals, Cambridge, MA*

SANGJUN LEE • *Department of Molecular Pathology, The University of Texas M.D. Anderson Cancer Center, Houston, TX*

MARCEL LEIST • *Department of Molecular Disease Biology, H. Lundbeck A/S, Copenhagen, Denmark*

PETER LISTON • *Molecular Genetics, Children's Hospital of Eastern Ontario Research Institute, Ottawa, Ontario, Canada*

SEAMUS J. MARTIN • *Molecular Cell Biology Laboratory, Department of Genetics, The Smurfit Institute, Trinity College, Dublin, Ireland*

TIMOTHY J. MCDONNELL • *Department of Molecular Pathology, The University of Texas M.D. Anderson Cancer Center, Houston, TX*

RON MITTLER • *Department of Botany, Iowa State University, Ames, IA*

BRONA M. MURPHY • *Molecular Cell Biology Laboratory, Department of Genetics, The Smurfit Institute, Trinity College, Dublin, Ireland*

SEAN L. O'CONNOR • *Department of Molecular Pathology, The University of Texas M.D. Anderson Cancer Center, Houston, TX*

YIGONG SHI • *Department of Molecular Biology, Lewis Thomas Laboratory, Princeton University, Princeton, NJ*

VLADIMIR SHULAEV • *Virginia Bioinformatics Institute, Virginia Polytechnic Institute and State University, Blacksburg, VA*

KEVIN B. SPURGERS • *Department of Molecular Pathology, The University of Texas M.D. Anderson Cancer Center, Houston, TX*

R. ANNE STETLER • *Department of Neuroscience, University of Pittsburgh School of Medicine, Pittsburgh, PA*

MANJERI A. VENKATACHALAM • *Department of Pathology, University of Texas Health Science Center, San Antonio, TX*

YI-CHUN WU • *Zoology Department, Graduate Institute of Molecular and Cellular Biology, National Taiwan University, Taipei, Taiwan*

JINZHAO WANG • *Department of Cellular Biology and Anatomy, Medical College of Georgia, Augusta, GA*

DING XUE • *Department of MCD Biology, University of Colorado, Boulder, CO*

XIAO-MING YIN • *Department of Pathology, University of Pittsburgh School of Medicine, Pittsburgh, PA*

YONGGE ZHAO • *Department of Pathology, University of Pittsburgh School of Medicine, Pittsburgh, PA*

I

Molecules and Pathways of Apoptosis

1

Caspases

Structure, Activation Pathways,
and Substrates

Brona M. Murphy and Seamus J. Martin

INTRODUCTION

Apoptosis, or programmed cell death (PCD), is an important counterpart to mitosis for the regulation of cell numbers during development, in homeostatic cell turnover in the adult, and in many other settings *(1)*. Apoptosis is characterized by a series of distinct morphological and biochemical alterations to the cell such as DNA fragmentation, chromatin condensation, cell shrinkage, and plasma-membrane blebbing *(2)*. Although it can be readily appreciated that large numbers of cells die during development—during the sculpting processes that shape organs or form cavities, for example—it is often less readily appreciated that cells die on a vast scale in mature organisms.

For example, billions of erythrocytes are released from the bone marrow every day, and it follows that a corresponding number must be eliminated to make way for these new arrivals. In a similar manner, millions of neutrophils are also replaced daily and the cells of the skin and gut epithelia are under constant renewal. Within the adaptive immune system, activated T and B lymphocytes clonally expand to produce many millions of effector cells upon each encounter with foreign antigen. Although a proportion of these cells are retained as "memory" cells to help fight subsequent infections, most are rapidly eliminated through apoptosis. For this reason, T- and B-cell numbers are maintained at relatively constant levels throughout our lives despite almost continuous clonal expansion of specific lymphocyte subpopulations in response to antigenic challenge.

The huge scale of this ongoing cell death would present a major problem for multicellular organisms were it not for the existence of a mechanism that prevents the escape of cellular constituents from dying cells and ensures safe removal of the dead cell corpses *(3)*. Significantly, the phenotypic changes that occur during numerous instances of programmed cell death, from nematodes to man, are strikingly similar. This suggests that the molecular machinery that controls this process has been preserved through evolution.

Studies conducted over the past 7–10 years have revealed a complex web of molecules (the death machinery) that regulates apoptosis. The death machinery can be activated by diverse stimuli and has as its central components a group of proteolytic enzymes, called caspases (cysteinyl aspartate-specific proteases). Upon activation, the cooperative actions of the caspases produce the alterations to the cell that we recognize as the apoptotic phenotype *(4–10)*. Here, we discuss the caspases, how they become activated during apoptosis, some of the controls placed upon them, and finally their cellular substrates.

From: *Essentials of Apoptosis: A Guide for Basic and Clinical Research*
Edited by: X-M. Yin and Z. Dong © Humana Press Inc., Totowa, NJ

DISCOVERY OF THE CASPASES

The discovery of a role for the caspases in apoptosis has its origins in observations made by Horvitz's group on the regulation of programmed cell death during development of the nematode worm *Caenorhabditis elegans (11)*. *C. elegans* is an ideal organism for the study of cell fate because these animals have relatively few cells and are transparent, making it a relatively simple task to track the appearance and disappearance of cells in the living animal. Initial studies established that, of the 1090 somatic cells formed during the development of an adult hermaphrodite worm, 131 of these died during the process and were engulfed by neighboring cells. Because these cell deaths occurred at precise locations and times during worm development, they were considered to be genetically programmed. A mutagenesis screen for genes involved in regulating programmed cell death in *C. elegans* yielded a series of 14 worm mutants (*ced-1* to *ced-10*, *nuc-1*, *ces-1*, *ces-2*, *egl-1*) that exhibited defects in various aspects of the cell death process (reviewed in *11*) (*see* Chapter 9).

Three of the genes that were discovered using this approach—*ced-3*, *ced-4*, and *ced-9*—were found to occupy a particularly central position in the programmed cell-death pathway. Mutant worms carrying defective *ced-3* or *ced-4* genes exhibited almost no developmentally related cell deaths and so possessed many extra cells. This suggested that *ced-3* and *ced-4* were positive regulators of the cell-death program—their genes either encoding toxic proteins or somehow activating molecules lethal to the cells in which they were expressed. In contrast, mutations that inactivated the *ced-9* gene were lethal owing to extensive death of cells that would have normally survived, suggesting that the *ced-9* gene product was a negative regulator of cell death. Further support for this interpretation was provided by observations on animals carrying a gain of function *ced-9* mutation in which all programmed cell deaths were blocked.

Sequence analysis of the *ced-3* and *ced-4* genes did not provide any major clues concerning their function until it was discovered in 1993 that *ced-3* was homologous to a then recently cloned human gene that encoded interleukin-1β (IL-1β)-converting enzyme (*ICE*), an aspartate-specific protease that was responsible for conversion of pro-IL-1ß to its mature form *(12,13)*. Although it was soon reported that ectopic expression of *Ice* in rodent cells resulted in apoptosis *(14)*, the significance of this observation was undermined by experiments that demonstrated that the introduction of almost any protease into cells resulted in a similar outcome *(15)*.

More convincing evidence for protease involvement in apoptosis was provided by observations by a number of groups that certain proteins were specifically proteolysed during apoptosis, typically after aspartate residues *(16–20)*. Further supportive evidence for a role for *ICE* in apoptosis derived from studies that utilized tetrapeptide inhibitors (YVAD) of this protease *(21–23)*. Exposure of cells to aldehyde or ketone derivatives of this tetrapeptide partially protected from apoptosis in a number of contexts. Moreover, a naturally occurring inhibitor of *ICE*—the poxvirus-derived CrmA protein—was also found to be a potent inhibitor of several forms of apoptosis *(24,25)*. However, considerable doubt was cast upon the view that *ICE* played a central coordinating role in apoptosis when it was found that *ICE* null mice were phenotypically normal, with few, if any, defects associated with cell death regulation *(26,27)*. These and other observations stimulated a search for other *ced-3/ICE* homologs (now called caspases) that played a more direct role in the regulation of mammalian apoptosis and several were soon discovered.

THE CASPASES: AGENTS OF CELLULAR DEMOLITION

Caspases are proteases that use cysteine as the nucleophilic group for substrate cleavage and cleave peptide bonds on the carboxyl side of aspartic acid residues *(6,8,9,28)*. To date, 11 human caspases have been identified and these proteases appear to comprise a complex proteolytic system, rather like the clotting or complement systems *(7,30)*. Stimuli that trigger apoptosis all appear to do so through caspase activation. Activated caspases can promote activation of other members of the caspase fam-

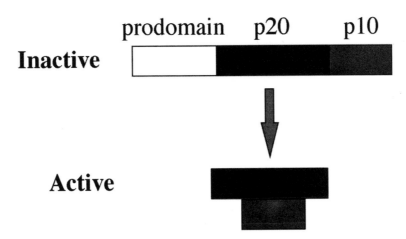

Fig. 1. Caspase domain structure. Caspases typically under proteolytic processing between their large and small subunits during activation.

ily and proteolysis of a diverse array of cellular proteins, all of which contributes to the controlled collapse of the cell.

Caspases are synthesized as inactive proenzymes that require proteolytic processing at internal Asp residues for full catalytic activity (Fig. 1). Structurally, the caspases are organized into a prodomain region, a large subunit, and a small subunit. Upon activation, the large and small caspase subunits are released from the proenzyme by cleaving an Asp-X bond between the prodomain and large subunit. Similarly, the large and small subunits are separated via a second cleavage event at an Asp-X bond between these two domains. The presence of Asp residues at the internal cleavage sites enables certain caspases to auto-activate or to be activated by other caspases as part of an amplification cascade (discussed in detail later). Active caspases are generally tetrameric, comprised of two large and two small subunits, and therefore possess two active sites. All caspases contain a pentapeptide surrounding the active site cysteine of general structure QACXG, where X is R, Q, or G.

The mammalian caspases can be divided into two broad families; those that are thought to play a central role in apoptosis (caspases-2, -3, -6, -7, -8, -9, -10, and -12) and those most closely related to caspase-1 (*ICE*), whose primary role seems to be in cytokine processing (caspases-1, -4, -5, and -11). The caspases involved in apoptosis can be further subdivided into either upstream/initiator caspases or downstream/effector caspases based on the length of their prodomains. The basic premise of this classification is that long prodomain caspases (such as caspases-2, -8, -9, and -10) are more likely to act as initiator caspases as they can be recruited to caspase-activating molecules (such as FADD or Apaf-1) through CARD (caspase recruitment domain) or CARD-like motifs in their pro-domain regions. Short prodomain caspases (such as caspases-3, -6, and -7) that do not possess CARD motifs are therefore less likely to be proximally activated in cell-death pathways. Such a scheme, however, oversimplifies caspase classification. For example, although caspases-8 and -9 do act as initiator caspases, caspase-2 has yet to be shown to play a decisive role as an upstream initiator caspase in any context. In fact, studies using *CASP-2*[-/-] mice have revealed that these mice develop normally and their cells die normally via apoptosis in response to various stimuli *(31)*. In addition, in the mitochondrial pathway of caspase activation, caspase-3 (a short prodomain caspase) plays a key role in disseminating and amplifying the caspase cascade *(32)*. Moreover, *CASP-3*[-/-] mice exhibit perinatal mortality owing to a gross accumulation of cells in the brains of these animals, arguing that caspase-

3 participates in an upstream "decision to die" capacity during brain development *(33)*. The latter observation highlights the importance that this "effector" caspase plays in embryonic development. Nonetheless this classification scheme does help when attempting to assign likely hierarchical positions to caspase molecules within uncharacterized activation pathways.

CASPASE ACTIVATION

The common strategy adopted to achieve apical caspase activation appears to be the formation of complexes containing several caspase zymogens *(34)*. This is achieved through recruitment of apical caspases into complexes by specific adaptor proteins. This appears to facilitate activation of the apical caspase, because "inactive" caspase zymogens possess low but detectable catalytic activity that is sufficient to process other caspases in circumstances where sustained close proximity between the zymogens is achieved. There are many contexts in which cells die by apoptosis and it is still unclear as to how caspases become activated in all of these situations. However, many stimuli that promote apoptosis appear to engage the caspases in one of three main ways, which we briefly discuss here.

The Mitochondrial Pathway to Caspase Activation

Mitochondria act as important sensors of cellular damage (*see* Chapter 6). Various forms of cellular stress such as DNA damage, heat shock, and oxidative stress result in an increase in the permeability of the outer mitochondrial membrane. This allows certain proteins, such as cytochrome *c*, to be released from the mitochondrial intermembrane space into the cytosol. Upon release into the cytosol, cytochrome *c* then binds to Apaf-1, a mammalian homolog of the *C. elegans* CED-4 protein *(35,36)*. The binding of cytochrome *c*, in association with ATP/dATP, results in a conformational change in Apaf-1 that promotes its oligomerization *(37,38)*. Procaspase-9 molecules can then bind to each of the Apaf-1 monomers via CARD-CARD interactions between both proteins. This high molecular-weight complex is called the apoptosome and its formation promotes caspase-9 activation due to the increase in the local concentration of caspase-9 zymogens. In addition, Apaf-1 appears to act as an allosteric regulator of caspase-9, because free proteolytically processed caspase-9 possesses little enzymatic activity. Recent evidence suggests that there are in fact seven Apaf-1 monomers in the apoptosome, each of which has a caspase-9 dimer bound to it *(38)*. In addition, it appears that only one monomer of each caspase-9 dimer is catalytically active *(38,39)* (*see* Chapter 4).

Upon activation within the apoptosome, caspase-9 then propagates the caspase cascade through activation of caspases-3 and -7. Caspase-3 in turn activates caspases-2 and -6 and the latter then promotes activation of caspases-8 and -10 downstream *(32)*. Caspase-3 also participates in a feedback amplification loop to further process caspase-9 *(32,40)*. Thus, it is evident that once caspase-9 is activated within the apoptosome, there is a rapid amplification of the death signal through activation of a panoply of other caspases (Fig. 2).

Ligation of Death Receptors

Members of the death receptor family include Fas/Apo-1/CD95, tumor necrosis factor-α (TNF-α), DR4 (TRAIL-R1), DR5 (TRAIL-R2), and DR6 *(41)* (*see* Chapter 5). The death receptors are a subset of the TNF/NGF receptor family that contain a conserved motif, termed the death domain (DD), within their cytoplasmic tails. These DD are responsible for recruiting adaptor molecules that, in turn, recruit caspases to the receptor complex. In most cases, the recruited caspase is caspase-8; however, caspase-10 may substitute in certain cases *(42,43)*. Once again, as in the apoptosome context, caspase-8 activation in the context of death-receptor signaling is thought to be owing to the increase in concentration of caspase-8 zymogens through aggregation *(44)*. For example, binding of Fas ligand to the Fas receptor leads to recruitment of the adaptor molecule FADD to the cytoplasmic

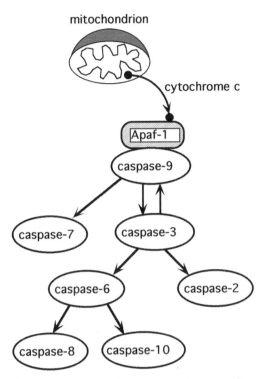

Fig. 2. Order of caspase activation events downstream of cytochrome *c* release into the cytosol (see text for further details).

tail of Fas through DD–DD interactions. This complex is referred to as the death-inducing signaling complex (DISC; ref. *45*). FADD in turn binds to the prodomain of procaspase-8, through interactions involving CARD-like death-effector domains (DED) within both proteins.

At this point, the death signal diverges in different cell types *(46)*. The precise pathway followed appears to be determined by the ability of caspase-8 to activate sufficient quantities of caspase-3 downstream. In type I cells caspase-8 can activate sufficient procaspase-3 and there is no requirement for mitochondrial participation. Active caspase-3 can then either cleave caspase-9 or directly process caspases-2 and -6 to further disseminate the caspase cascade. In type II cells, caspase-8 activation appears insufficient to directly activate procaspase-3. Instead caspase-8 cleaves the BH3– only protein, Bid, to generate an active ~15kDa C-terminal fragment termed tBid *(47,48)*. tBid appears to engage the mitochondrial pathway by stimulating cytochrome *c* release via Bax and/or Bak oligo-merization and insertion into mitochondrial membranes *(49)*. Once in the cytoplasm, cytochrome *c* activates caspases as previously described for the apoptosome pathway.

Granzyme B-Induced Caspase Activation

Cytotoxic T lymphocytes (CTL) and natural killer (NK) cells contain granules that are released onto the surface of target cells, such as transformed or virally infected cells *(50)*. These granules contain a variety of cytotoxic enzymes, but of particular importance are granzyme B and perforin. Perforin is a pore-forming protein whose likely function is to facilitate the entry of other granule components into target cells. Granzyme B is a serine protease that cleaves following aspartate residues, making caspases attractive targets. Caspase-3 was the first caspase identified as a substrate for granzyme B *(51,52)*. Subsequently, several other caspases have been identified as substrates for this

Table 1
Caspase Substrates Possessing DXXD Cleavage Sites

Substrate	*Cleavage Site*
PARP	**DEVD**
SREBP	**DEPD**
αII-Fodrin	**DETD**
βII-Fodrin	**DEVD**
Hsp90	**DEED**
Rb	**DEAD**
Gelsolin	**DQTD**
IκB-α	**DRGD**
DFF45/ICAD site I	**DETD**
DFF45/ICAD site II	**DAVD**
Bcl-2	**DAGD**
Acinus	**DELD**

CTL granule component in vitro *(53)*. However, only caspases-3 and -8 have been shown to be major substrates of granzyme B in intact cells *(54,55)*.

In addition to caspases, granzyme B can also cleave Bid into active gtBid, which can then translocate to mitochondria and provoke cytochrome *c* release *(56–58)*. Indeed, from in vitro comparisons of granzyme-B-mediated cleavage of Bid, caspase-8, and caspase-3, Bid appears to be the preferred substrate of this CTL protease *(58,59)*. Therefore, even though granzyme B is capable of activating caspases directly, its physiological means of triggering apoptosis may be through activation of the mitochondrial pathway, via Bid.

CASPASE SUBSTRATE PROTEINS

Active caspases are responsible for producing the distinctive morphological changes associated with apoptosis through targeting several proteins within the cell for activation or inactivation. Caspases typically recognize tetrapeptide (P_4–P_3–P_2–P_1) motifs in their substrates (e.g., DEVD, YVAD, DEAD), the scissile bond occurring between the P1 amino acid residue and the adjacent C-terminal amino acid in the peptide chain *(8–10)*. Caspases have an absolute requirement for aspartate at the P_1 position, however, it is the residue at the P_4 position that is the most critical in determining the substrate specificity of the individual caspases. A list of caspase substrates with DXXD cleavage sites is presented in Table 1. However, many substrates lacking this motif are also known (Table 2).

The list of identified substrates of the caspase family has grown rapidly and is likely to contain several hundred proteins upon completion *(see* ref. *28* for a recent review). These substrates represent an eclectic group of proteins and it is likely that only a relatively minor subset of caspase substrates are specifically targeted during apoptosis. Thus it is likely that the majority of caspase substrates are "innocent bystanders" and that their proteolysis contributes little to the process. For this reason, the consequences of many of the substrate cleavage events that take place during apoptosis are still the subject of speculation at present. However, evidence is accumulating to link specific caspase-mediated proteolytic events with the stereotypical destructive changes that have long been known to take place in the cell during apoptosis. The overall picture is one where the cell is coordinately dismantled in preparation for recognition and removal by phagocytic cells. Some of the caspase substrates that have been identified are discussed below.

Several cytoskeletal proteins such as gelsolin, fodrin, nuclear lamins A and B, Gas 2, keratin 18, and ß-catenin are well-established caspase substrate proteins *(60–66)*. Cleavage of such proteins may

Table 2
Caspase Substrates Possessing Non-DXXD Cleavage Sites

Substrate	Cleavage Site
STAT1	**MELD**
Lamin B	**VEVD**
NF-κB	**VFTD**
Bcl-x_L	**HLAD**
Bid	**LQTD**
Actin	**ELPD**
Lamin A	**VEID**

contribute to the dramatic reorganization of the cell body that takes place during apoptosis. Some of these proteolytic event may also contribute to the production of intact cell fragments (apoptotic bodies) that is seen during apoptosis.

DFF45 (DNA fragmentation factor 45 kDa) or ICAD (inhibitor of caspase-activated DNase) is a particularly well-studied caspase substrate *(67,68)*. The latter protein is an inhibitor of a DNase termed DFF40/CAD (caspase-activated DNase). During apoptosis, ICAD/DFF45 is cleaved, which results in CAD activation and subsequent degradation of DNA, a hallmark of apoptosis.

Many other caspase substrates are known but preliminary proteomic analyses of apoptotic cells suggest that the total number may well exceed 500. Clearly much work remains done before the final demolition phase of apoptosis is fully understood. In addition, the specific array of substrates that each caspase is capable of targeting is still largely undefined. However, studies using cell-free extracts depleted of specific caspases suggests that caspase-3 is the primary executioner caspase with relatively few physiological substrates for caspases-6 and -7 thus far identified *(69)*.

CONCLUSION

A major focus in the cell-death field at present is on understanding the complex mechanisms of caspase activation and regulation within mammalian cells. Many key questions still remain to be answered, however. For example, it is still not clear how important triggers of caspase activation, such as p53 or c-Myc, activate the cell-death machinery. We do not understand how caspases trigger the membrane changes that ultimately result in removal of apoptotic cells by phagocytes. It remains unclear what proportion of the cellular proteome is targeted by caspases during demolition phase of apoptosis. Answers to these and other questions will shed considerable light upon this fundamental biological process. Furthermore, understanding the molecular control of apoptosis will undoubtedly yield a new generation of drug targets for the manipulation of apoptosis in disease contexts.

ACKNOWLEDGMENTS

We thank the Health Research Board (RP61–2000), Science Foundation Ireland (PI.1/B038), and The Wellcome Trust (047580) for support of some of the work discussed in this review.

REFERENCES

1. Jacobson, M. D., Weil, M., and Raff, M. C. (1997) Programmed cell death in animal development. *Cell* **88,** 347–354.
2. Wyllie, A. H., Kerr, J. F. R., and Currie, A. R. (1980) Cell death: the significance of apoptosis. *Int. Rev. Cytol.* **68,** 251–306.
3. Savill, J. and Fadok, V. (2000) Corpse clearance defines the meaning of cell death. *Nature* **407,** 784–788.
4. Martin, S. J. and Green, D. R. (1995) Protease activation during apoptosis: death by a thousand cuts? *Cell* **82,** 349–352.
5. Nicholson, D. W. (1996) *ICE*/CED-3–like proteases as therapeutic targets for the control of inappropriate apoptosis. *Nature Biotechnol.* **14,** 297–301.

6. Cohen, G. M. (1997) Caspases: the executioners of apoptosis. *Biochem. J.* **326,** 1–16.
7. Salvesen, G. S. and Dixit V. M. (1997) Caspases: intracellular signaling by proteolysis. *Cell* **91,** 443–446.
8. Villa, P., Kaufmann, S. H. and Earnshaw, W. C. (1997) Caspases and caspase inhibitors. *Trends Biochem. Sci.* **22,** 388–393.
9. Earnshaw, C. W., Martins, M. L., and Kaufmann, H. S. (1999) Mammalian caspases: structure, activation, substrates and functions during apoptosis. *Annu. Rev. Biochem.* **68,** 383–424.
10. Nicholson, D. W. and Thornberry, N. A. (1997) Caspases: killer proteases. *Trends Biochem. Sci.* **8,** 299–306.
11. Ellis, R. E., Yuan, J., and Horvitz, H. R. (1991) Mechanisms and function of cell death. *Annu. Rev. Cell Biol.* **7,** 663–698.
12. Yuan, J., Shaham, S., Ledoux, S., Ellis, H. M, and Horvitz, H.R. (1993) The C. elegans cell death gene ced-3 encodes a protein similar to mammalian interleukin-1β-converting enzyme. *Cell* **75,** 641–652.
13. Xue, D., Shaham, S., and Horvitz, H. R. (1996) The *Caenorhabditis elegans* cell death protein CED-3 is a cysteine protease with substrate specificities similar to those of human CPP32 protease. *Genes Dev.* **10,** 1073–1083.
14. Miura, M., Zhu, H., Rotello, R., Hartweig, E. A., and Juan, J. (1993) Induction of apoptosis in fibroblasts by IL-1β-converting enzyme, a mammalian homolog of the C. elegans cell death gene ced-3. *Cell* **75,** 653–660.
15. Williams, M. S. and Henkart, P. A. (1994) Apoptotic cell death induced by intracellular proteolysis. *J. Immunol.* **153,** 4247–4255.
16. Kaufmann, S. H., Desnoyers, S., Ottaviano, Y., Davidson, N. E., and Poirier, G. G. (1993) Specific proteolytic cleavage of poly(ADP-ribose) polymerase: an early marker of chemotherapy-induced apoptosis. *Cancer Res.* **53,** 3976–3985.
17. Lazebnik, Y. A., Kaufmann, S. H., Desnoyers, S., Poirier, G. G., and Earnshaw, W. C. (1994) Cleavage of poly (ADP-ribose) polymerase by a proteinase with properties like *ICE*. *Nature* **371,** 346–347.
18. Casciola-Rosen, L. A., Miller, D. K., Anhalt, G. J., and Rosen, A. (1994) Specific cleavage of the 70-kDa protein component of the U1 small nuclear ribonucleoprotein is a characteristic biochemical feature of apoptotic cell death. *J. Biol. Chem.* **269,** 30757–30760.
19. Neamati, N., Fernandez, A., Wright, S., Kiefer, J., and McConkey, D. J. (1995) Degradation of lamin B1 precedes oligonucleosomal DNA fragmentation in apoptotic thymocytes and isolated thymocyte nuclei. *J. Immunol.* **154,** 3788–3795.
20. Martin, S. J., O'Brien, G. A., Nishioka, W. K., McGahon, A. J., Saido, T., and Green D. R. (1995) Proteolysis of Fodrin nonerythroid spectrin during apoptosis. *J. Biol. Chem.* **270,** 6425–6428.
21. Enari, M., Hug, H., and Nagata, S. (1995) Involvement of an *ICE*-like protease in Fas-mediated apoptosis. *Nature* **375,** 78–81.
22. Los, M., Van de Craen, M., Penning, L. C., Schenk, H., Westendorp, M., Baeuerle, P. A., et al. (1995) Requirement for an *ICE*/CED-3 protease for Fas/APO-1-mediated apoptosis. *Nature* **375,** 81–83.
23. Martin, S. J., Newmeyer, D. D., Mathias, S., Farschon, D., Wang, H. G., Reed, J. C., et al. (1995) Cell-free reconstitution of Fas-, UV radiation- and ceramide-induced apoptosis. *EMBO J.* **14,** 5191–5200.
24. Gagliardini, V., Fernandez, P. A., Lee, R. K., Drexler, H. C. A., Rotello, R. J., Fishman, M. C., and Yuan, J. (1994) Prevention of vertebrate neuronal death by the CrmA gene. *Science* **263,** 826–828.
25. Tewari, M. and Dixit, V. M. (1995) Fas- and tumor necrosis factor-induced apoptosis is inhibited by the poxvirus CrmA gene product. *J. Biol. Chem.* **270,** 3255–3260.
26. Li, P., Allen, H., Banerjee, S., Franklin, S., Herzog, L., Johnston, C., et al. (1995) Mice deficient in IL-1β-converting enzyme are defective in production of mature IL-1β and resistant to endotoxic shock. *Cell* **80,** 401–411.
27. Kuida, K., Lippke, J. A., Ku, G., Harding, M. W., Livingston, D. J., Su, M. S., and Flavell, R. A. (1995) Altered cytokine export and apoptosis in mice deficient in interleukin-1 beta converting enzyme. *Science* **267,** 2000–2003.
28. Nicholson, D. W. (1999) Caspase structure, proteolytic substrates, and function during apoptotic cell death. *Cell Death Differ.* **6,** 1028–1042.
29. Yuan, J., Shaham, S., Ledoux, S., Ellis, H. M., and Horvitz, H. R. (1993) The *C. elegans* cell death gene ced-3 encodes a protein similar to mammalian interleukin-1 beta-converting enzyme. *Cell* **75,** 641–652.
30. Slee, E. A., Adrain, C., and Martin, S. J. (1999) Serial killers: ordering caspase activation events in apoptosis. *Cell Death Differ.* **6,** 1067–1074.
31. Bergeron, L., Perez, G. I., Macdonald, G., Shi, L., Sun, Y., Jurisicova, A., et al. (1998) Defects in regulation of apoptosis in caspase-2 deficient mice. *Genes Dev.* **12,** 1304–1314.
32. Slee, E. A., Harte, M. T., Kluck, R. M., Wolf, B. B., Casiano, C. A., Newmyer, D. D., et al. (1999) Ordering the cytochrome c-initiated caspase cascade: hierarchial activation of caspases-2, -3, -6, -7, -8 and –10 in a capsase-9 dependent manner. *J. Cell Biol.* **144,** 281–292.
33. Kluida, K., Zheng, T. S., Na, S., Kuan, C. Y., Yang, D., Karasuyama, H., et al. (1996) Decreased apoptosis in the brain and premature lethality in CPP32 deficient mice. *Nature* **384,** 368–372.
34. Salvesen, G. S. and Dixit, V. M. (1999) Caspase activation: the induced-proximity model. *Proc. Natl. Acad. Sci. USA* **96,** 10964–10967.
35. Liu, X., Kim, C. N., Yang, J., Jemmerson R., and Wang X. (1996) Induction of apoptotic program in cell-free extracts: requirement for dATP and cytochrome c. *Cell* **86,** 147–157.
36. Li, P., Nijhawan, D., Budihardjo, I., Srinivasula, S. M., Ahmad, M., Alnemri, E. S., and Wang, X. (1997) Cytochrome c and dATP-dependent formation of Apaf-1/caspase-9 complex initiates an apoptotic protease cascade. *Cell* **91,** 479–489.
37. Adrain, C. and Martin, S. J. (2001) The mitochondrial apoptosome: a killer unleashed by the cytochrome seas. *Trends Biochem. Sci.* **26,** 390–397.

38. Acehan, D., Jiang, X., Morgan, D. G., Heuser, J. E., Wang, X., and Akey, C. W. (2002) Three-dimensional structure of the apoptosome. Implications for assembly, procaspase-9 binding and activation. *Mol. Cell* **9,** 423–432.
39. Renatus, M., Stennicke, H. R., Scott, F. L., Liddington, R. C., and Salvesen, G. (2001) Dimer formation drives the activation of the cell death protease caspase-9. *PNAS* **98,** 14250–14255.
40. Srinivasula, S. M., Ahmad, M., Fernandes-Alnemri, T., and Alnemri, E. S. (1998) Autoactivation of procaspase-9 by Apaf-1 mediated oligomerization. *Mol. Cell* **1,** 949–957.
41. Ashkenzai, A. and Dixit, V. M. (1998) Death receptors: signalling and modulation. *Science* **281,** 1305–1308.
42. Kischkel, F. C., Lawerence, D. A., Tinel, A., LeBlanc, H., Virmani, A., Schow, P., et al. (2001) Death receptor recruitment of endogenous caspase-10 and apoptosis in the absence of caspase-8. *J. Biol. Chem.* **276,** 46639–46646.
43. Wang, J., Chun, H. J., Wong, W., Spencer, D. M., and Lenardo, M. J. (2001) Caspase-10 is an initiator caspase in death receptor signalling. *Proc. Natl. Acad. Sci. USA* **98,** 13884–13888.
44. Muzio, M., Stockwell, B. R., Stennike, H. R., Salvesen, G. S., and Dixit, V. M. (1998) An induced proximity model for caspase-8 activation. *J. Biol. Chem.* **273,** 2926–2930.
45. Kischkel, F. C., Hellbardt, S., Behrmann, I., Germer, M., Pawlita, M., Krammer, P. H., and Peter, M. E. (1995) Cytotoxicity-dependent APO-1 (Fas/CD95)-associated proteins form a death-inducing signaling complex (DISC) with the receptor. *EMBO J.* **14,** 5579–5588.
46. Scaffidi, C., Schmitz, I., Zha, J., Korsmeyer, S. J., Krammer, P. H., and Peter, M. E. (1999) Differential modulation of apoptosis sensitivity in CD95 type I and type II cells. *J. Biol. Chem.* **274,** 22532–22538.
47. Lou, X., Budihardjo, I., Zou, H., Slaughter, C., and Wang, X. (1998) Bid, a Bcl2 interacting protein, mediates cytochrome c release from mitochondria in response to activation of cell surface death receptors. *Cell* **94,** 481–490.
48. Li, H., Zhu, H., Xu, C. J., and Yuan, J. (1998) Cleavage of Bid by caspase-8 mediates the mitochondrail damage in the Fas pathway of apoptosis. *Cell* **94,** 491–501.
49. Wei, M. C., Zong, W. X., Cheng, E. H., Lindsten, T., Panoutsakopoulou, V., Ross, A. J., et al. (2001) Proapoptotic BAX and Bak: a requisite gateway to mitochondrial dysfunction and death. *Science* **292,** 727–730.
50. Froelich, C. J., Dixit, V. M., and Yang, X. (1998) Lymphocyte granule-mediated apoptosis: matters of viral mimicry and deadly proteases. *Immunol. Today* **19,** 30–36.
51. Darmon, A. J., Nicholson, D. W., and Bleackley R. C. (1995) Activation of the apoptotic protease CPP32 by cytotoxic T-cell-derived granzyme B. *Nature* **377,** 446–448.
52. Martin, S. J., Amarante-Mendes, G. P., Shi, L., Chuang, T. H., Casiano, C. A., O'Brien, G. A., et al. (1996) The cytotoxic cell protease granzyme B initiates apoptosis in a cell-free system by proteolytic processing and activation of the *ICE*/CED-3 family protease, CPP32, via a novel two-step mechanism. *EMBO J.* **15,** 2407–2416.
53. Van de Craen, M., Van den Brande, I., Declercq, W., Irmler, M., Beyaert, R., et al. (1997) Cleavage of caspase family members by granzyme B: a comparative study in vitro. *Eur. J. Immunol.* **27,** 1296–1299.
54. Medema, J. P., Toes, R. E., Scaffidi, C., Zheng, T. S., Flavell, R. A., Melief, C. J., et al. (1997) Cleavage of FLICE (caspase-8) by granzyme B during cytotoxic T lymphocyte-induced apoptosis. *Eur. J. Immunol.* **27,** 3492–3498.
55. Atkinson, E. A., Barry, M., Darmon, A. J., Shostak, I., Turner, P.C., Moyer, R. W., and Bleackley, R. C. (1998) Cytotoxic T lymphocyte-assisted suicide. Caspase-3 activation is primarily the result of the direct action of granzyne B. *J. Biol. Chem.* **273,** 21261–21266.
56. Barry, M., Heiben, J. A., Pinkoski, M. J., Lee, S. F., Moyer, R. W., Green, D. R., and Bleackley, R. C. (2000) Granzyme B short-circuits the need for caspase-8 activity during granule-mediated cytotoxic T-lymphocyte killing by directly cleaving Bid. *Mol. Cell Biol.* **20,** 3781–3794.
57. Heiben, J. A., Goping, I. S., Barry, M., Pinkoski, M. J., Shore, G. C., Green, D. R., and Bleackley, R. C. (2000) Granzyme B-mediated cytochrome c release is regulated by the Bcl-2 family members Bid and Bax. *J. Exp. Med.* **192,** 1391–1402.
58. Sutton, V. R., Davis, J. E., Cancilla, M., Johnstone, R. W., Ruefi, A. A., Sedelies, K., et al. (2000) Initiation of apoptosis by granzyme B requires direct cleavage of bid, but not direct granzyme B-mediated caspase activation. *J. Exp. Med.* **192,** 1403–1414.
59. Pinkoski, M. J., Waterhouse, N. J., Heiben, J. A., Wolf, B. B., Kuwana, T., Goldstein, J. C., et al. (2001) Granzyme B-mediated apoptosis proceeds predominantly through a Bcl-2 inhibitable mitochondrial pathway. *J. Biol. Chem.* **276,** 12060–12067.
60. Kothakota, S., Azuma, T., Reinhard, C., Klippel, A., Tang, J., Chu, K., et al. (1997) Caspase-3-generated fragment of gelsolin: effector of morphological change in apoptosis. *Science* **278,** 294–298.
61. Martin, S. J., O'Brien, G. A., Nishioka, W. K., McGahon, A. J., Saido, T., and Green, D. R. (1995) Proteolysis of Fodrin nonerythroid spectrin during apoptosis. *J. Biol. Chem.* **270,** 6425–6428.
62. Takahashi, A., Alnemri, E. S., Lazebnik, Y. A., Fernandes-Alnemri, T., Litwack, G., Moir, R. D., et al. (1996) Cleavage of lamin A by Mch2α but not CPP32: multiple interleukin 1β-converting enzyme-related proteases with distinct substrate recognition properties. *Proc. Natl. Acad. Sci. USA* **93,** 8395–8400.
63. Neamati, N., Fernandez, A., Wright, S., Kiefer, J., and McConkey, D. J. (1995) Degradation of lamin B1 precedes oligonucleosomal DNA fragmentation in apoptotic thymocytes and isolated thymocyte nuclei. *J. Immunol.* **154,** 3788–3795.
64. Brancolini, C., Bendetti, M., and Schneider, C. (1995) Microfilament reorganization during apoptosis: the role of Gas2, a possible substrate for *ICE*-like proteases. *EMBO J.* **14,** 5179–5190.
65. Caulin, C., Salvesen, G. S., and Oshima, R. G. (1997) Caspase cleavage of keratin 18 and reorganization of intermediate filaments during epithelial cell apoptosis. *J. Cell Biol.* **138,** 1379–1394.

66. Brancolini, C., Lazarevic, D., Rodriguez, J., and Schneider, C. (1997) Dismantling cell-cell contacts during apoptosis is coupled to a caspase-dependent proteolytic cleavage of beta-catenin. *J. Cell Biol.* **139,** 759–771.
67. Liu, X., Zou, H., Slaughter, C., and Wang, X. (1997) DFF, a heterodimeric protein that functions downstream of caspase-3 to trigger DNA fragmentation during apoptosis. *Cell* **90,** 405–413.
68. Enari, M., Sakahira, H., Yokoyama, H., Okawa, K., Iwamatsu, A., and Nagata, S. (1998) A caspase-activated DNase that degrades DNA during apoptosis, and its inhibitor ICAD. *Nature* **391,** 43–50.
69. Slee, E. A., Adrain, C., and Martin, S. J. (2001) Executioner caspases–3,-6 and –7 perform distinct, non-redundant roles during the demolition phase of apoptosis. *J. Biol. Chem.* **276,** 7320–7326.

2

Bcl-2 Family Proteins
Master Regulators of Apoptosis

Xiao-Ming Yin, Wen-Xing Ding, and Yongge Zhao

INTRODUCTION

Apoptosis is an active process of cellular self-destruction with distinctive morphological and bio-chemical features *(1)*. Two major apoptotic pathways have been defined in mammalian cells: the death-receptor pathway and the mitochondrial pathway. In the mitochondrial pathway, the Bcl-2 family proteins are perhaps the most important regulators. These proteins can also be responsible for bridging signals from the death-receptor pathway to the mitochondrial pathway. The Bcl-2 family of proteins consists of both anti-apoptosis and pro-apoptosis members. While the pro-apoptosis members serve as sensors to death signals and executors of the death program, the anti-apoptosis members inhibit the initiation of the death program. The Bcl-2 family proteins are evolutionarily conserved, but may accomplish these tasks by different mechanisms in different species. In addition, multiple cellular signals can modify the activities and locations of these proteins, thus forming an intracellular signaling network that maintains the delicate balance between life and death.

EVOLUTIONARY CONSERVATION OF BCL-2 FAMILY PROTEINS

Bcl-2, the prototype of the Bcl-2 family proteins, was the first defined molecule involved in apoptosis. It was initially cloned from the t(14;18) breakpoint in human follicular lymphoma *(2–4)*. Although its role as a proto-oncogene was quickly realized, its biological function as an anti-apoptosis gene was not realized until some years later *(5,6)*. A number of proteins were soon discovered that share sequence homology with Bcl-2, but only some of those engage in anti-apoptosis activities; others actually promote apoptosis *(see* Table 1).

Notably, this family of proteins is evolutionarily conserved. A number of viruses encode Bcl-2 homologs, including most, if not all, gamma herpes viruses *(7)*. Most of these viral homologs are anti-apoptotic, probably because viruses need to keep the infected cells alive for latent and persistent infection *(7,8)*. The nematode *Caenorhabditis elegans* has its own sequence and functional homologs for a death antagonist, *CED-9 (9)*, and a BH3-only death agonist, EGL-1 *(10)*. On the other hand, only prodeath homologs (dBorg-1/Drob-1/Debcl and dBorg-2/Buffy) have been described in the *Drosophila (11)*. These homologs are discussed in details in Chapters 9 and 10, respectively.

From: *Essentials of Apoptosis: A Guide for Basic and Clinical Research*
Edited by: X-M. Yin and Z. Dong © Humana Press Inc., Totowa, NJ

Table 1
The Bcl-2 Family Proteins

Function	Organisms	Members	BH domains[a]
Anti-apoptosis	Mammals	Bcl-2, Bcl-x$_L$, Bcl-w, Mcl-1, A1/Bfl-1, Boo/Diva, Bcl-B/Bcl-2L-10/Nrh	Multidomain
	C. elegans	*CED-9*	
	Virus	E1B-19K (Adenovirus), BHRF-1(EBV), KS-Bcl-2 (HHV8), ORF16 (HSV), LMW5-HL (ASFV)	
	Xenopus	XR1, XR11	
Pro-apoptosis	Mammals	Bax, Bak, Bok/Mtd, Bcl-x$_S$, Bcl-G$_L$	Multidomain
		Bad, Bid, Bik/Nbk, Blk, Hrk/DP5, Bim/Bod, Bmf, Nip3/BNIP3, Nix/Bnip3L, Noxa/APR, PUMA, MAP-1, Bcl-G$_S$	BH3-only
	C. elegans	EGL-1	BH3-only
	Drosophila	dBorg^{-1}/Drob^{-1}Debcl, dBorg^{-2}/Buffy	Multidomain

[a]Multidomain family proteins may contain more than one BH domain, whereas BH3-only members contain the BH3 domain only. Domain structures of representative Bcl-2 family proteins are depicted in Fig. 1.

One of the key features of the Bcl-2 family proteins is that members share sequence homology in four domains—the BH1, 2, 3, and 4 domains—although not all members have all the domains *(12–15)* (*see* Table 1 and Fig. 1). Mutagenesis studies have revealed that these domains are important for the various molecular functions and for protein interactions among the family members. The BH1 and BH2 domains are necessary for the death-repression function of the anti-apoptosis molecules, whereas the BH3 domain is required for the death-promotion function of the prodeath molecules *(16,17)*. In addition, the BH4 domain, which is present mainly in the anti-apoptosis molecules, is also important for death-inhibition functions (14, 15).

The prodeath molecules can be further divided into those with only the BH3 domain and those with multiple domains. It seems that the so-called "BH3-only" molecules, such as Bid, Bim, and Bad, are sensors for the peripheral death signals and are able to activate the "multidomain" executioner molecules, Bax or Bak *(17–20)*. This process in some ways resembles the caspase cascade, in which the initiator caspases activate the effector caspases.

BCL-2 FAMILY PROTEIN INTERACTIONS AND THEIR FUNCTIONAL SIGNIFICANCE

Protein Interactions Among Bcl-2 Family Members

The Bcl-2 family proteins can interact with each other and also with several other proteins. In fact, the first pro-apoptosis Bcl-2 family protein, Bax, was cloned based on its interaction with Bcl-2 *(21)*. Many other Bcl-2 family proteins were also cloned based on this type of interaction. A number of methods have been utilized to determine such interactions, including yeast two-hybrid, co-immuno-precipitation, and GST pull-down assay. Interpretation of the protein–protein interactions observed in vitro sometimes can be complicated. For example, it was found that the in vitro interaction of Bax with Bcl-2 occurred only in the presence of detergent, which caused a conformational change of Bax *(22)*. It is not certain whether or not interactions among other Bcl-2 family proteins in vitro will be affected by the detergent. However, the authenticity of such interactions may be verified by an in vivo interaction system, such as the yeast two-hybrids, which do not involve the use of detergent *(23,24)*. In addition, interactions among the family members, such as Bcl-2 and Bax, likely occur on membranes in vivo, where a proper conformation enabling them to interact may be formed as the result of activation by death signals *(22,25–28)*.

Based on these analyses, several interaction types can be defined. The most common type is the interaction between the antideath and prodeath members, such as Bcl-2 vs Bax *(21)* or Bid *(29)*. This interaction can result in antagonistic action of the two types of molecules and thus could set a rheostat control of the death program *(30)*. Interestingly, not all antideath molecules can interact with all prodeath molecules. It seems that some members of one group will preferentially bind to some members of the other group. For example, the antideath molecule Bcl-B binds to Bax, but not to Bak *(31)*. On the other hand, the prodeath molecule Bok/Mtd binds to Mcl-1 and to BHRF-1, but not to Bcl-2, Bcl-x_L, or Bcl-w *(32)*. Correspondingly, these molecules may only antagonize the function of those molecules to which they bind *(31,32)*. This type of selectivity suggests that specific amino acids required for particular interactions may only exist in some but not all of the family members. In addition, it may also suggest that in certain tissues and for certain death stimuli, a specific set of Bcl-2 family proteins is critically involved.

The second type of interaction occurs between two prodeath members, usually one BH3-only molecule and one multidomain molecule, such as Bid to Bax *(29)* or Bak *(33)*, MAP-1 to Bax *(34)*, and Bim_S or BimAD to Bax *(35)*. Such interactions could be important for the activation of the multidomain executioner molecules, Bax or Bak, by the BH3-only sensor molecules *(28,33)*. A further discussion of the functional significance of this type of interaction is given below.

The third type of interaction is multimerization of the same molecule. This has been observed in both antideath molecules, such as Bcl-2 or Bcl-x_L *(12,23,24,36)*, and prodeath molecules, such as Bax, Bak, and MAP-1 *(25,33,34,37)*. The ability of Bax or Bak to oligomerize has been considered an important factor in their role as a mitochondrial channel for releasing mitochondrial apoptotic factors such as cytochrome *c (33,37)*. A more detailed discussion can be found in Chapter 6.

The Importance of BH Domains in Bcl-2 Family Protein Interactions

BH domains are critically involved in protein interactions among family members. The BH1 and BH2 domains of the antideath molecules seem to interact with the BH3 domain of the prodeath molecules *(12,13,29)*, with the latter serving as the donor for the "pocket site" of the former, as suggested by the structural studies *(see* ref. *16* and next section). Similarly, the domains involved in the interactions between two prodeath proteins, such as Bid and Bax, are the BH3 domain of the BH3-only molecule and the BH1 and/or BH2 domain of the multi-domain molecules *(29)*. Mutations at one of the domains can usually disrupt such interactions. However, it may be necessary to introduce mutations in all the BH domains of a particular antideath molecule to disrupt its interaction with a prodeath molecule *(34)*. At other times, regions outside of the BH domain may be required for interactions, such as the interaction between BNIP1 and Bcl-x_L *(38)*. Critical amino acids have been defined in each BH domain, such as Gly[145] in the BH1 domain, Trp[188] in the BH2 domain of Bcl-2, and Gly[94] in the BH3 domain of Bid *(12,29)*.

Interestingly, certain amino acids are used selectively in one molecule for interacting with different partners, and this selective use is of functional significance. For example, although Bcl-x_L can bind to both BH3-only molecules (such as Bid and Bad) and multidomain molecules (such as Bax and Bak), certain amino acids (Phe[131] and Asp[133]) seem to be important for binding to BH3-only molecules, but not to Bax *(20,39)*. The mutant (F131V, D133A) will bind to Bid, Bad, or Bim$_L$, but not to Bax. The mutant, however, retains the antideath function, and it may block the activity of the BH3-only molecules *(20)*. In addition, variations in certain key amino acids could result in different affinities in binding to the same molecule, which occurs with two Bcl-2 isoforms in binding to Bak or Bad-derived BH3 peptides *(40)*.

Interactions Between Bcl-2 Family Proteins and Other Proteins

The Bcl-2 family proteins can also interact with several other proteins. On one hand, some of these proteins can sequester the BH3–only molecules in the cytosol; specifically, 14–3–3ε binds to phosphorylated Bad and prevents it from translocating to the mitochondria *(41)*. In addition, the dynein light chain complex 1 or 2 retain Bim_{EL}/Bim_L or Bmf in the microtubular dynein motor complex or actin filamentous myosin V motor complex, respectively *(42,43)*. On the other hand, some of the binding partners can be regulated by the Bcl-2 family proteins for their functions. These may include: *CED-4*, as inhibited by CED-*9 (44)*; calcineurin, as inhibited by Bcl-2 *(45)*; calcineurin, as inhibited by Bcl-x_L *(46)*; and VDAC, as activated by Bax *(47)*. In these cases, the Bcl-2 family proteins prevent caspase (CED-*3*) activation *(44)*, inhibit cell-cycle progression *(45)*, prevent FasL transcription *(46)*, or induce mitochondrial permeability transition *(47)*, respectively. Finally, a group of proteins may serve as bridges between the Bcl-2 family proteins and other cellular proteins. For example, Bcl-x_L may bind to BAR to regulate the activity of caspase-8 *(48)*, and may bind to Aven to regulate the activity of Apaf-1 *(49)*. Bcl-2 or Bcl-x_L may also bind to Bap 31 at the site of the endoplasmic reticulum to regulate the activity of caspase-8 *(50)*.

THE CRYSTAL AND SOLUTION STRUCTURE OF BCL-2 FAMILY PROTEINS

The crystal and solution structures of several Bcl-2 family proteins, Bcl-x_L, Bcl-2, Bax and Bid, have been defined (*see* refs. *40,51–56* and Chapter 4). One of the common structural features is that these proteins are all composed of alpha helices and assume a similar conformation (Fig. 1). The alpha helical structures consist of central hydrophobic helix (helices) surrounded by multiple amphipathic helices. Such an arrangement of alpha helices is similar to that of the membrane-translocation domain of bacterial toxins, in particular diphtheria toxin and the colicins, which suggests that the Bcl-2 family proteins may have a pore-forming function. Indeed, all four proteins tested showed channel activities in vitro on lipid bilayers or liposomes. This activity may relate to the function of these molecules in regulating mitochondrial permeability (*see* Chapter 6).

One major structural difference between the multi-domain proteins, Bcl-2, Bcl-x_L and Bax, and the BH3-only protein, Bid, is that the BH1, BH2, and BH3 domains form a hydrophobic pocket in the multi-domain proteins, but not in the BH3-only proteins. This suggests that the BH3 domain of Bid or any other BH3-only protein may function as a "donor" in its interaction with the multi-domain proteins, whose hydrophobic pocket can serve as an "acceptor." Indeed, when a BH3 peptide derived from Bak is in complex with Bcl-x_L, it binds to this hydrophobic pocket *(52)*.

Differences in the structural topology and electrostatic potential of the hydrophobic pocket exist between Bcl-2 and Bcl-x_L, despite their overall similarity in solution structure. This is consistent with the finding that the two molecules have different affinities to various interacting molecules *(40,51,56)*. Such a difference even exits between two different isoforms of human Bcl-2 *(40)*, and between the human Bcl-2 and its viral homolog, KSHV Bcl-2 *(56)*. These variations within an overall conserved structure are compatible with the conserved anti-apoptosis function, but are also indicative of emphasis on different strategies to achieve this function.

The multidomain prodeath molecule, Bax, has a structure very similar to that of Bcl-x_L *(55)*. It is not clear how Bax functions in opposition to the antideath molecules. One clue from the structural study is that the full-length Bax actually has a conformation similar to that of C-terminal-truncated Bcl-x_L binding to a Bak BH3 peptide. The transmembrane domain of Bax is actually occupying its own hydrophobic pocket. It is known that Bax needs to be activated through conformation change for its pro-apoptotic function *(22,28)*, and it is likely that the solution structure of an activated Bax would be quite different from that of a quiescent Bax. The transmembrane domain of Bax may be released when Bax changes its conformation, thus freeing the hydrophobic pocket for interaction with other Bcl-2 family proteins, such as Bid, and/or exposing the BH3 domain to exercise the prodeath function. The latter possibility is further confirmed in Bid, which has assumed an overall conserved structure before activation *(53,54)*. However, activation of Bid by proteolysis can cause the exposure of its BH3 domain for its killing function *(16,54)*. In contrast, the anti-apoptosis molecules usually have their BH3 domain buried, which may explain why they are not apoptotic. It can be postulated that if their BH3 domain is ever exposed, these molecules may also assume pro-apoptotic functions. Indeed, when Bcl-2 or Bcl-x_L are cleaved by caspase-3 to remove the N-termini, they are endowed with apoptotic abilities (*see* below) *(57,58)*.

REGULATION OF BCL-2 FAMILY PROTEINS

Regulation of Expression

Because of the potent effects of Bcl-2 family proteins on the balance of life and death, cells impose strict regulations on the expression and activity of these molecules. Whereas certain antideath or prodeath molecules are expressed constitutively in cells, others are expressed only following death stimuli. This is particularly true for a number of pro-apoptosis molecules. For example, DNA damage can induce the expression of Noxa and PUMA in a p53-dependent manner *(59–61)*. Upregulation of prodeath molecules can also be developmentally regulated; an example is EGL-1, which is required for the death of the HSN neurons in the male *C. elegans (10)* (*see* Chapter 9). Similarly, deprivation of nutritional factors can also induce expression of pro-apoptosis molecules. For example, Hrk/DP5 or Bim_{EL} can be induced in cultured sympathetic neurons following NGF withdrawal *(62)*. Whereas the death signals for the upregulation of BH3-only molecules may be specific, those that can upregulate the multidomain molecule, Bax, are often more diverse, suggesting that Bax is involved in the common death pathway (*see* below).

Expression of anti-apoptosis molecules can be induced by survival signals or inflammatory signals, which may occur in a cell-specific or time-specific manner. For example, Mcl-1 can be upregulated by GM-CSF in myeloid cells *(63,64)*, and A1 can be induced in endothelial cells or neutrophils in response to phorbol esters, LPS, tumor necrosis factor-α (TNF-α), interleukin-1 (IL-1), or G-CSF *(65–67)*. Expression of Bcl-x_L and Bcl-2 in thymocytes and mature T-cells is a good example of how homeostasis can be maintained by differential expression of these genes in a temporal-specific manner *(68)*. Thus Bcl-x_L, but not Bcl-2, is preferentially expressed in immature CD4/CD8 double-positive cells. On the other hand, Bcl-2, but not Bcl-x_L, is expressed in mature CD4 or CD8 single-positive cells. However, expression of Bcl-x_L is upregulated in activated mature T-cells. This probably allows the activated cells to survive for their immune functions *(69)*.

Regulation Through Alternative Splicing

A puzzling fact of the regulation of the Bcl-2 family protein is alternative splicing. A number of these proteins, including pro- and antideath members, can be expressed in different forms. For some, the different spliced forms can have opposite functions, such as Bcl-x_L vs Bcl-x_S *(70)*, and Mcl-1_L vs Mcl-1_S *(71)*. For others, alternative splicing does not alter the prodeath or antideath nature of the product, only its potency. The long form of Bcl-2, Bcl-2α, is more potent than the short form, Bcl-2ß *(72)*; however, the short form of Bim, Bim_S, is much more potent than the long form, Bim_L, or the extra-long form, Bim_{EL} *(73)*. Bim_S and another newly defined Bim splicing variant, BimAD, may also activate the mitochondria by different mechanisms from other forms of Bim (*see* below). In no case is it clear how the alternative splicing is regulated. Tissue-specific or signal-specific mechanisms may be involved. For example, Bcl-G_L is widely expressed, but Bcl-G_S is only found in testis *(74)*. Thus, it is possible that splicing variants could regulate apoptosis in a temporal- and/or spatial-specific way.

Regulation Through Post-Translational Modifications

Post-translational modification is probably the most significant mechanism regulating the activities of the Bcl-2 family proteins. This is particularly important for those prodeath molecules that are normally expressed in a healthy cell. These modifications often occur in response to death or survival signals and mainly include proteolytic cleavage and phosphorylation. In addition, protein conformation changes or degradation could be also induced as a post-translational event in response to the signals.

Subcellular Translocation of Prodeath Molecules Resulting from Activation by Post-Translational Modifications

One of the main outcomes of the post-translational modifications is the translocation of the modified death agonists to the mitochondria, as in the case of Bax, Bid, Bim, and Bad. In such cases, the Bcl-2 family proteins serve as sensors to the external death signals and transmit those signals to the mitochondria.

The first type of post-translational modification is conformation change, which, for Bax, is the first step in response to death signals *(22,27,74a)*. This change may be due to an elevated cytosolic pH *(26)*. Cellular alkalinization may alter the ionization of key amino acid residues at the N- and C-termini, thus breaking the intra-molecular interactions maintained by the ionic force. The conformation change allows the exposure of the two termini, and the availability of the hydrophobic C-terminus now confers on the molecule the ability to target the mitochondrial membrane *(26)*. The insertion of Bax seems to be also greatly facilitated by the presence of another prodeath molecule, Bid, which may induce further conformational change of the molecule *(37,75)*.

Translocation of Bid is dependent on caspase cleavage, which is the second type of post-translational modification. Bid is a BH3-only pro-apoptosis Bcl-2 family protein normally present in the cytosol *(29)*. Bid is activated by caspase-8, usually following the activation of Fas or TNF-R1 *(76–78)*. The cleavage occurs at the so-called "loop region" (aa 43–77) *(53)* (Fig. 1), which is also susceptible to cleavage by granzyme B (Asp[75]) and lysosomal enzymes (Arg[65]) *(79)*. The 15 kD carboxy-terminal caspase-8-cleaved fragment of Bid (aa 60–195), called tBid, can be further myristoylated at Gly[60] near the N-terminus *(80)*. The modified Bid can now efficiently translocate to mitochondria. This newly acquired ability may be owing to the appearance of a large hydrophobic surface that was previously buried but revealed by the protease cleavage *(16,54)*. The changes in hydrophobic exposure and the related surface charges, together with the myristoylation, contribute to the translocation and integration of tBid into the mitochondrial membranes.

The third type of post-translational modification that results in subcellular translocation is phosphorylation and dephosphorylation. This is exemplified by Bad, which is also a BH3-only prodeath molecule *(81)*. The subcellular localization of Bad depends on its phosphorylation status. In the presence of a growth factor, such as IL-3, Bad can be phosphorylated *(41)*. There are two main phosphorylation sites in Bad. While Ser[136] seems to be mainly phosphorylated by Akt/PKB, a serine-threonine kinase *(82,83)*, the site of Ser[112] is phosphorylated by a cAMP-dependent kinase *(84)*. When phosphorylated, Bad binds to a cytosolic protein, 14–3–3, and is trapped in the cytosol. Subsequent to a death stimulus, such as IL-3 deprivation or calcium influx, dephosphorylation occurs through certain phosphatases, such as calcineurin *(85)*. De-phosphorylated Bad disassociates from 14–3–3 and translocates to mitochondria, contributing to cell death *(41)*.

There are other mechanisms to induce translocation of Bcl-2 family proteins. For example, translocation of Bim$_{EL}$/Bim$_L$ from the microtubule-associated dynein motor complex to the mitochondria can be induced by cytokine withdrawal, chemicals such as taxol, or ultraviolet (UV)-irradiation *(43)*. Another BH3-only molecule, Bmf, can be activated by anoikis or UV-irradiation. Bmf is released from myosin V motor complex and translocated to the mitochondria *(42)*. In both cases, it seems that some noncaspase proteases are involved to release these molecules from their normal location in cells.

Post-Translational Modification Inactivating Bcl-2 Family Proteins

There are no reports indicating that post-translational modifications can affect the subcellular locations of the antideath Bcl-2 family proteins, which are usually present either exclusively or par-

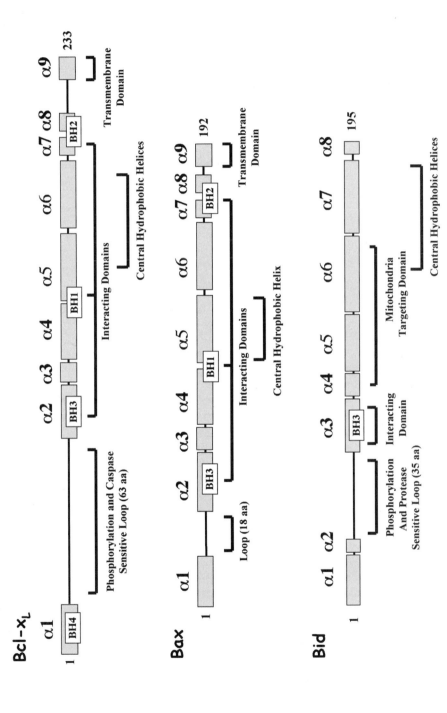

Fig. 1. Schematic representation of Bcl-2 family proteins. The structural features of three Bcl-2 family members, Bcl-x_L, Bax, and Bid are shown. They represent the three subgroups of the family proteins: the multidomain anti-apoptosis molecules, the multidomain pro-apoptosis molecules, and the BH3-only pro-apoptosis molecules, respectively (*see* Table 1). These molecules are composed of alpha helical structure with 1–2 central hydrophobic helix (helices) surrounded by 6–8 amphipathic helices (*51,53–55*). The α7 helix of Bcl-x_L (*55*) is also called α6' helix (*56*). A large, flexible loop is present in all three molecules, with a length of 18–67 amino acids. The loop is known to be a regulatory domain, sensitive to protease or kinase effects. The Bcl-2 homology domains 1–4 are distributed either in one alpha helix or across two alpha helices. They are involved in protein-protein interactions. Other alpha helices are involved in membrane binding (the TM domain in Bcl-x_L or Bax), membrane targeting (the alpha helices 4–6 in Bid), or pore-forming (the alpha helices 5–6 in Bcl-x_L). Overall the structural features are conserved among the Bcl-2 family members, but variations are present as to the number of alpha helices, the arrangement of the helices, the length of the loop, and the number of BH domains. See text for details.

tially in intracellular organelles. But modifications do occur in response to death stimuli, which can result in inactivation of the molecules or even convert them to prodeath molecules.

Both Bcl-2 and Bcl-x_L can be phosphorylated by death stimuli. The phosphorylation seems to occur on serine residuals in the so-called loop region, which is between the first and second alpha helices (Fig. 1) *(51,86–88)*. The phosphorylation results in decreased anti-apoptotic activities of either Bcl-2 or Bcl-x_L *(88)*, e.g., by reducing their ability to interact with Bax *(89)*. The chemotherapeutic drug taxol is well-known for its ability to inactivate Bcl-2 by inducing its phosphorylation *(86,88)*.

Recently, Hardwick and colleagues found that both Bcl-2 and Bcl-x_L could be also subject to caspase cleavage *(57,58)*. Caspase-3 is the main caspase that cleaves these molecules, again at the loop region. The cleavage does more than merely inactivate the function of these molecules; it actually confers on them an ability to induce apoptosis. Thus, the C-terminal fragment of Bcl-2 or Bcl-x_L (tBcl-2 or tBcl-x_L) is able to induce apoptosis. Furthermore, cleavage-resistant molecules, which are engineered through site-directed mutagenesis to delete the caspase-3 recognition site, exhibit stronger anti-apoptotic activity *(57,58)*. It seems that the altered tBcl-2 or tBcl-x_L may not contribute to cell death at the early initiation stage, because the modification occurs only after caspase activation. However, these truncated molecules can further accelerate the death process.

The prodeath molecules can also be inactivated by post-translational modifications. Bid can be phosphorylated by casein kinase I or II at the loop region (Ser[61] and Ser[64]), which results in the resistance of Bid to being cleaved by caspase-8 *(90)*. Phosphorylation-resistant mutant (S61A, S64A) is more cytotoxic than the wild-type molecule, reflecting that this type of phosphorylation could probably be physiologically relevant. In another example, phosphorylation of Bad causes the molecule not only to be trapped in the cytosol by 14–3–3 (*see* above) but also to be inactivated. The latter is accomplished through phosphorylation at Serine[155], which is in the middle of the BH3 domain. This can render Bad unable to interact with Bcl-x_L, thus losing its activity *(91,92)*.

CONTROL OF APOPTOSIS BY BCL-2 FAMILY PROTEINS

Mitochondria as the Major Targets

The Bcl-2 family proteins can be localized in the mitochondria either constitutively or by induction *(16,17)*. This is consistent with the current knowledge that the Bcl-2 family proteins regulate apoptosis mainly via their effects on mitochondria. Certainly, Bcl-2 molecules have also been found in other subcellular locations such as endoplasmic reticulum (ER) and thus can regulate the contributions of these organelles to apoptosis *(93)*. However, Bcl-2 proteins seem to be most important for the mitochondria apoptosis pathway.

The activation of the mitochondria pathway is signified by the release of mitochondrial apoptotic proteins and by mitochondrial dysfunction *(16,94)*. Both processes are inhibited by the death antagonists (Bcl-2, Bcl-x_L, etc.) but promoted by the death agonists (Bax, Bak, Bid, Bad, etc.). A detailed discussion of the mitochondria activation and mechanisms can be found in Chapter 6. Briefly, the release of the mitochondrial apoptotic proteins is due to an increase in outer membrane permeability, which may be owing to the opening of the mitochondria permeability transition pore, the pore formed by the Bcl-2 family proteins, such as Bax or Bak, or a pore made of the components of the two. However, another possibility is that the Bcl-2 family proteins may regulate the transportation of key metabolites, such as ADP and ATP, across the mitochondria and thus indirectly regulate the mitochondrial functions *(95)*.

Activation of Multidomain Bax and/or Bak by BH3-Only Molecules to Induce Mitochondrial Apoptosis

The multidomain prodeath proteins, Bax or Bak, are responsible for the induction of mitochondrial apoptosis. Deletion of both Bax and Bak (but not of just one of them) renders the cell completely resistant to all major mitochondrial death signals, including DNA damage, growth factor deprivation, and ER stress, and to the extrinsic pathway signals mediated by Bid *(18–20,96)*. Mice deficient in both Bax and Bak also have defects in developmentally regulated apoptosis (*[96]*; also *see* Chapter 11). It should be noted, however, that Bax and Bak may not be completely redundant in serving their functions. For example, in the human colon carcinoma cell line HCT116, which has mismatch repair deficiency, apoptosis induced by nonsteroidal anti-inflammatory drugs (NSAIDs; sulindac or indomethacin) or tumor necrosis factor-related apoptosis-inducing ligand (TRAIL) is dependent on Bax, but not Bak *(97,98)*. In addition, genetic deletion of Bax alone is sufficient to render sympathetic neurons resistant to neuronal growth factor (NGF)-deprivation-induced apoptosis *(99)*. One may keep in mind that Bax is usually localized in the cytosol in healthy cells and translocated to the mitochondria in response to death stimuli, whereas Bak seems to constitutively reside in the mitochondria. This distinction may contribute to the differential stimulation of Bax and Bak in certain cases.

It now seems clear that Bax and Bak are activated by the BH3-only molecules, which are activated by the peripheral death signals *(18–20)*. Two models, a direct activation model and an indirect activation model, have been proposed to explain how the BH3-only molecules can activate Bax or Bak, based on whether they interact with each other or not.

In the first model, the BH3-only molecules can directly activate the multidomain molecules, Bax and Bak, to initiate the mitochondrial events. This model is mainly supported by the observation that Bid induces Bax insertion into the mitochondrial membranes and its oligomerization *(28,37)*, as well as Bak oligomerization *(33)*. Bid can interact with Bax, Bak, and Bcl-2 *(28,29,33)*; however, its apoptosis-inducing capability is dependent on its ability to interact with the prodeath multidomain molecules but not with the antideath Bcl-2 or Bcl-x_L *(29)*. Conversely, although Bcl-x_L can interact with both Bid and Bax/Bak, its effects seem to be more dependent on its binding to Bid than to Bax, based on the use of Bcl-x_L mutants that can differentially bind to Bid and Bax *(20,39)*. Bim$_L$ was also found to be able to induce Bak oligomerization just as Bid *(20)*. Similarly, it seems that Bim$_L$ and Bad can also be sequestered by Bcl-x_L and lose their ability to bind to Bak or Bax *(20)*.

According to the second model, the BH3-only molecules, once translocated to the mitochondria, may not affect Bax/Bak directly, but rather bind to the antideath Bcl-2 family proteins to antagonize their function or to convert them to Bax/Bak-like molecules for oligomerization with Bax or Bak- *(17,19)*. Perhaps the best evidence for this hypothesis is that in *C. elegans*, the BH3-only molecule, EGL-1, binds competitively to the antideath molecule, CED-9, so that CED-4 is released from the binding with CED-9 to activate the caspase homolog, CED-*3 (44)*. Whether such a mechanism operates in the mammalian system is still unknown, although it has been speculated based on the interaction pattern of some of the molecules. From this aspect, it is found that Bim$_{EL}$ *(35)* or Bmf *(42)* does not bind to Bax, although whether they bind to Bak is not known. Bim$_L$ may be able to interact with Bak, because it can cause Bak oligomerization *(20)*. In addition, it is not known whether Bad or Noxa is able to interact with Bax or Bak. However, all these molecules are able to interact with Bcl-2 or Bcl-x_L, and their ability to induce apoptosis seems to depend on Bax and/or Bak *(19,20,42,59,73,100)*. Therefore, it is possible that these molecules may work through the second mechanism. Interestingly, the different spliced forms of Bim may work through different mechanisms. Thus, both Bim$_S$, which is more potent than Bim$_{EL}$ and Bim$_L$ and does not bind to the dynein motor complex, and another newly defined Bim spliced variant, BimAD, can actually bind to Bax and cause Bax conformation change *(35)*. Although both Bim$_S$ and BimAD can also interact with Bcl-2, mutants of BimAD that bind to Bax but not to Bcl-2 are still apoptogenic, suggesting that BimAD can directly activate Bax without interacting with Bcl-2 *(35)*.

THE PHYSIOLOGICAL ROLES OF THE BCL-2 FAMILY PROTEINS IN DEVELOPMENT AND HOMEOSTASIS

Because of the critical functions that Bcl-2 family proteins perform in the regulation of apoptosis induced by a variety of death signals, these proteins are important to organisms for their physiological functions in both embryonic development and homeostasis in adult life.

Bcl-2 family proteins play a critical role in programmed cell death. The term programmed cell death (PCD), initially defined by developmental biologists, describes the temporally and spatially controlled death of cells during development *(101)*. The genetic pathway of PCD was first systemically characterized in the nematode *C. elegans* by Horvitz and colleagues *(102)*. The antideath molecule CED-*9* is essential to the normal development of the worm, so that the loss of function mutation of this molecule causes normally surviving cells to die, which results in embryonic lethality *(103)*. This essential role of antideath molecules has also been observed in mammals. Inactivation of some mammalian anti-apoptosis genes, *bcl-x$_L$* or *mcl-1*, leads to embryonic lethality. Whereas *bcl-x$_L$* seems to be important for the development of the neuronal and hematopoietic systems *(104)*, *mcl-1* is critical to the development of trophectoderm, important for the implantation of embryos to the uterus *(105)*. Although deletion of Bcl-2 from the mouse genome only results in partial lethality so that some mice survive through the term, these mice nevertheless have significant developmental defects, including thymic atrophy, polycystic kidney disease, and melanocyte maturation arrest that leads to hypopigmentation *(106)*. A number of nonlethal developmental defects have also been observed in other genetic models where the Bcl-2 family genes were deleted by gene-targeting techniques. For example, both *bcl-lw-* and *bax*-deficient male mice, although surviving through the term, are infertile owing to abnormal spermatogenesis *(107,108)*. More examples are given in Chapter 11.

Another key function of apoptosis for multicellular organisms is to maintain cellular homeostasis during adult life. The Bcl-2 family proteins are particularly important in this aspect. For example, many types of stem cells express Bcl-2 for long-term survival *(109)*. Long-lived plasma cells and memory B cells also express a high level of Bcl-2 so that they will be available to counteract any future invasion of the organism *(110)*. This ability of Bcl-2 to maintain cell survival over a long period may be dangerous if it is not under tight control; indeed, the abnormal expression of this gene can lead to oncogenesis. A chromosomal translocation (14;18) that results in a deregulated expression of Bcl-2 may be responsible for the etiology of 85% of follicular lymphomas and 20% of diffuse B-cell lymphomas in humans *(111)*. This finding leads to the definition of a new type of proto-oncogene, represented by *bcl-2*, which does not cause abnormal cellular proliferation but affects cell death *(112)*. This type of proto-oncogene, when collaborating with the traditional proto-oncogenes such as *myc*, could lead to highly malignant, rapidly progressive tumors as demonstrated in several murine models *(113,114)*. Conversely, the prodeath Bcl-2 family proteins, such as Bax or Noxa, could serve an anti-tumor function by inducing cell death. Indeed, both Bax and Noxa are transcriptional targets of the tumor-suppressor gene p53 and could mediate its tumor-suppressing function *(59,115–117)*. A recent study has also demonstrated that induction of apoptosis is the only mechanism required for p53 to suppress tumorigenesis *(118)*. The concept that cell death genes serve a role in neoplasm is also helpful in dealing with cancers that have developed a resistance to chemotherapy. Many of these cancers express high levels of anti-apoptosis Bcl-2 family proteins that inhibit apoptosis induced by the drugs *(111)*.

The Bcl-2 family proteins are important in enabling organisms to perform their physiological functions in their responses to certain endogenous or exogenous stimuli. One example is that short-lived T-cells, when activated by foreign antigens, begin to express Bcl-x$_L$, which allows the activated T-cells to survive for a sufficient period of time to mature for their immune functions *(119)*. An expression of a prodeath gene, *bim*, may ensure the death of these cells later on to avoid autoimmunity *(120)*. Mice with the deletion of the *bim* gene actually manifest autoimmune symptoms *(121)*. In fact, Bim has been shown to be important to both central and peripheral tolerance of T-cells and B-cells *(122)*.

A similar process occurs in myeloid cells, where the anti-apoptosis molecule Mcl-1 is induced when a differentiation process starts *(63,64)*, and in some TNF-sensitive cells, where another anti-apoptosis molecule, A1, is induced to prevent the toxicity of TNF *(123)*. This temporal regulation of the expression of selected antideath genes allows the cells to perform their specific functions without committing suicide. On the other hand, constitutional expression of certain prodeath proteins, such as Bax in sympathetic neurons *(99)*, and Bid in hepatocytes *(75,124)*, may ensure that an effective death program is exercised when the environment becomes hostile. However, the physiological benefits of such an arrangement to the host are not quite clear.

Finally, because Bcl-2 family proteins can interact with other proteins, they may participate in other kind of cellular functions. One notable example is the regulation of cell-cycle progression, which can be inhibited by the antideath molecules Bcl-2 or Bcl-x_L *(68)* but promoted by the prodeath molecules Bax *(125)* or Bad *(126)*.

CONCLUDING REMARKS

The Bcl-2 family proteins include both anti-apoptosis and pro-apoptosis members. Despite their significant sequence diversities, they share homology in certain BH domains. This group of proteins is important to a number of physiological functions during normal embryonic development and during maintenance of homeostasis. The molecular mechanisms of these proteins in regulation of apoptosis could be as diverse as the members of the family. However, it seems that their central function is to regulate the mitochondrial physiology to inhibit or promote cell death. In this role, the pro- and antideath molecules may directly or indirectly antagonize each other's function. In addition, other functions have been indicated beyond apoptosis regulation for some of the family members. The role of Bcl-2 family proteins as a molecular sensor in many different pathophysiological contexts has yet to be fully explored. Thus, the future work with these proteins will focus on these critical issues.

REFERENCES

1. Kerr, J. F., Wyllie, A. H., and Currie, A. R. (1972) Apoptosis: a basic biological phenomenon with wide-ranging implications in tissue kinetics. *Br. J. Cancer* **26(4)**, 239–257.
2. Tsujimoto, Y., Finger, L. R., Yunis, J., Nowell, P. C., and Croce, C. M. (1984) Cloning of the chromosome breakpoint of neoplastic β cells with the t(14;18) chromosome translocation. *Science* **226(4678)**, 1097–1099.
3. Cleary, M. L. and Sklar, J. (1985) Nucleotide sequence of a t(14;18) chromosomal breakpoint in follicular lymphoma and demonstration of a breakpoint-cluster region near a transcriptionally active locus on chromosome 18. *Proc. Natl. Acad. Sci. USA* **82(21)**, 7439–7443.
4. Bakhshi, A., Jensen, J. P., Goldman, P., Wright, J. J., McBride, O. W., Epstein, A. L., and Korsmeyer, S. J. (1985) Cloning the chromosomal breakpoint of t(14;18) human lymphomas: clustering around JH on chromosome 14 and near a transcriptional unit on 18. *Cell* **41(3)**, 899–906.
5. Vaux, D. L., Cory, S., and Adams, J. M. (1988) Bcl-2 gene promotes haemopoietic cell survival and cooperates with c-myc to immortalize pre-β cells. *Nature* **335(6189)**, 440–442.
6. Nunez, G., London, L., Hockenbery, D., Alexander, M., McKearn, J. P., and Korsmeyer, S. J. (1990) Deregulated Bcl2 gene expression selectively prolongs survival of growth factor-deprived hemopoietic cell lines. *J. Immunol.* **144(9)**, 3602–3610.
7. Hardwick, J. M. (1998) Viral interference with apoptosis. *Semin. Cell Dev. Biol.* **9(3)**, 339–349.
8. Gangappa, S., van Dyk, L. F., Jewett, T. J., Speck, S. H., and Virgin, H. W. (2002) IV, Identification of the in vivo role of a viral Bcl2. *J. Exp. Med.* **195(7)**, 931–940.
9. Hengartner, M. O. and Horvitz, H. R. (1994) *C. elegans* cell survival gene CED-9 encodes a functional homolog of the mammalian proto-oncogene Bcl-2. *Cell* **76(4)**, 665–676.
10. Conradt, B. and Horvitz, H. R. (1998) The *C. elegans* protein EGL-1 is required for programmed cell death and interacts with the Bcl2-like protein *ced-9*. *Cell* **93(4)**, 519–529.
11. Vernooy, S. Y., Copeland, J., Ghaboosi, N., Griffin, E. E., Yoo, S. J., and Hay, B. A. (2000) Cell death regulation in drosophila: Conservation of mechanism and unique insights. *J. Cell Biol.* **150(2)**, F69–F76.
12. Yin, X. M., Oltvai Z. N., and Korsmeyer, S. J. (1994) BH1 and BH2 domains of Bcl-2 are required for inhibition of apoptosis and heterodimerization with bax (see comments). *Nature* **369(6478)**, 321–323.

13. Chittenden, T., Flemington, C., Houghton, A. B., Ebb, R. G., Gallo, G. J., Elangovan, B., Chinnadurai, G., and Lutz, R. J. (1995) A conserved domain in Bak, distinct from BH1 and BH2, mediates cell death and protein binding functions. *EMBO J.* **14(22)**, 5589–5596.

14. Hunter, J. J., Bond, B. L., and Parslow, T. G. (1996) Functional dissection of the human Bcl-2 protein: Sequence requirements for inhibition of apoptosis. *Mol. Cell Biol.* **16(3)**, 877–883.

15. Huang, D. C., Adams, J. M., and Cory, S. (1998) The conserved n-terminal BH4 domain of Bcl-2 homologues is essential for inhibition of apoptosis and interaction with CED-4. *EMBO J.* **17(4)**, 1029–1039.

16. Gross, A., McDonnell, J. M., and Korsmeyer, S. J. (1999) Bcl-2 family members and the mitochondria in apoptosis. *Genes Dev.* **13**, 1899–1911.

17. Puthalakath, H. and Strasser, A. (2002) Keeping killers on a tight leash: transcriptional and post-translational control of the pro-apoptotic activity of BH3-only proteins. *Cell Death Differ.* **9(5)**, 505–512.

18. Wei, M. C., Zong, W. X., Cheng, E. H., Lindsten, T., Panoutsakopoulou, V., Ross, A. J., et al. (2001) Proapoptotic bax and Bak: a requisite gateway to mitochondrial dysfunction and death. *Science* **292(5517)**, 727–730.

19. Zong, W. X., Lindsten, T., Ross, A. J., MacGregor, G. R., and Thompson, C. B. (2001) BH3-only proteins that bind pro-survival Bcl-2 family members fail to induce apoptosis in the absence of Bax and Bak. *Genes Dev.* **15(12)**, 1481–1486.

20. Cheng, E. H., Wei, M. C., Weiler, S., Flavell, R. A., Mak, T. W., Lindsten, T., and Korsmeyer, S. J. (2001) Bcl-2, Bcl-x$_L$ sequester BH3 domain-only molecules preventing Bax- and Bak-mediated mitochondrial apoptosis. *Mol. Cell.* **8(3)**, 705–711.

21. Oltvai, Z. N., Milliman, C. L., and Korsmeyer, S. J. (1993) Bcl-2 heterodimerizes in vivo with a conserved homolog, Bax, that accelerates programmed cell death. *Cell* **74(4)**, 609–619.

22. Hsu, Y. T. and Youle, R. J. (1998) Bax in murine thymus is a soluble monomeric protein that displays differential detergent-induced conformations. *J. Biol. Chem.* **273(17)**, 10777–10783.

23. Sedlak, T. W., Oltvai, Z. N., Yang, E., Wang, K., Boise, L. H., Thompson, C. B., and Korsmeyer, S. J. (1995) Multiple Bcl-2 family members demonstrate selective dimerizations with Bax. *Proc. Natl. Acad. Sci. USA* **92(17)**, 7834–7838.

24. Hanada, M., Aime-Sempe, C., Sato, T., and Reed, J. C. (1995) Structure-function analysis of Bcl-2 protein. Identification of conserved domains important for homodimerization with Bcl-2 and heterodimerization with Bax. *J. Biol. Chem.* **270(20)**, 11962–11969.

25. Gross, A., Jockel, J., Wei, M. C., and Korsmeyer, S. J. (1998) Enforced dimerization of Bax results in its translocation, mitochondrial dysfunction and apoptosis. *EMBO J.* **17(14)**, 3878–3885.

26. Khaled, A. R., Kim, K., Hofmeister, R., Muegge, K., and Durum, S. K. (1999) Withdrawal of IL-7 induces Bax translocation from cytosol to mitochondria through a rise in intracellular pH. *Proc. Natl. Acad. Sci. USA* **96(25)**, 14476–14481.

27. Nechushtan, A., Smith, C. L., Hsu, Y. T., and Youle, R. J. (1999) Conformation of the Bax c-terminus regulates subcellular location and cell death. *EMBO J.* **18(9)**, 2330–2341.

28. Desagher, S., Osen-Sand, A., Nichols, A., Eskes, R., Montessuit, S., Lauper, S., et al. (1999) Bid-induced conformational change of Bax is responsible for mitochondrial cytochrome c release during apoptosis. *J. Cell Biol.* **144(5)**, 891–901.

29. Wang, K., Yin, X. M., Chao, D. T., Milliman, C. L., and Korsmeyer, S. J. (1996) Bid: A novel BH3 domain-only death agonist. *Genes Dev.* **10(22)**, 2859–2869.

30. Oltvai, Z. N. and Korsmeyer, S. J. (1994) Checkpoints of dueling dimers foil death wishes (comment). *Cell* **79(2)**, 189–192.

31. Ke, N., Godzik, A., and Reed, J. C. (2001) Bcl-b, a novel Bcl-2 family member that differentially binds and regulates Bax and Bak. *J. Biol. Chem.* **276(16)**, 12481–12484.

32. Hsu, S. Y., Kaipia, A., McGee, E., Lomeli, M., and Hsueh, A. J. (1997) Bok is a pro-apoptotic Bcl-2 protein with restricted expression in reproductive tissues and heterodimerizes with selective anti-apoptotic Bcl-2 family members. *Proc. Natl. Acad. Sci. USA* **94(23)**, 12401–12406.

33. Wei, M. C., Lindsten, T., Mootha, V. K., Weiler, S., Gross, A., Ashiya, M., et al. (2000) tBid, a membrane-targeted death ligand, oligomerizes Bak to release cytochrome c. *Genes Dev.* **14(16)**, 2060–2071.

34. Tan, K. O., Tan, K. M., Chan, S. L., Yee, K. S., Bevort, M., Ang, K. C., and Yu, V. C. (2001) Map-1, a novel proapoptotic protein containing a BH3-like motif that associates with Bax through its Bcl-2 homology domains. *J. Biol. Chem.* **276(4)**, 2802–2807.

35. Marani, M., Tenev, T., Hancock, D., Downward, J., and Lemoine, N. R. (2002) Identification of novel isoforms of the BH3 domain protein Bim which directly activate Bax to trigger apoptosis. *Mol. Cell. Biol.* **22(11)**, 3577–3589.

36. Diaz, J. L., Oltersdorf, T., Horne, W., McConnell, M., Wilson, G., Weeks, S., et al. (1997) A common binding site mediates heterodimerization and homodimerization of Bcl-2 family members. *J. Biol. Chem.* **272(17)**, 11350–11355.

37. Eskes, R., Desagher, S., Antonsson, B., and Martinou, J. C. (2000) Bid induces the oligomerization and insertion of Bax into the outer mitochondrial membrane. *Mol. Cell Biol.* **20(3)**, 929–935.

38. Yasuda, M. and Chinnadurai, G. (2000) Functional identification of the apoptosis effector BH3 domain in cellular protein bnip1. *Oncogene* **19(19)**, 2363–2367.

39. Cheng, E. H., Levine, B., Boise, L. H., Thompson, C. B., and Hardwick, J. M. (1996) Bax-independent inhibition of apoptosis by Bcl-x$_L$. *Nature* **379(6565)**, 554–556.

40. Petros, A. M., Medek, A., Nettesheim, D. G., Kim, D. H., Yoon, H. S., Swift, K., et al. (2001) Solution structure of the antiapoptotic protein Bcl-2. *Proc. Natl. Acad. Sci. USA* **98(6)**, 3012–3017.

41. Zha, J., Harada, H., Yang, E., Jockel, J., and Korsmeyer, S. J. (1996) Serine phosphorylation of death agonist bad in response to survival factor results in binding to 14-3-3 not Bcl-x$_L$. *Cell* **87(4)**, 619–628.

42. Puthalakath, H., VILlunger, A., O'ReILly, L. A., Beaumont, J. G., Coultas, L., Cheney, R. E., et al. (2001) Bmf: a proapoptotic BH3-only protein regulated by interaction with the myosin V actin motor complex, activated by anoikis. *Science* **293(5536)**, 1829–1832.

43. Puthalakath, H., Huang, D. C., O'ReILly, L. A., King, S. M., and Strasser, A. (1999) The proapoptotic activity of the Bcl-2 famILy member bim is regulated by interaction with the dynein motor complex. *Mol. Cell* **3(3)**, 287–296.

44. Chen, F., Hersh, B. M., Conradt, B., Zhou, Z., Riemer, D., Gruenbaum, Y., and Horvitz, H. R. (2000) Translocation of *C. elegans* CED-4 to nuclear membranes during programmed cell death. *Science* **287(5457)**, 1485–1489.

45. Shibasaki, F., Kondo, E., Akagi, T., and McKeon, F. (1997) Suppression of signalling through transcription factor NF-AT by interactions between calcineurin and Bcl-2. *Nature* **386(6626)**, 728–731.

46. Biswas, R. S., Cha, H. J., Hardwick, J. M., and Srivastava, R. K. (2001) Inhibition of drug-induced FAS ligand transcription and apoptosis by Bcl-x$_L$. *Mol. Cell Biochem.* **225(1–)**, 7–20.

47. Shimizu, S., Narita, M., and Tsujimoto, Y. (1999) Bcl-2 famILy proteins regulate the release of apoptogenic cytochrome c by the mitochondrial channel VDAC. *Nature* **399(6735)**, 483–487.

48. Zhang, H., Xu, Q., Krajewski, S., Krajewska, M., Xie, Z., Fuess, S., et al. (2000) BAR: an apoptosis regulator at the intersection of caspases and Bcl-2 family proteins. *Proc. Natl. Acad. Sci. USA* **97(6)**, 2597–2602.

49. Chau, B. N., Cheng, E. H., Kerr, D. A., and Hardwick, J. M. (2000) Aven, a novel inhibitor of caspase activation, binds Bcl-x$_L$ and Apaf-1. *Mol. Cell* **6(1)**, 31–40.

50. Ng, F. W., Nguyen, M., Kwan, T., Branton, P. E., Nicholson, D. W., Cromlish, J. A., and Shore, G. C. (1997) p28 Bap31, a Bcl-2/Bcl-x$_L$- and procaspase-8-associated protein in the endoplasmic reticulum. *J. Cell Biol.* **139(2)**, 327–338.

51. Muchmore, S. W., Sattler, M., Liang, H., Meadows, R. P., Harlan, J. E., Yoon, H. S., et al. (1996) X-ray and NMR structure of human Bcl-x$_L$, an inhibitor of programmed cell death. *Nature* **381(6580)**, 335–341.

52. Sattler, M., Liang, H., Nettesheim, D., Meadows, R. P., Harlan, J. E., Eberstadt, M., et al. (1997) Structure of Bcl-x$_L$/Bak peptide complex: recognition between regulators of apoptosis. *Science* **275(5302)**, 983–986.

53. Chou, J., Li, H., Salvesen, G., Yuan, J., and Wagner, G. (1999) Solution structure of Bid, an intracellular amplifier of apoptotic signaling. *Cell* **96(5)**, 615–624.

54. McDonnell, J., Fushman, D., MILliman, C., Korsmeyer, S., and Cowburn, D. (1999) Solution structure of the proapoptotic molecule Bid: a structural basis for apoptotic agonist and antagonists. *Cell* **96(5)**, 625–634.

55. Suzuki, M., Youle, R. J., and Tjandra, N. (2000) Structure of Bax: coregulation of dimer formation and intracellular localization. *Cell* **103(4)**, 645–654.

56. Huang, Q., Petros, A. M., Virgin, H. W., Fesik, S. W., and Olejniczak, E. T. (20002) Solution structure of a Bcl-2 homolog from kaposi sarcoma virus. *Proc. Natl. Acad. Sci. USA* **99(6)**, 3428–3433.

57. Cheng, E. H., Kirsch, D. G., Clem, R. J., Ravi, R., Kastan, M. B., Bedi, A., et al. (1997) Conversion of Bcl-2 to a Bax-like death effector by caspases. *Science* **278(5345)**, 1966–1968.

58. Clem, R. J., Cheng, E. H., Karp, C. L., Kirsch, D. G., Ueno, K., Takahashi, A., et al. (1998) Modulation of cell death by Bcl-x$_L$ through caspase interaction. *Proc. Natl. Acad. Sci. USA* **95(2)**, 554–559.

59. Oda, E., Ohki, R., Murasawa, H., Nemoto, J., Shibue, T., Yamashita, T., et al. (2000) Noxa, a BH3-only member of the Bcl-2 famILy and candidate mediator of p53-induced apoptosis. *Science* **288(5468)**, 1053–1058.

60. Yu, J., Zhang, L., Hwang, P. M., Kinzler, K. W., and Vogelstein, B. (2001) PUMA induces the rapid apoptosis of colorectal cancer cells. *Mol. Cell* **7(3)**, 673–682.

61. Nakano, K. and Vousden, K. H. (2001) PUMA, a novel proapoptotic gene, is induced by p53. *Mol. Cell.* **7(3)**, 683–694.

62. Sanchez, I. and Yuan, J. (2001) A convoluted way to die. *Neuron* **29(3)**, 563–566.

63. Chao, J. R., Wang, J. M., Lee, S. F., Peng, H. W., Lin, Y. H., Chou, C. H., et al. (1998) Mcl-1 is an immediate-early gene activated by the granulocyte- macrophage colony-stimulating factor (GM-CSF) signaling pathway and is one component of the GM-CSF viabILity response. *Mol. Cell Biol.* **18(8)**, 4883–4898.

64. Moulding, D. A., Quayle, J. A., Hart, C. A., and Edwards, S. W. (1998) Mcl-1 expression in human neutrophILs: regulation by cytokines and correlation with cell survival. *Blood* **92(7)**, 2495–2502.

65. Karsan, A., Yee, E., Kaushansky, K., and Harlan, J. M. (1996) Cloning of human Bcl-2 homologue: inflammatory cytokines induce human A1 in cultured endothelial cells. *Blood* **87(8)**, 3089–3096.

66. Chuang, P. I., Yee E., Karsan A., Winn R. K., and Harlan, J. M. (1998) A1 is a constitutive and inducible Bcl-2 homologue in mature human neutrophILs. *Biochem. Biophys. Res. Commun.* **249(2)**, 361–365.

67. Hu, X., Yee, E., Harlan, J. M., Wong, F., and Karsan, A. (1998) Lipopolysaccharide induces the antiapoptotic molecules, A1 and A2o, in microvascular endothelial cells. *Blood* **92(8)**, 2759–2765.

68. Chao, D. T. and Korsmeyer, S. J. (1998) Bcl-2 famILy: regulators of cell death. *Annu. Rev. Immunol.* **16**, 395–419.

69. Boise, L. H., Minn, A. J., Noel, P. J., June, C. H., Accavitti, M. A., Lindsten, T., and Thompson, C. B. (1995) CD28 costimulation can promote T cell survival by enhancing the expression of Bcl-x$_L$. *Immunity* **3(1)**, 87–98.

70. Boise, L. H., Gonzalez-Garcia, M., Postema, C. E., Ding, L., Lindsten, T., Turka, L. A., et al. (1993) Bcl-x, a Bcl-2-related gene that functions as a dominant regulator of apoptotic cell death. *Cell* **74(4)**, 597–608.

71. Bingle, C. D., Craig, R. W., Swales, B. M., Singleton, V., Zhou P., and Whyte, M. K. B. (2000) Exon skipping in Mcl-1 results in a Bcl-2 homology Domain 3 only gene product that promotes cell death. *J. Biol. Chem.* **275(29)**, 22136–22146.

72. Hockenbery, D. M., Oltvai, Z. N., Yin, X. M., Milliman, C. L., and Korsmeyer, S. J. (1993) Bcl-2 functions in an antioxidant pathway to prevent apoptosis. *Cell* **75(2)**, 241–251.

73. O'Connor, L., Strasser, A., O'Reilly, L. A., Hausmann, G., Adams, J. M., Cory, S., and Huang, D. C. (1998) Bim: a novel member of the Bcl-2 family that promotes apoptosis. *EMBO J.* **17(2)**, 384–395.

74. Guo, B., Godzik, A., and Reed, J. C. (2001) Bcl-G, a novel pro-apoptotic member of the Bcl-2 famILy. *J. Biol. Chem.* **276(4),** 2780–2785.

74a. Griffiths, G. J., Dubrez, L., Morgan, C. P., Jones, N. A., Whitehouse, J., Corfe, B., et al. (1999) Cell damage-induced conformational changes of the pro-apoptotic protein Bak in vivo precede the onset of apoptosis. *J. Cell Biol.* **144(5),** 903-914.

75. Zhao, Y., Li, S., ChILds, E. E., Kuharsky, D. K., and Yin, X. M. (2001) Activation of prodeath Bcl-2 family proteins and mitochondria apoptosis pathway in tnfa-induced liver injury. *J. Biol. Chem.* **276(29),** 27432–27440.

76. Li, H., Zhu, H., Xu, C. J., and Yuan, J. (1998) Cleavage of Bid by caspase 8 mediates the mitochondrial damage in the fas pathway of apoptosis. *Cell* **94(4),** 491–501.

77. Luo, X., Budihardjo, I., Zou, H., Slaughter, C., and Wang, X. (1998) Bid, a Bcl-2 interacting protein, mediates cytochrome c release from mitochondria in response to activation of cell surface death receptors. *Cell* **94(4),** 481–490.

78. Gross, A., Yin, X. M., Wang, K., Wei, M. C., Jockel, J., Milliman, C., et al. (1999) Caspase cleaved Bid targets mitochondria and is required for cytochrome c release, while Bcl-x$_L$ prevents this release but not Tumor Necrosis Factor-r1/Fas death. *J. Biol. Chem.* **274(2),** 1156–1163.

79. Stoka, V., Turk, B., Schendel, S. L., Kim, T.-H., Cirman, T., Snipas, S. J., et al. (2001) Lysosomal protease pathways to apoptosis. Cleavage of Bid, not pro-caspases, is the most likely route. *J. Biol. Chem.* **276(5),** 3149–3157.

80. Zha, J., Weiler, S., Oh, K. J., Wei, M. C., and Korsmeyer, S. J. (2000) Posttranslational N-myristoylation of Bid as a molecular switch for targeting mitochondria and apoptosis. *Science* **290(5497),** 1761–1765.

81. Zha, J., Harada, H., Osipov, K., Jockel, J., Waksman, G., and Korsmeyer, S. J. BH3 domain of Bad is required for heterodimerization with Bcl-x$_L$ and pro-apoptotic activity. *J. Biol. Chem.* **272(39),** 24101–24104.

82. Datta, S. R., Dudek, H., Tao, X., Masters, S., Fu, H., Gotoh, Y., and Greenberg, M. E. (1997) Akt phosphorylation of Bad couples survival signals to the cell-intrinsic death machinery. *Cell* **91(2),** 231–241.

83. del Peso, L., Gonzalez-Garcia, M., Page, C., Herrera, R., and Nunez, G. (1997) Interleukin-3-induced phosphorylation of Bad through the protein kinase AKT. *Science* **278(5338),** 687–689.

84. Harada, H., Becknell, B., WILm, M., Mann, M., Huang, L. J., Taylor, S. S., et al. (1999) Phosphorylation and inactivation of Bad by mitochondria-anchored protein kinase A. *Mol. Cell* **3(4),** 413–422.

85. Wang, H. G., Pathan, N., Ethell, I. M., Krajewski, S., Yamaguchi, Y., Shibasaki, F., et al. (1999) Ca^{2+}-induced apoptosis through calcineurin dephosphorylation of Bad. *Science* **284(5412),** 339–343.

86. Fang, G., Chang, B. S., Kim, C. N., Perkins, C., Thompson, C. B., and Bhalla, K. N. (1998) "Loop" domain is necessary for taxol-induced mobILity shift and phosphorylation of Bcl-2 as well as for inhibiting taxol-induced cytosolic accumulation of cytochrome c and apoptosis. *Cancer Res.* **58(15),** 3202–3208.

87. Poruchynsky, M. S., Wang, E. E., Rudin, C. M., Blagosklonny, M. V., and Fojo, T. (1998) Bcl--x$_L$ is phosphorylated in malignant cells following microtubule disruption. *Cancer Res.* **58(15),** 3331–3338.

88. Yamamoto, K., Ichijo, H., and Korsmeyer, S. J. (1999) Bcl--2 is phosphorylated and inactivated by an ASK1/JUN N-terminal protein kinase pathway normally activated at G(2)/M. *Mol. Cell Biol.* **19(12),** 8469–8478.

89. Haldar, S., Chintapalli, J., and Croce, C. M. (1996) Taxol induces Bcl-2 phosphorylation and death of prostate cancer cells. *Cancer Res.* **56(6),** 1253–1255.

90. Desagher, S., Osen-Sand, A., Montessuit, S., Magnenat, E., Vilbois, F., Hochmann, A., et al. (2001) Phosphorylation of Bid by casein kinases I and II regulates its cleavage by caspase 8. *Mol. Cell* **8(3),** 601–611.

91. Zhou, X. M., Liu, Y., Payne, G., Lutz, R. J., and Chittenden, T. (2000) Growth factors inactivate the cell death promoter Bad by phosphorylation of its BH3 domain on ser155. *J. Biol. Chem.* **275(32),** 25046–25051.

92. Datta, S. R., Katsov, A., Hu, L., Petros, A., Fesik, S. W., Yaffe, M. B., and Greenberg, M. E. (2000) 14–3–3 proteins and survival kinases cooperate to inactivate Bad by BH3 domain phosphorylation. *Mol. Cell.* **6(1),** 41–51.

93. Akao, Y., Otsuki, Y., Kataoka, S., Ito, Y., and Tsujimoto, Y. (1994) Multiple subcellular localization of Bcl-2: Detection in nuclear outer membrane, endoplasmic reticulum membrane, and mitochondrial membranes. *Cancer Res.* **54(9),** 2468–2471.

94. Green, D. R. and Reed, J. C. (1998) Mitochondria and apoptosis. *Science* **281(5381),** 1309–1312.

95. Vander Heiden, M. G., Chandel, N. S., Schumacker, P. T., and Thompson, C. B. (1999) Bcl--xl prevents cell death following growth factor withdrawal by facILitating mitochondrial ATP/ADP exchange. *Mol. Cell* **3(2),** 159–167.

96. Lindsten, T., Ross, A. J., King, A., Zong, W., Rathmell, J. C., Shiels, H. A., et al. (2000) The combined functions of proapoptotic Bcl-2 famILy members Bak and Bax are essential for normal development of multiple tissues. *Mol. Cell.* **6(6),** 1389–1399.

97. Zhang, L., Yu, J., Park, B. H., Kinzler, K. W., and Vogelstein, B. (2000) Role of Bax in the apoptotic response to anticancer agents. *Science* **290(5493),** 989–992.

98. LeBlanc, H., Lawrence, D., Varfolomeev, E., Totpal, K., Morlan, J., Schow, P., et al. (2002) Tumor-cell resistance to death receptor-induced apoptosis through mutational inactivation of the proapoptotic Bcl-2 homolog Bax. *Nat. Med.* **8(3),** 274–281.

99. Deckwerth, T. L., Elliott, J. L., Knudson, C. M., Johnson, Jr., E. M., Snider, W. D., and Korsmeyer, S. J. (1996) Bax is required for neuronal death after trophic factor deprivation and during development. *Neuron* **17(3),** 401–411.

100. Yang, E., Zha, J., Jockel, J., Boise, L. H., Thompson, C. B., and Korsmeyer, S. J. (1995) Bad, a heterodimeric partner for Bcl-x$_L$ and Bcl-2, displaces Bax and promotes cell death. *Cell* **80(2),** 285–291.

101. Saunders, J. W., Jr. (1966) Death in embryonic systems. *Science* **154(749),** 604–612.

102. Horvitz, H. R., Shaham, S., and Hengartner, M. O. (1994) The genetics of programmed cell death in the nematode *Caenorhabditis elegans*. Cold Spring Harb. Symp. Quant. Biol. **59,** 377–385.

103. Hengartner, M. O., Ellis, R. E., and Horvitz, H. R. (1992) *Caenorhabditis elegans* gene *ced-9* protects cells from programmed cell death. *Nature* **356(6369),** 494–499.
104. Motoyama, N., Wang, F., Roth, K. A., Sawa, H., Nakayama, K., Negishi, I., et al. (1995) Massive cell death of immature hematopoietic cells and neurons in Bcl-x-deficient mice. *Science* **267(5203),** 1506–1510.
105. Rinkenberger, J. L., Horning, S., Klocke, B., Roth, K., and Korsmeyer, S. J. (2000) Mcl-1 deficiency results in peri-implantation embryonic lethality. *Genes Dev.* **14(1),** 23–27.
106. Veis, D. J., Sorenson, C. M., Shutter, J. R., and Korsmeyer, S. J. (1993) Bcl--2-deficient mice demonstrate fulminant lymphoid apoptosis, polycystic kidneys, and hypopigmented hair. *Cell* **75(2),** 229–240.
107. Print, C. G., Loveland, K. L., Gibson L., Meehan T., Stylianou A., Wreford N., et al. (1998) Apoptosis regulator Bcl-w is essential for spermatogenesis but appears otherwise redundant. *Proc. Natl. Acad. Sci. USA* **95(21),** 12424–12431.
108. Knudson, C. M., Tung, K. S., Tourtellotte, W. G., Brown, G. A., and Korsmeyer, S. J. (1995) Bax-deficient mice with lymphoid hyperplasia and male germ cell death. *Science* **270(5233),** 96–99.
109. Hockenbery, D. M., Zutter, M., Hickey, W., Nahm, M., and Korsmeyer, S. J. (1991) Bcl-2 protein is topographically restricted in tissues characterized by apoptotic cell death. *Proc. Natl. Acad. Sci. USA* **88(16),** 6961–6965.
110. Nunez, G., Hockenbery, D., McDonnell, T. J., Sorensen, C. M., and Korsmeyer, S. J. (1991) Bcl--2 maintains β cell memory. *Nature* **353(6339),** 71–73.
111. Yang, E. and Korsmeyer, S. J. (1996) Molecular thanatopsis: a discourse on the Bcl2 famILy and cell death. *Blood* **88(2),** 386–401.
112. Korsmeyer, S. J. (1992) Bcl--2: a repressor of lymphocyte death. *Immunol. Today* **13(8),** 285–288.
113. Strasser, A., Harris, A. W., Bath, M. L., and Cory, S. (1990) Novel primitive lymphoid tumours induced in transgenic mice by cooperation between myc and Bcl-2. *Nature* **348(6299),** 331–333.
114. McDonnell, T. J. and Korsmeyer, S. J. (1991) Progression from lymphoid hyperplasia to high-grade malignant lymphoma in mice transgenic for the t(14; 18). *Nature* **349(6306),** 254–256.
115. Miyashita, T. and Reed, J. C. (1995) Tumor suppressor p53 is a direct transcriptional activator of the human Bax gene. *Cell* **80(2),** 293–299.
116. Selvakumaran, M., Lin, H. K., Miyashita, T., Wang, H. G., Krajewski, S., Reed, J. C., et al. (1994) Immediate early up-regulation of bax expression by p53 but not TGF beta 1: A paradigm for distinct apoptotic pathways. *Oncogene* **9(6),** 1791–1798.
117. Schuler, M., Bossy-Wetzel, E., Goldstein, J. C., Fitzgerald, P., and Green, D. R. (2000) p53 induces apoptosis by caspase activation through mitochondrial cytochrome c release. *J. Biol. Chem.* **275(10),** 7337–7342.
118. Schmitt, C. A., Fridman, J. S., Yang, M., Baranov, E., Hoffman, R. M., and Lowe, S. W. (2002) Dissecting p53 tumor suppressor functions in vivo. *Cancer Cell* **1(3),** 289–298.
119. Boise, L. H., Minn, A. J., and Thompson, C. B. (1995) Receptors that regulate T-cell susceptibILity to apoptotic cell death. *Ann. NY Acad. Sci.* **766,** 70–80.
120. Hildeman, D. A., Zhu, Y., Mitchell, T. C., Kappler, J., and Marrack, P. (2002) Molecular mechanisms of activated T cell death in vivo. *Curr. Opin. Immunol.* **14(3),** 354–359.
121. Bouillet, P., Metcalf, D., Huang, D. C., Tarlinton, D. M., Kay, T. W., Kontgen, F., et al. (1999) Proapoptotic Bcl-2 relative Bim required for certain apoptotic responses, leukocyte homeostasis, and to preclude autoimmunity. *Science* **286(5445),** 1735–1738.
122. BouILlet, P., Purton, J. F., Godfrey, D. I., Zhang, L. C., Coultas, L., Puthalakath, H., et al. (2002) BH3–only Bcl-2 family member Bim is required for apoptosis of autoreactive thymocytes. *Nature* **415(6874),** 922–926.
123. Zong, W. X., Edelstein, L. C., Chen, C., Bash, J. and Gelmas C. (1999) The prosurvival Bcl-2 homolog Bfl-1/A1 is a direct transcriptional target of NF-kappaβ that blocks TNfalpha-induced apoptosis. *Genes Dev.* **13(4),** 382–387.
124. Yin, X. M., Wang, K., Gross, A., Zhao, Y., Zinkel, S., Klocke, B., Roth, K. A., and Korsmeyer, S. J. (1999) Bid-deficient mice are resistant to Fas-induced hepatocellular apoptosis. *Nature* **400(6747),** 886–891.
125. Knudson, C. M., Johnson, G. M., Lin, Y., and Korsmeyer, S. J. (2001) Bax accelerates tumorigenesis in p53-deficient mice. *Cancer Res.* **61(2),** 659–665.
126. Chattopadhyay, A., Chiang, C. W., and Yang, E. (2001) BAD/Bcl-x$_L$ heterodimerization leads to bypass of G0/G1 arrest. *Oncogene* **20(33),** 4507–4518.

3

Inhibitors of Apoptosis
Proteins

Peter Liston, Wai Gin Fong, and Robert G. Korneluk

INTRODUCTION

The preceding decade has witnessed a genuine revolution in our understanding of the regulation of cell homeostasis in health and disease. Apoptosis has emerged as the fundamental mechanism of cell loss in most, if not all, genetic disorders and injury states, and has also resulted in a fundamental change in our understanding of the molecular basis of cancer. At the molecular level, the evolving description of mitochondrial dysfunction and caspase activation as central elements of apoptosis have transpired to challenge our underlying assumptions of cellular function. Of the many break-through discoveries over the preceding decade, the discovery and characterization of the caspases represents a milestone in our understanding of cellular processes. Consequently, the control of caspase activity now holds tremendous therapeutic potential and is under intense investigation in academic and pharmaceutical company laboratories worldwide.

Activation and control of the caspase cascade is central to several cellular processes, extending beyond classical cell death to include proliferation and terminal cellular differentiation. Controlled, limited caspase activation has been demonstrated to be essential for the proliferation of T lymphocytes (1,2), as well as in the generation of mature erythrocytes (3), monocytes (4), and epidermal cells (5). The maintenance of caspase control is thus essential for both homeostasis and cellular differentiation. The pivotal role of caspase activation in virtually all cell death events suggests that strategies that prevent or limit their activation will find utility in the treatment of numerous disease and injury processes. The inhibitors of apoptosis (IAPs) are proteins that function as intrinsic regulators of the caspase cascade. Cellular proteins have been identified that inhibit specific "upstream" or initiator caspases, but the IAPs are the only known endogenous proteins that regulate the activity of both initiator (caspase-9) and effector caspases (caspases-3 and -7). Controlled expression of the IAPs has been shown to influence cell death in a variety of contexts, including in vivo models of neurodegenerative disorders, as well as in hyper-proliferative disorders, such as cancer and autoimmune diseases.

From: *Essentials of Apoptosis: A Guide for Basic and Clinical Research*
Edited by: X-M. Yin and Z. Dong © Humana Press Inc., Totowa, NJ

Table 1

IAP	Alternative names	Domains	Caspase specificity	Other binding proteins	Chromosome location
NAIP	BIRC1	BIR(3) NOD LRR	Caspase-3, -7	Hippocalcin	5q13.1
XIAP	BIRC4 API3 MIHA ILP-1	BIR(3) RING	Caspase-3, -7, -9	Smac/DIABLO Omi/HtrA2 XAF1 TAB1 NRAGE	Xq25
c-IAP1	BIRC2 API1 MIHB HIAP2	BIR(3) CARD RING	Caspase-3, -7, -9	Smac/DIABLO Omi/HtrA2 TRAF1 TRAF2	11q22
c-IAP2	BIRC3 API2 MIHC HIAP2	BIR(3) CARD RING	Caspase -3, -7, -9	Smac/DIABLO Omi/HtrA2 TRAF1 TRAF2 Bcl10	11q22
Survivin	BIRC5 API4 TIAP (mouse)	BIR Coiled-coil	Caspase-3, -7	β-tubulin Smac	17q25
Livin	BIRC7 KIAP ML-IAP	BIR RING	Caspase-3,-7, -9	Smac	20q13.3
Ts-IAP	BIRC8 ILP-2	BIR RING	Caspase-9		19q11.3

DISCOVERY OF THE IAPS

The IAPs are a family of proteins characterized by one or more 70–80 amino acid Baculoviral IAP Repeat (BIR) domains. The BIR domain is a characteristic cysteine and histidine-rich protein folding domain that chelates zinc and forms a compact globular structure consisting of four or five alpha helices and a variable number of anti-parallel B-pleated sheets. The core of a BIR domain consists of the variable consensus sequence $C_{X2}C_{X6}$ $W_{X3}D_{X5}H_{X6}C$, where X is any amino acid. The first IAPs were identified in baculoviruses (reviewed in ref. *6*), but have since been extended to *Drosophila* and numerous vertebrate species. The characterization of IAP proteins indicates a role for these proteins as endogenous caspases inhibitors, as well as in participating in cell-cycle regulation and the modulation of receptor-mediated signal transduction. The subsequent identification of BIR-containing proteins in unicellular organisms, such as yeast, that do not possess a cell-death program, has necessitated a more strict definition of an IAP than the mere presence of a BIR domain(s). Members of the larger family are now termed BIRPs, for BIR containing proteins, of which two major subfamilies have been described (reviewed in ref. *7*). The IAP group consists of those members possessing one or more BIRs and to which have been ascribed an anti-apoptotic activity. Other members of the BIRP family generally contain only a single BIR domain, and appear to function in cytokinesis and chromatin segregation. This chapter reviews the structural and biochemical features of the IAPs, their regulation, their role in signal-transduction pathways, and their potential as therapeutic targets in a variety of disease states.

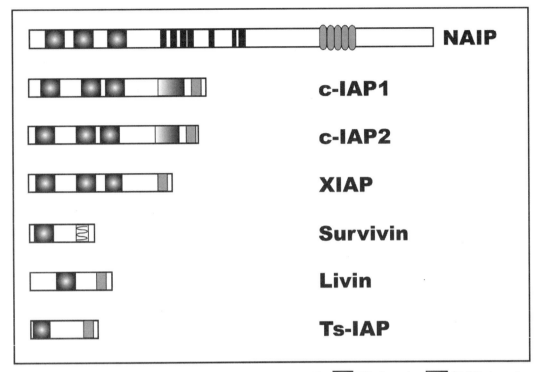

Proteins were drawn indicating various important motifs. ■ BIR domain, ▢ CARD domain, ▮ RING zinc finger, ▐▌▌▐ NOD motif, ▒ Leucine rich region and ▤ Coiled-coil domain

Fig. 1. The human IAP family.

STRUCTURE AND FUNCTIONAL DISSECTION OF THE IAPS

The concomitant identification of the IAPs in independent laboratories has led to multiple names being conferred on virtually all of the family members, and is summarized in Table 1. The most frequently cited nomenclature has been used in this chapter. The first mammalian IAP homolog to be identified was Neuronal Apoptosis Inhibitory Protein (NAIP), which was isolated during a positional cloning effort to identify the causative gene for spinal muscular atrophy (SMA, ref. *8*). In contrast to the baculoviral IAPs, which possess two BIR domains and a carboxy terminal RING zinc finger, NAIP has three BIR domains and a very large and unique carboxy terminus with homology to members of the NACHT subfamily of NOD (Nucleotide binding, Oligomerization Domain) proteins (reviewed in ref. *9*). SMA is a neuromuscular degenerative disease characterized by progressive loss of motor neurons leading to wasting of the voluntary muscles. Despite the fact that there are at least three phenotypic variants of SMA, (types I, II, and III), distinguished by the severity of symptoms and age of onset, a single genetic locus at chromosome 5q13 *(10)* had been implicated in all cases. The loss of an apoptotic inhibitor at this locus is in accordance with pathological continuation of neuronal apoptosis beyond that normally required for the generation of a functional neuromusculature. Concurrent with the isolation of *naip*, a second candidate gene, Survival Motor Neuron (*smn*; *11*), was identified in the immediate chromosomal vicinity. Although *naip* deletion has since been ruled out as the primary mutation event in SMA, deletions of the neighboring *smn* gene that encompass *naip* appear to lead to greater disease severity, suggesting that NAIP may have a modulating role (reviewed in ref. *12*). Subsequent to the identification of NAIP, the IAP family was rapidly expanded with the identification of c-IAP1 (cellular IAP1), c-IAP2, and XIAP (X-linked IAP), all of which contained three BIR domains and a carboxy terminal RING finger *(13–16)*. The family continued to expand with the identification of Survivin (single BIR, carboxy-terminal coiled-coil domain; *17*), Livin (single BIR and carboxy terminal RING finger; *18–20*), and Ts-IAP (Testis-specific IAP; single BIR and carboxy terminal RING finger; *21,22*). Table 1 summarizes the names, domain structure, and interactions of the IAP family members, which are illustrated schematically in Fig. 1.

Function of the BIR Domains

Cell-culture transfection experiments initially indicated that the IAPs could suppress or delay apoptosis in a wide variety of cell-death paradigms (reviewed in refs. *23,24*). Without any understanding of mechanism, results showed that both intrinsic and extrinsic pathways of apoptosis were inhibited. At the same time, the importance of the caspases was emerging with the identification and characterization of the multiple mammalian homologs of the *Caenorhabditis elegans* ICE-like protease CED-3. It was not until 1997 that the two stories were reconciled with the demonstration that the IAPs could directly inhibit some of the caspases *(25,26)*. Subsequent work established that the BIR domains alone could inhibit the caspases *(27)*, and in some cases, deletion of the carboxy terminal RING finger actually improved the potency of the protein.

The BIR domains of the IAPs are the most characterized functional units of the IAPs. Each BIR domain folds into a functionally independent structure that chelates a zinc ion, and consists of a globular head and an unstructured tail derived from the amino terminal "linker" region located upstream of the individual BIR domains (*see* Chapter 4). The majority of IAP interactions have been mapped to the BIR domains, including inhibition of caspases, binding of c-IAP1 and 2 to tumor necrosis factor (TNF) receptor associated factors (TRAFs), and XIAP interaction with the TAB1 protein. Specific interactions with initiator (–9) and effector (-3 and -7) caspases have been mapped to individual BIR domains. In general, IAPs with multiple BIR motifs use the third BIR domain to inhibit caspase-9, and the second BIR domain functions to inhibit caspases-3 and -7 *(26–28)*. Within the single BIR-containing proteins, Ts-IAP inhibits caspase-9 *(21)*, whereas the single Survivin BIR domain inhibits caspases-3 and -7 *(29)*. Interestingly, the single BIR domain in Livin has been reported to inhibit caspases -3, -7, and -9 *(18,20)*, and thus appears to have a broader range of activity than any other single BIR domain.

Despite the overall similarity of the BIR domains, the mechanisms of caspase inhibition differ significantly. Most of these studies have focused on XIAP, but are believed to hold true for the other IAPs as well. X-ray crystallographic *(30–32)* as well as mutagenesis studies *(33,34)* on XIAP BIR2 complexed with caspases-3 or -7 revealed that the linker region proximal to BIR2 is far more important than the BIR domain itself. The linker region is stretched across the active site of caspases-3 or -7 and precludes substrate entry. Furthermore, the peptide is positioned in the reverse orientation relative to peptide inhibitors and natural substrates of the caspases, and is thus very distinct from other known protease inhibitors. Although caspases-3 and -7 are very closely related, and the linker region inhibits both enzyme catalytic sites, there are differences. The linker region accounts entirely for the inhibition of caspases-3, and BIR2 can be replaced by an irrelevant protein such as GST *(31,32)*. In contrast, the XIAP BIR2 domain makes contact with the amino terminus of capase-7 and stabilizes the interaction of the linker in the catalytic groove. As a consequence, XIAP-mediated inhibition of caspase-3 is competitive, whereas inhibition of caspase-7 occurs by both a competitive and noncompetitive mechanism *(35)*. The domain structure and interactions of XIAP are shown schematically in Fig. 2.

The BIR3 domains of XIAP, c-IAP1, c-IAP2 *(36,37)*, and the single BIR domains in Livin *(20)* and Ts-IAP *(21)* have been demonstrated to bind and inhibit caspase-9. The mechanism of caspase-9 inhibition has again been best-characterized for the XIAP protein, and is very distinct from the bind-

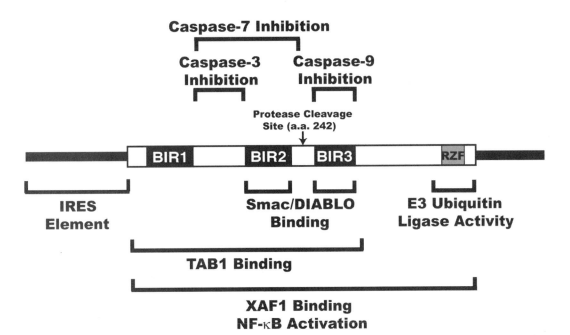

Fig. 2. Functional map of XIAP interactions.

ing to caspases-3 or -7. Caspase-9 becomes catalytically active through a conformational change when bound by Apaf1 (Apoptosis Protease Activating Factor-1), and thus appears to be distinct among the caspases in that it lacks a requirement for proteolytic activation. In addition to its ability to proteolytically process caspase-3, caspase-9 can undergo a self-cleavage event in the linker region between the p20 and p10 subunits at Asp_{315}. The XIAP BIR3 domain directly engages caspase-9 by binding to this newly exposed amino terminus, thereby occluding the caspase active site. A second cleavage event can be catalyzed by caspase-3 at Asp_{330}, 15 amino acids from the initial cleavage site. This feedback proteolysis does not alter the enzymatic activity of caspase-9, but rather removes the peptide sequence that binds XIAP, thereby preventing IAP inhibition (38).

There are thus distinctly different mechanisms of inhibition for each of the three known caspase targets of the IAPs. Caspase-3 is inhibited by the BIR1–2 linker region exclusively, caspase-7 is inhibited by a combination of the BIR1–2 linker and the BIR2 domain, and caspase-9 is inhibited by BIR3, without the apparent contribution of the upstream linker region. Finally, it should be noted that the BIR1 domains of XIAP, c-IAP1, c-IAP2, and NAIP, do not display any caspase-inhibiting activity, and are the least conserved of the three BIR domains. Nevertheless, their presence in each of these IAPs suggests that some unknown function remains to be identified.

Function of the RING Finger

RING fingers are a subclass of zinc-finger domains that chelate two zinc ions in a characteristic cross-brace arrangement, and are usually found at the amino terminus of proteins that function as E3 ubiquitin ligases. RING finger-containing proteins function as adapters, recruiting target proteins to a multi-component complex containing an E2 enzyme, and provide specificity for ubiquitin-conjugating activity, and hence proteosomal degradation (reviewed in ref. 39). Initial studies on the RING-finger domains of XIAP and c-IAP1 suggest that this domain provokes the ubiquitination and degradation of IAP proteins in response to apoptotic stimuli (Fig. 2). Treatment of cells with glucocorticoids or etoposide results in the rapid degradation of these IAPs, which can be blocked by proteosome inhibitors (40). Later work established that c-IAP2 and XIAP can trigger the ubiquitination of caspases-3 and -7 *(41,42)*, suggesting that targeting of caspases to the proteosome may be one of the anti-apoptotic mechanisms of the IAPs. It remains unclear whether the RING finger domain actually contributes to the anti-apoptotic activity of the protein, or whether it in fact antagonizes this activity, and conflicting results have been obtained using RING-finger deletion mutants of the IAPs *(40,42,43)*. A third possibility is that the RING-finger functions to suppress apoptosis in conditions of low apoptotic stimulus (via ubiquitination of caspases), whereas higher levels of apoptotic stress trigger self-degradation of the IAPs and hence apoptosis. Other potential targets of RING mediated ubiquitination, such as the XAF1, Smac, Omi, or the TNF receptors (*see* below) have not been investigated.

Function of CARD, NOD, and Coiled-Coil Domains in Some IAP Family Members

The CARD (Caspase Recruitment Domain) domain is a type of protein fold that typically mediates oligomerization with other CARD-containing proteins, as well as mediating homodimerization (reviewed in refs. *44,45*). Both c-IAP1 and c-IAP2 have a CARD domain between the BIR domains and the RING zinc finger (Fig. 1). The location of the CARD domain in the middle of these proteins is unusual. CARDs, death domains (DDs) and death effector domains (DEDs), which are all structurally related, are almost universally located in the amino terminus of the protein. To date, no function has been ascribed to the CARD domains of either c-IAP1 or c-IAP2.

The unique carboxy terminus of NAIP consists of a nucleotide-binding oligomerization domain (NOD) followed by a distal cluster of five leucine rich repeats (LRR) of 20–29 amino acids each. NOD-containing proteins are organized and function in a similar manner: the centrally located NOD domain mediates self-association, triggering induced proximity of proteins bound by the amino terminal domain (reviewed in ref. *46*). Activation of the NOD cassette is controlled by a carboxy terminal "sensor" domain. Apaf1 is the prototype of these proteins, in which cytochrome *c* binding to the WD40 repeats in the "sensor" domain triggers NOD-mediated oligomerization and a conformational change that exposes the amino terminal CARD domain for recruitment of caspase-9. Additional NOD proteins have been identified that recruit caspases (NOD1/CARD4, Ipaf/CLAN/CARD12) and/or activated the NF-κB pathway (Ipaf, NOD2/CARD15). Like NAIP, the "sensor" domains of NOD1, NOD2, and Ipaf1 are LRRs, and are structurally related to a series of plant pathogen resistance (R) proteins. All of these LRR proteins bind intracellular lipopolysaccharide (LPS) secreted by bacterial pathogens, and are involved in host responses such as cytokine production and NF-κB activation. The mouse *naip* gene cluster, which consists of at least six copies of *naip (47)*, maps to the legionella-susceptibility locus *(48)*, again providing intriguing indications that NAIP may be unique among the IAPs in playing a role in the host cell response to intracellular bacterial infection. We propose that NAIP binds LPS via the LRR domain, which triggers NOD oligomerization and exposure/activation of the BIR domains. Although this model accounts for the conservation of the carboxy terminal domain of NAIP, oligomerization and BIR domain activation in response to LPS has yet to be demonstrated.

The coiled-coil domain found in the carboxy terminus of Survivin projects from the "bow tie" structure of the dimeric Survivin protein *(49)*. Survivin is expressed at the G_2/M point in the cell cycle and associates with the mitotic spindle apparatus. This association is believed to be mediated by coiled-coil domain interaction with β-tubulin, and may recruit additional proteins to the kinetochore *(50)*. Single BIR containing proteins in *C. elegans* (BIR-1) as well as in yeast appear to play critical roles in cell-cycle progression through the G_2/M checkpoint. Interference with yeast BIR proteins results in aberrant mitosis and cytokinesis, resulting in multinucleated cells. In mammalian cells, inhibition of Survivin expression also results in polyploidy, aberrant mitosis, and defective cytokinesis. It was originally proposed that survivin monitors the success of mitosis and chromatin segregation, functioning to inhibit a default program that would otherwise activate caspases and apoptosis *(50)*. Although this is consistent with the embryonic lethal phenotype of the knockout mouse, and the expression of Survivin in many human cancers (reviewed in refs. *51,52*), it does not account for the chromosomal defects that occur in tissue-culture cells when Survivin expression is ablated. One would have predicted quite the opposite, in that removing a caspase inhibitor should result in massive apoptosis as soon as cell-cycle checkpoints are reached, and not the accumulation of multinucleated and polyploid cells. There are obviously many questions that need to be addressed in terms of caspase activation during cell-cycle progression and the role of nuclear caspases and IAPs.

PROTEOLYTIC PROCESSING

In addition to inhibiting caspases, XIAP and c-IAP1 can also serve as substrates for these proteases. Apoptosis triggered by a variety of mechanisms results in the appearance of proteolytic fragments of XIAP *(53)*. In vitro protease reactions suggest that a number of caspases, including caspases-3, -6, -7, and -8 can cleave XIAP. The initial processing event occurs at Asp_{242}, located in the linker region between BIR2 and BIR3, and generates two products; a BIR1–2 fragment, and a BIR3-RING fragment. The BIR1–2 product, when synthesized in bacteria, retains the ability to inhibit caspase-3, and the corresponding coding region expressed from a plasmid will inhibit Fas-induced apoptosis. This inhibition does not appear to be as effective as full-length XIAP protein, and the BIR1–2 fragment is further degraded to small peptide fragments during apoptosis. The BIR3-RING fragment appears to be longer-lived, and retains the ability to inhibit caspase-9 and Bax-induced apoptosis. The physiological significance of XIAP processing remains uncertain. A model was proposed in which the modular nature of the XIAP protein was capitalized on by separating the two functional regions of the protein, allowing independent targeting of caspases *(53)*. However, in all instances, the proteolytic fragments have only been observed in apoptotic cells *(53–55*; H. Gibson and R.G. Korneluk, unpublished observations), suggesting that overwhelming caspase activation resulting in XIAP cleavage is a way of eliminating these apoptotic inhibitors. Cells exposed to a sublethal apoptotic stress do not appear to accumulate XIAP fragments, suggesting that cleavage is not a significant pathway in enhancing XIAP-mediated protection. It is also noteworthy that XIAP fragments appear late in the apoptotic process, as much as 8 h after the initiation of apoptosis using anti-Fas antibody *(53)*. At this point, all of the morphological and biochemical characteristics of apoptosis are well-progressed, and the cell is past the point of no return.

c-IAP1 can also be cleaved by caspase-3, at Asp_{351}, just distal to the BIR3 domain. The evidence to date suggests that the carboxy terminal fragment, consisting of the CARD domain and RING finger, is pro-apoptotic *(43)*, whereas the amino terminal fragment consisting of the three BIR domains is rapidly degraded. Precedent for caspase cleavage converting an anti-apoptotic protein into a pro-apoptotic protein is well-established with the Bcl-2 and Bcl-X$_L$ proteins. Both the mechanism by which the CARD-RING fragment of c-IAP1 triggers apoptosis and the physiological significance of caspase cleavage is unknown.

INHIBITING THE INHIBITORS: NEGATIVE REGULATORS OF IAP FUNCTION

Currently, there are three proteins that have been identified that bind IAPs and suppress their activity. These proteins have been termed XAF1 (XIAP-Associated Factor1), Smac (Second mitochondrial Activator of Caspases, a.k.a Diablo), and Omi (a.k.a HtrA2). The XAF1 protein was identified by two-hybrid screening with XIAP, and encoded a novel, zinc finger-rich protein. In vitro experiments using purified, recombinant proteins demonstrate that XAF1 can directly bind XIAP and interfere with XIAP-mediated caspase-3 inhibition *(56)*. Cell-culture experiments using recombinant adenoviruses demonstrate that XAF1 reverses XIAP-mediated protection against chemotherapeutic drugs such as etoposide or cisplatin. XAF1 protein accumulates in the nucleus, whereas XIAP is predominantly cytosolic. However, XAF1 can trigger the re-localization of XIAP from the cytosol to the nucleus, perhaps as a means of sequestering XIAP. Interestingly, XAF1 is ubiquitously expressed in normal tissues, but found at extremely low levels (less than 1% of control normal tissues) in the majority of the NCI 60 cell-line panel of cancer cells *(57)*. This loss of XAF1 expression in transformed cells has been proposed to contribute to apoptosis suppression by allowing unrestricted IAP activity.

Unlike XAF1, the Smac protein resides in the mitochondria of healthy cells, and is released upon apoptotic stress with similar kinetics to cytochrome *c*. In the process of being released from the mitochondria, the 55 amino acid mitochondrial localization-signal peptide is proteolytically removed *(58,59)*. The mechanism of Smac release has not been entirely resolved. Treatment of apoptotic cells with caspase inhibitors allowed cytochrome *c* release, but blocked release of Smac, suggesting egress by separate mechanisms *(60)*. The relatively small cytochrome *c* (approx 12 kDa) protein has been proposed to escape through Bax- or Bak-formed membrane pores. By comparison, released of the Smac dimer (approx 100 kDa) may require further development of mitochondrial pathology secondary to caspase activation.

Smac has been demonstrated to bind all of the IAPs tested to date, including XIAP, c-IAP1, c-IAP2, Survivin *(58)*, and Livin *(61)*. Although Smac can bind to either BIR2 or BIR3 of XIAP, thereby interfering with either caspases-3 and -7 or caspase-9 inhibition, the binding is considerably stronger with BIR3 *(62,63)*. The crystal structure of Smac revealed that the protein forms a long, bridge-like structure consisting of three extended alpha helices bundled together, and an unstructured amino terminus *(64)* (*see* Chapter 4). Smac homodimers form via a large hydrophobic interface, and this homodimerization appears to be required for activity *(65)*. The unstructured, newly generated amino terminus of Smac makes critical contacts with XIAP BIR3 and mediates XIAP inhibition. Additional contacts with the helical bundles of Smac are predicted from the crystal structure, and are more significant in interactions with the BIR2 domain *(63)*. Co-crystalization of Smac and XIAP BIR3 established that the amino terminal tetrapeptide sequence of Smac (Ala-Val-Pro-Ile) fits within a surface groove of the BIR3 domain, with the alanine residue bound within a hydrophobic pocket. Some, but not all of Smac's ability to inhibit XIAP BIR3 function can be reconstituted with short peptides. The surface contacts that Smac makes with XIAP BIR3 overlapped completely with a surface map of BIR3/caspase-9 contacts *(34)*. Furthermore, the similarity to the cleavage site in the linker region of caspase-9 (Ala_{316}-Thr-Pro-Phe) suggested a competition model in which Smac competes for or displaces XIAP from caspase-9. A key component of this model was established when Asp_{316} cleavage site mutants of caspase-9 were established. Mutant caspase-9 is fully active, but cannot undergo proteolytic processing between the large and small subunits. XIAP cannot inhibit this enzyme, demonstrating that a critical feature of XIAP-mediated inhibition is interaction with the amino terminus of the linker region of partially processed caspase-9 *(38,66)*.

The identification of Smac also provided a critical insight into the function of the pro-apoptotic *Drosophila* proteins Reaper, Hid, and Grim. These proteins control virtually all cell death in the fly and can all be antagonized by DIAP1, but share little or no sequence conservation (reviewed in ref. *67*). However, the amino termini of these proteins all share homology with the Smac tetrapeptide motif, and have been shown to bind a similar groove on the surface of BIR2 of DIAP1 *(68)*.

Subsequent to the identification of Smac, a second mitochondrial IAP binding protein, called Omi or HtrA2, was identified by several groups *(69–73)*. Like Smac, the Omi protein is released from the mitochondria of apoptotic cells and is processed to generate a Smac-like tetrapeptide motif at the amino terminus. Direct binding to XIAP, and inhibition of XIAP-caspase interaction, appears to be only one of Omi's pro-apoptotic activities. The serine protease activity of Omi also contributes to cell death in a noncaspase-dependent manner, though the cellular targets of the protease activity have not been identified *(70,71,73)*. By analogy to the *Escherichia coli* htrA2 protein, it was predicted that the human Omi/HtrA2 protein would be involved in the proteolytic degradation of misfolded proteins under conditions of cellular stress. Although no cytosolic targets have been identified, proteolytic removal of the 155 amino acid mitochondrial targeting peptide of Omi is a self-catalyzed event *(74)*, raising the possibility that other mitochondrial leader sequences, including that of Smac, may be processed by Omi.

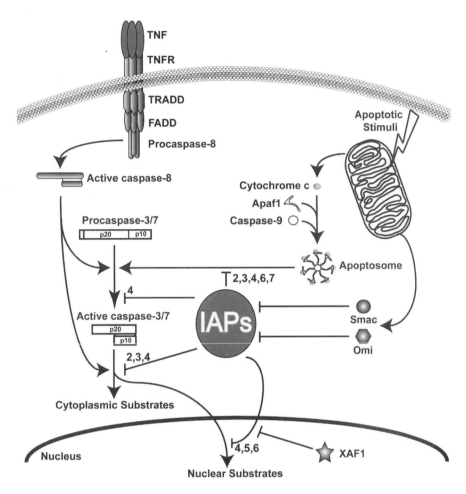

Fig. 3. Role of IAPs in apoptotic pathways. The known caspase inhibiting abilities of the various IAPs are shown. 1–NAIP, 2–c-IAP1, 3–c-IAP2, 4–XIAP, 5–Survivin, 6–Livin, 7–Ts-IAP.

In a simplified version of the overall process of cell death as it relates to caspase activation and the IAPs, the following model (Fig. 3) is proposed. Apoptotic stresses acting through the intrinsic pathway triggers the expression and/or activation of pro-apoptotic Bcl-2 family members (*see* Chapters 2 and 6). The balance of pro- and anti-apoptotic Bcl-2 proteins constitutes the first decision point. Given sufficient activation, the pro-apoptotic Bcl-2 proteins trigger the release of cytochrome *c*, presumably via channel formation in the outer mitochondrial membrane. The release of cytochrome *c* triggers a second decision point, in which levels of the IAPs determine the outcome. Pre-exposure to apoptotic stress (*see* below) or pre-existing high levels of IAP expression may suppress newly activated caspase-9 and any effector caspases that have become activated, and the ubiquitin ligase activity triggers the disposal of activated caspases via the proteosome. However, if sufficient caspase-9 activation occurs, caspases-3 and/or -7 are activated and removes the binding site for the IAPs on caspase-9. In addition, activated caspases trigger the further permeabilization of the mitochondrial membranes through the activation of the permeability transition pore, which allows the release of Apoptosis-Inducing Factor (AIF), Smac, and Omi. AIF transits to the nucleus and triggers chromatin condensation, while Smac and Omi bind and inhibit any further participation of the IAPs, thereby allowing unrestricted caspase activity to proceed. In addition, the IAPs are themselves degraded by the caspases (i.e., XIAP) or processed into pro-apoptotic fragments (i.e., c-IAP1).

SPECIALIZATION AND REGULATION: IAPS ARE NOT AS REDUNDANT AS THEY APPEAR

With the discovery that XIAP could inhibit only a subset of the caspases, the logical prediction was that the other IAPs would display high affinities for the remaining members of the caspase family. Quite the reverse has held true, and all IAPs appear to be restricted to inhibiting caspases-3 and -7, and/or -9, although their respective affinities do vary. This apparent redundancy ignores critical aspects of IAP expression and subcellular localization that make each IAP distinct. Some of the unique attributes of individual IAPs have been defined, while others no doubt remain to be discovered.

The most distantly related IAP, Survivin, is restricted to expression at the G_2/M point in the cell cycle, and is the only IAP to associate with chromatin structures *(50)*. Survivin thus appears to play a unique role in monitoring the success of chromosome replication and the suppression of caspase activity in the nucleus. Livin has also been reported to localize in the nucleus, but does not appear to be cell cycle-regulated nor does it associate with particular structures within the nucleus *(18)*. One possibility is that Livin serves a "housekeeping" function and prevents accidental caspase activation in the nucleus.

Ts-IAP is an autosomal, retrotransposed, intronless copy of XIAP, and is expressed solely in the testis *(21,22)*. There are examples of other X chromosome-linked genes having retrotransposed autosomal copies. The PGK gene is one such example, in which the chromosome 19 copy contains no introns and is expressed under the control of a promoter unrelated to the X-linked ancestral gene. Like Ts-IAP, the PGK-2 gene is active only in testis, where it is required in order to compensate for X chromosome inactivation during spermatogenesis *(75)*. The distribution of Ts-IAP within the testis has not been determined, and it will be interesting to determine if it is indeed expressed in spermatogonia.

Less clear is the need for multiple copies of the triple-BIR containing IAPs, NAIP, c-IAP1, c-IAP2, and XIAP. The unique carboxy terminus of NAIP suggests that it may function in some way related to host defense against intracellular parasites, but much remains to be discovered. Although the remaining IAPs are all cytoplasmic and display similar activities, they have different tissue distribution. XIAP mRNA is expressed in all tissues at a relatively constant level, but protein synthesis is controlled by a unique mechanism. The XIAP transcript is approx 9 kb, yet the coding region accounts for only 1.5 kb of this. The 5' untranslated region (UTR) is at least 1.5 kb, as it has not been completely characterized. Such extraordinarily long 5' UTRs are exceedingly rare in eukaryotic mRNAs, and are predicted to constitute an insurmountable obstacle to normal scanning ribosome initiation. Subcloning of the XIAP 5' UTR into bicistronic expression vectors has demonstrated that the UTR functions as an Internal Ribosome Initiation Site (IRES) element *(76)*.

IRES elements were first identified in picornaviruses, which shut down host protein synthesis by inactivating a key initiation factor necessary for cap-dependent translation. The IRES element recruits ribosomes internally at the beginning of the ORF and thereby allows viral polyprotein synthesis to continue. Cellular IRES elements are rare, but have been identified in a number of oncogenes and growth-factor genes. IRES-containing transcripts thus continue to direct protein synthesis under a number of cellular stress conditions in which cap-dependent translation is shut down. There are many cellular stresses that trigger shut-down of scanning ribosome initiation, including serum starvation, chemotherapeutic drugs, γ-irradiation, heat shock, viral infection, and stress in the endoplasmic reticulum (ER) (reviewed in ref. *77*). In some cases, the explanation is clear as to why the cell would do this. For example, in heat-shocked cells unfolded proteins are not exported to the golgi and accumulate in the ER. When this happens, ER stress triggers the shut down of scanning ribosome initiated protein synthesis, thereby halting the further accumulation of proteins in the ER. The cell then tries to use this "time out" to correct the problem before resuming protein synthesis. Proteins that are initiated by an IRES element and are thus resistant to this shutdown include chaperone proteins like BiP, which helps in the refolding of misfolded proteins, and XIAP, which helps the cell to survive this period of cell stress. Thus the regulation of XIAP is distinct from that of the other IAPs, and may at least partially explain the requirement for more than one IAP gene.

Two of the IAPs, c-IAP1 and c-IAP2, are components of the protein complex that forms on the cytoplasmic tail of TNF-α receptor 2. This binding is not direct, and is mediated by two additional proteins, TRAF1 and TRAF2 *(13)*. In addition, c-IAP1 can also complex with TRAF2 and TRADD on the TNFR1 cytoplasmic domain *(78)*. Signal transduction from the TNF receptors is extremely complex, resulting in either apoptosis or proliferation depending on the cell type and environmental cues. In the simplest scenario, recruitment of TRADD and FADD results in the formation of a death-inducing signaling complex (DISC) that recruits and activates caspase-8 by an induced proximity activation mechanism. The formation of alternative complexes involving the TRAF and IAP proteins favors NF-κB activation, which in turn transcriptionally activates several pro-survival genes, including c-IAP1, 2 and XIAP (reviewed in ref. *79*). In what appears to be a positive feedback loop, c-IAP2 and XIAP appear to be able to trigger the activation of NF-κB *(55,80,81)*. The role of the IAPs in TNF receptor complexes is not fully understood, and may be unrelated to direct caspase inhibition. Interestingly, c-IAP1 and -2 are not as potent as XIAP in both biochemical assays of caspase inhibition *(26)*, and in most cell-death models. The one exception to this is TNF-α-mediated cell death, in which c-IAP1 outperforms the other IAPs *(82)*, again suggesting specialized roles for each of the IAPs.

THERAPEUTIC OPPORTUNITIES

Neuronal cell death in most neurodegenerative disorders, as well as in traumatic brain injury and spinal-cord damage, exhibits most of the hallmarks of apoptosis (reviewed in refs. *83–85*). Limiting the extent of caspase activation in the target neuronal population may be therapeutically relevant in slowing the progression of Alzheimer's, Parkinson's, ALS, and Huntington's diseases, as well as in retinal degenerations and in the injured central nervous system (CNS) (*see* Chapter 14). Given that the IAPs have been demonstrated to suppress apoptosis initiated by virtually every trigger tested to date in tissue-culture cells (reviewed in ref. *23*), their utility has been explored in in vivo model systems. Stereotactic injection of adenoviral expression vectors has been used to determine the protective effect of NAIP and XIAP in the rat hippocampus in the four-vessel occlusion global ischemia model. Suppression of caspase activation shortly after the ischemic event, as well as long-term histological preservation of the vulnerable CA1 neurons, was observed. Perhaps most significantly, functional rescue of memory and learning ability was demonstrated *(86,87)*. Adenoviral vectors encoding NAIP, c-IAP1, and c-IAP2 have been shown to suppress apoptosis in the sciatic-nerve axotomy model *(88,89)*, and adeno-XIAP in an optic-nerve axotomy model *(90)*. Finally, NAIP overexpression has been shown to be protective both histologically and functionally in the 6-hydroxy dopamine model of Parkinson's disease *(91)*. These studies raise the hope that neurodegenerative disease intervention may be possible using more advanced gene-therapy vectors encoding the IAPs.

Cancer is a disease that is extremely heterogeneous, in which many different tumor types can arise from virtually any tissue. Numerous proto-oncogenes have been identified, which, when mutated or aberrantly expressed, can contribute to the transformed phenotype. Despite the complexity of cancer genetics, there are fundamental characteristics shared by all cancers, regardless of tumor origin. The deregulation of most growth-promoting oncogenes triggers apoptosis in an otherwise normal cell (reviewed in ref. *92*) (*see* Chapter 12). As a consequence, the suppression of apoptosis is a fundamental and requisite change in all cancer cells, regardless of origin *(93)*. Apoptosis is also the primary means by which radio- and chemotherapy modalities kill cancer cells (reviewed in ref. *94*). The discovery that the *Bcl-2* oncogene functions as an inhibitor of apoptosis revolutionized cancer biologists' concepts of tumor initiation and progression, identifying apoptotic inhibition as a key event in cancer formation, progression, and resistance to therapy.

Given the central role of the IAPs in controlling apoptosis, it is not surprising that expression studies have revealed elevated IAP levels in a wide variety of cancer-cell lines and primary tumor-biopsy samples *(17,20,23,57,95,96)*. Survivin-expression patterns provide the most dramatic example of this, with expression largely restricted to embryonic tissues and many different tumor types, but absent in most adult tissues (reviewed in refs. *51,52*). Furthermore, patient biopsy samples displaying Survivin expression indicate a poor prognosis, increased rates of treatment failure, and relapse (reviewed in ref. *97*). Detection of Survivin protein in urine is being developed as a diagnostic and prognostic test for bladder carcinoma *(98)*, and blood testing for other tumor types may be feasible (reviewed in ref. *52*). The prognostic significance of IAP overexpression is less clear for some of the other IAPs. For example, XIAP protein levels correlate with disease severity and prognosis in acute myelogenous leukemia (AML) *(95)*, but not in nonsmall cell lung carcinoma (NSCLC) *(99–101)*.

Beyond these correlative expression studies, direct genetic evidence is emerging that the IAPs can function as oncogenes. Chromosome amplification of the 11q21–q23 region, which encompasses both c-IAP1 and c-IAP2, has been observed in a variety of malignancies, including medulloblastomas, renal-cell carcinomas, glioblastomas, and gastric carcinomas. Esophageal squamous-cell carcinomas frequently display this amplification, and transcriptional profiling has identified c-IAP1 as the sole target gene that is consistently overexpressed in these tumors *(102)*.

Additional direct genetic evidence for an oncogenic role of the IAPs is found in extranodal marginal zone mucosa-associated lymphoid tissue (MALT) B-cell lymphomas. Two recurrent translocation events have been documented in MALT lymphomas: t(11;18)(q21;q21), and t(1;14)(p22;q32), both of which involve NF-κB activation and c-IAP2. Wild-type MALT1 binds Bcl-10 via a CARD-CARD interaction, and this oligomerization triggers NF-κB activation *(103,104; see* Fig. 4). Phosphorylation of Bcl-10 also regulates the interchangeable interaction of TRAF2 and c-IAP2 in this complex *(105,106)*. Hyperphosphorylation of Bcl-10 triggers release of TRAF2 and association of c-IAP2, and correlates with induction of apoptosis. The model presented in Fig. 4 proposes direct interaction of TRAF2 with MALT1 that in turn binds to Bcl-10 and is responsible for NF-κB activation and the generation of pro-survival signals. We propose that this is due to CARD-CARD interactions between Bcl-10 and c-IAP2, which in some way interferes with c-IAP2 activity.

The more frequent t(11;18) translocations that occur in up to 50% of extranodal MALT lymphomas *(107)*, are unusual in that they invariably encode an in-frame chimeric protein consisting of the c-IAP2 BIR domains (minus the CARD and RING domains), fused to the carboxy terminus of MALT1 *(103;* Fig. 4). Significantly, the majority of gastric MALT lymphomas that do not respond to antibiotic therapy display the c-IAP2–MALT1 translocation *(108)*. This appears to be owing to a feedback mechanism, in which the c-IAP2–MALT1 fusion protein triggers NF-κB activation, which in turn transcriptionally upregulates the NF-κB responsive *c-iap2* promoter *(80,109,110;* Fig. 4B). Remarkably, Bcl-10 translocation to the immunoglobulin heavy chain locus is the basis of the less frequent t(1;14) translocation event observed in MALT lymphomas, and again results in an NF-κB-inducing gene being expressed under the control of an NF-κB-responsive promoter (Fig. 4C). These positive feedback loops result in constitutive high-level expression of *c-iap2* or the translocated gene encoding the fusion protein, as well as other NF-κB inducible genes, several of which are anti-apoptotic. A positive feedback loop also explains why tumor regression does not occur with antibiotic therapy, because there is no longer a requirement for exogenous pro-inflammatory signaling to achieve chronic NF-κB activation.

Inhibition of IAP expression or function therefore has clear therapeutic potential in cancer therapy. In addition to tumor types with underlying genetic alterations to the IAPs, a generalized approach for increasing the apoptotic sensitivity of cancer cells using IAP inhibitors may prove effective. Adenoviral vectors encoding an antisense XIAP cDNA sensitize chemo-resistant ovarian-carcinoma cell lines *(96,111)*, as well as increasing the radiation sensitivity of NSCLC cells *(112)*. A variety of strategies have been employed to interfere with survivin expression or function, including antisense cDNAs and oligonucleotides, and nonphosphorylatable mutants. Inhibition of Survivin induces

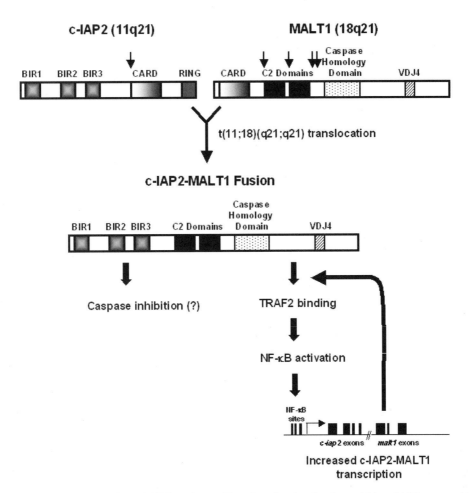

Fig. 4. The role of c-IAP2 in MALT lymphoma. Translocation breakpoints within c-IAP2 occur within the intron between exon 7 and exon 8, preserving the coding region proximal to the CARD domain (indicated by the arrow). The MALT1 break points map to several positions distal to the MALT1 CARD domain and proximal to the caspase homology domain (arrows). The t(11;18)(q21;q21) translocation generates a fusion protein that is both anti-apoptotic and NF-κB activating through TRAF2 binding. A positive feedback amplification loop exists in which the fused gene is expressed under the control of the *c-IAP2* promoter, which contains several NF-κB sites.

apoptosis, decreases colony formation in soft agar, and sensitizes various tumor cells to chemotherapeutic drugs *(17,113–118)*. In addition, in vivo experiments using human breast-cancer xenograft models suggest that survivin targeting is as effective as paclitaxel in suppressing tumor progression *(97)*. Together, these lines of evidence indicate that targeting IAP expression and/or function, alone or in combination with conventional anti-cancer therapeutics, will enter clinical trials in the very near future.

CONCLUSIONS

The process of apoptosis is controlled at multiple steps, each of which is influenced by both pro- and anti-apoptotic proteins. The equilibrium between the cell death inducing caspase cascade and the IAPs constitutes one such fundamental decision point. An ongoing debate in apoptosis research concerns the so-called "point of no return," after which the cell cannot be rescued. Many researchers have proposed that mitochondrial permeability changes irreversibly commit the cell to die, and that inhibition of downstream events, including caspase activation, will only delay apoptosis or result in secondary necrotic cell death. However, there is evidence that some cell types can recover and resume proliferation despite the transient activation of the caspase cascade and the appearance of apoptotic morphology (reviewed in ref. *119*). The recurrent upregulation of IAP expression in cancer-cell lines and tumors also indicates that this decision point is crucial in determining overall cell fate. Experimentally, overexpression of the IAPs blocks apoptosis and results in functional recovery in a number of neurodegenerative model systems, again suggesting that the window for apoptosis intervention may be larger than was previously assumed.

The IAPs not only control cell death, but also influence signal-transduction pathways, protein turnover, and progression through cell cycle, although many aspects of IAP function in all of these processes remain to be clarified. With the recognition of apoptosis as a fundamental aspect of so many human disease states, IAPs and other anti-apoptotic proteins are now acknowledged as being outstanding therapeutic targets.

REFERENCES

1. Kennedy, N. J., Kataoka, T., Tschopp, J., and Budd, R. C. (1999) Caspase activation is required for T cell proliferation. *J. Exp. Med.* **190,** 1891–1896.
2. Alam, A., Cohen, L. Y., Aouad, S., and Sekaly, R. P. (1999) Early activation of caspases during T lymphocyte stimulation results in selective substrate cleavage in nonapoptotic cells. *J. Exp. Med.* **190,** 1879–1890.
3. Bratosin, D., Estaquier, J., Petit, F., Arnoult, D., Quatannens, B., Tissier, J. P., et al. (2001) Programmed cell death in mature erythrocytes: a model for investigating death effector pathways operating in the absence of mitochondria. *Cell Death Differ.* **8,** 1143–1156.
4. Pandey, P., Nakazawa, A., Ito, Y., Datta, R., Kharbanda, S., and Kufe, D. (2000) Requirement for caspase activation in monocytic differentiation of myeloid leukemia cells. *Oncogene* **10,** 3941–3947.
5. Eckhart, L., Declercq, W., Ban, J., Rendl, M., Lengauer, B., Mayer, C., et al. (2000) Terminal differentiation of human keratinocytes and stratum corneum formation is associated with caspase-14 activation. *J. Invest. Dermatol.* **115,** 1148–1151.
6. Clem, R. J. (2001) Baculoviruses and apoptosis: the good, the bad, and the ugly. *Cell Death Differ.* **8,** 137–143.
7. Miller, L. K. (1999) An exegesis of IAPs: salvation and surprises from BIR motifs. *Trends Cell Biol.* **9,** 323–328.
8. Roy, N., Mahadevan, M. S., McLean, M., Shutler, G., Yaraghi, Z., Farahani, R., et al. (1995) The gene for neuronal apoptosis inhibitory protein is partially deleted in individuals with spinal muscular atrophy. *Cell* **80,** 167–178.
9. Koonin, E. V. and Aravind, L. (2000) The NACHT family: a new group of predicted NTPases implicated in apoptosis and MHC transcription activation. *Trends Biochem. Sci.* **25,** 223–224.
10. Melki, J., Abdelhak, S., Sheth, P., Bachelot, M.F., Burlet, P., Marcadet, A., et al. (1990) Gene for chronic proximal spinal muscular atrophies maps to chromosome 5q. *Nature* **344,** 767–768.
11. Lefebvre, S., Burglen, L., Reboullet, S., Clermont, O., Burlet, P., Violet, L., et al. (1995) Identification and characterization of a spinal muscular atrophy-determining gene. *Cell* **80,** 155–165.
12. Gendron, N. H. and MacKenzie, A. E. (1999) Spinal muscular atrophy: molecular pathophysiology. *Curr. Opin. Neurol.* **12,** 137–142.
13. Rothe, M., Pan, M.-G., Henzel, W. J., Ayres, T. M., and Goeddel, D. V. (1995) The TNFR2–TRAF signaling complex contains two novel proteins related to baculoviruial inhibitor or apoptosis proteins. *Cell* **83,**1243–1252.
14. Liston, P., Roy, N., Tamai, K., Lefebvre, C., Baird, S., Cherton-Horvat, G., et al. (1996) Suppression of apoptosis in mammalian cells by NAIP and a related family of IAP genes. *Nature* **379,** 349–353.
15. Duckett, C. S., Nava, V. E., Gedrich, R. W., Clem, R. J., Van Dongen, J. L., Gilfilan, M. C., et al. (1996) A conserved family of cellular genes related to the baculovirus iap gene and encoding apoptosis inhibitors. *EMBO J.* **15,** 2685–2694.
16. Uren, A. G., Pakusch, M., Hawkins, C. J., Puls, K. L., and Vaux, D. L. (1996) Cloning and expression of apoptosis inhibitory protein homologs that function to inhibit apoptosis and/or bind tumor necrosis factor receptor-associated factors. *Proc. Natl. Acad. Sci. USA* **93,** 4974–4978.
17. Ambrosini, G., Adida, C., and Altieri, D. (1997) A novel anti-apoptosis gene, survivin, expressed in cancer and lymphoma. *Nat. Med.* **3,** 917–921.
18. Kasof, G. M. and Gomes, B. C. (2001) Livin, a novel inhibitor-of-apoptosis (IAP) family member. *J. Biol. Chem.* **276,** 3238–3246.
19. Lin, J. H., Deng, G., Huang, Q., and Morser, J. (2000) KIAP, a novel member of the inhibitor of apoptosis protein family. *Biochem. Biophys. Res. Commun.* **279,** 820–831.
20. Vucic, D., Stennicke, H. R., Pisabarro, M. T., Salvesen, G. S., and Dixit, V. M. (2000) ML-IAP, a novel inhibitor of apoptosis that is preferentially expressed in human melanomas. *Curr. Biol.* **10,** 1359–1366.

21. Richter, B. W., Mir, S. S., Eiben, L. J., Lewis, J., Reffey, S. B., Frattini, A., et al. (2001) Molecular cloning of ILP-2, a novel member of the inhibitor of apoptosis protein family. *Mol. Cell. Biol.* **21,** 4292–4301.

22. Lagace, M., Xuan, J. Y., Young, S. S., McRoberts, C., Maier, J., Rajcan-Separovic, E., and Korneluk, R. G. (2001) Genomic organization of the x-linked inhibitor of apoptosis and identification of a novel testis-specific transcript. *Genomics* **77,** 181–188.

23. Lacasse, E. C., Baird, S., Korneluk, R. G., and MacKenzie, A. E. (1998) The inhibitors of apoptosis (IAPs) and their emerging role in cancer. *Oncogene* **17,** 3247–3259.

24. Deveraux, Q. L. and Reed, J. C. (1999) IAP family proteins-suppressors of apoptosis. *Genes Dev.* **13,** 239–252.

25. Deveraux, Q. L., Takahashi, R., Salvesen, G. S., and Reed, J. C. (1997) X-linked IAP is a direct inhibitor of cell-death proteases. *Nature* **388,** 300–304.

26. Roy, N., Deveraux, Q. L., Takahashi, R., Salvesen, G. S., and Reed, J. C. (1997) The c-IAP-1 and c-IAP-2 proteins are direct inhibitors of specific caspases. *EMBO J.* **16,** 6914–6925.

27. Takahashi, R., Deveraux, Q., Tamm, I., Welsh, K., Assa-Munt, N., Salvesen, G. S., and Reed J. C. (1998) A single BIR domain of XIAP sufficient for inhibiting caspases. *J. Biol. Chem.* **273,** 7787–7790.

28. Maier, J. K. X., Lahoua, Z. Gendron, N. H., Fetni, R., Johnston, A., Nicholson, D. W., and MacKenzie, A. E. (2002) The neuronal apoptosis inhibitory protein is a direct inhibitor of caspases 3 and 7. *J. Neurosci.* **22,** 2035–2043.

29. Shin, S., Sung, B. J., Cho, Y. S., Kim, H. J., Ha, N. C., Hwang, J. I., et al. (2001) An anti-apoptotic protein human survivin is a direct inhibitor of caspase-3 and -7. *Biochemistry* **40,** 1117–1123.

30. Riedl, S. J., Renatus, M., Schwarzenbacher, R., Zhou, Q., Sun, C., Fesik, S. W., et al. (2001) Structural basis for the inhibition of caspase-3 by XIAP. *Cell* **104,** 791–800.

31. Huang, Y., Park, Y. C., Rich, R. L., Segal, D., Myszka, D. G., and Wu, H. (2001) Structural basis of caspase inhibition by XIAP: differential roles of the linker versus the BIR domain. *Cell* **104,** 781–790.

32. Chai, J., Shiozaki, E., Srinivasula, S. M., Wu, Q., Datta, P., Alnemri, E. S., et al. (2001) Structural basis of caspase-7 inhibition by XIAP. *Cell* **104,** 769–780.

33. Sun, C., Cai, M., Gunasekera, A. H., Meadows, R. P., Wang, H., Chen, J., et al. (1999) NMR structure and mutagenesis of the inhibitor-of-apoptosis protein XIAP. *Nature* **40,** 818–822.

34. Sun, C., Cai, M., Meadows, R. P., Xu, N., Gunasekera, A. H., Herrmann, J., et al. (2000) NMR structure and mutagenesis of the third Bir domain of the inhibitor of apoptosis protein XIAP. *J. Biol. Chem.* **275,** 33777–33781.

35. Suzuki, Y., Nakabayashi, Y., Nakata, K., Reed, J. C., and Takahashi, R. (2001) X-linked inhibitor of apoptosis protein (XIAP) inhibits caspases-3 and -7 in distinct modes. *J. Biol. Chem.* **276,** 27058–27063.

36. Deveraux, Q. L., Roy, N., Stennicke, H. R., Van Arsdale, T., Zhou, Q., Srinivasula, S. M., et al. (1998) IAPs block apoptotic events induced by caspase-8 and cytochrome c by direct inhibition of distinct caspases. *EMBO J.* **17,** 2215–2223.

37. Bratton, S. B., Walker, G., Srinivasula, S. M., Sun, X. M., Butterworth, M., Alnemri, E. S., and Cohen, G. M. (2001) Recruitment, activation and retention of caspases-9 and -3 by Apaf-1 apoptosome and associated XIAP complexes. *EMBO J.* **20,** 998–1009.

38. Srinivasula, S. M., Hegde, R., Saleh, A., Datta, P., Shiozaki, E., Chai, J., et al. (2001) A conserved XIAP-interaction motif in caspase-9 and Smac/DIABLO regulates caspase activity and apoptosis. *Nature* **410,** 112–116.

39. Pickart, C. M. (2001) Mechanisms underlying ubiquitination. *Ann. Rev. Biochem.* **70,** 503–533.

40. Yang, Y., Fang, S., Jensen, J. P., Weissman, A. M., and Ashwell, J. D. (2000) Ubiquitin protein ligase activity of the IAPs and their degradation in proteosomes in response to apoptotic stimuli. *Science* **288,** 874–877.

41. Huang, H.-K., Joazeiro, C. A. P., Bonfoco, E., Kamada, S., Leverson, J. D., and Hunter, T. (2000) The inhibitor of apoptosis, cIAP2, functions as a ubiquitin protein ligase and promotes in vitro monoubiquitination of caspase 3 and 7. *J. Biol. Chem.* **275,** 26661–26664.

42. Suzuki, Y., Nakabayashi, Y., and Takahashi, R. (2001) Ubiquitin-protein ligase activity of X-linked inhibitor of apoptosis protein promotes proteosomal degradation of caspase-3 and enhances its anti-apoptotic effect in Fas-induced cell death. *Proc. Natl. Acad. Sci. USA* **98,** 8662–8667.

43. Clem, R. J., Sheu, T.-T., Richter, B. W. M., He, W.-W., Thornberry, N. A., Duckett, C. S., and Hardwick, J. M. (2001) c-IAP1 is cleaved by caspases to produce a proapoptotic C-terminal fragment. *J. Biol. Chem.* **276,** 7602–7608.

44. Weber, C. H. and Vincenz, C. (2001) The death domain superfamily: a tale of two interfaces? *Trends Biochem. Sci.* **26,** 475–481.

45. Martin, S. J. (2001) Dealing the CARDs between life and death. *Trends. Cell. Biol.* **11,** 188–189.

46. Inohara, N. and Nunez, G. (2001) The NOD: a signaling module that regulates apoptosis and host defense against pathogens. *Oncogene* **20,** 6473–6481.

47. Yaraghi, Z., Korneluk, R. G., and MacKenzie, A. (1998) Cloning and characterization of the multiple murine homologues of NAIP (neuronal apoptosis inhibitory protein). *Genomics* **51,** 107–113.

48. Growney, J. D. and Dietrich, W. F. (2000) High-resolution genetic and physical map of the Lgn1 interval in C57BL/6J implicates Naip2 or Naip5 in Legionella pneumophila pathogenesis. *Genome Res.* **10,** 1158–1171.

49. Chantalat, L., Skoufias, D. A., Kleman, J. P., Jung, B., Dideberg, O., and Moargolis, R. L. (2000) Crystal structure of human surviving reveals a bow tie-shaped dimer with two unusual alpha-helical extensions. *Mol. Cell* **6,** 183–189.

50. Li, F., Ambrosini, G., Chu, E. Y., Plescia, J., Tognin, S., Marchisio, P. C., and Altieri, D. C. (1998) Control of apoptosis and mitotic spindle checkpoint by survivin. *Nature* **396,** 580–583.

51. Reed, J. C. and Bischoff, J. R. (2000) BIRinging chromosomes through cell division-and survivin' the experience. *Cell* **102,** 545–548.

52. Reed, J. C. (2001) The survivin saga goes in vivo. *J. Clin. Invest.* **108**, 965–969.

53. Deveraux, Q. L., Leo, E., Stennicke, H. R., Welsh, K., Salvesen, G. S. and Reed, J. C. (1999) Cleavage of human inhibitor of apoptosis protein XIAP results in fragments with distinct specificities for caspases. *EMBO J.* **18**, 5242–5251.

54. Johnson, D. E., Gastman, B. R., Wieckowski, E., Wang, G. Q., Amoscato, A., Delach, S. M., and Rabinowich, H. (2000) Inhibitor of apoptosis protein hILP undergoes caspase-mediated cleavage during T lymphocyte apoptosis. *Cancer Res.* **60**, 1818–1823.

55. Levkau, B., Garton, K. J., Kerri, N., Kloke, K., Nofer, J. R., Baba, H. A., et al. (2001) XIAP induces cell-cycle arrest and activates nuclear factor-kappaB: new survival pathways disabled by caspase-mediated cleavage during apoptosis of human endothelial cells. *Circ. Res.* **88**, 282–290.

56. Liston, P., Fong, W. G., Kelly, N. L., Toji, S., Miyazaki, T., Conte, D., et al. (2001) Identification of XAF1 as an antagonist of XIAP anti-Caspase activity. *Nat. Cell. Biol.* **3**, 128–133.

57. Fong, W. G., Liston, P., Rajcan-Separovic, E., St. Jean, M., Craig, C., and Korneluk, R. G. (2000) Expression and genetic analysis of XIAP-associated factor 1 (XAF1) in cancer cell lines. *Genomics* **70**, 113–122.

58. Du, C., Fang, M., Li, Y., Li, L., and Wang, X. (2000) Smac, a mitochondrial protein that promotes cytochrome c-dependent caspase activation by eliminating IAP inhibition. *Cell* **102**, 33–42.

59. Verhagen, A. M., Ekert, P. G., Pakusch, M., Silke, J., Connolly, L. M., Reid, G. E., et al. (2000) Identification of DIABLO, a mammalian protein that promotes apoptosis by binding to and antagonizing IAP proteins. *Cell* **102**, 43–53.

60. Adrian, C., Creagh, E. M., and Martin, S. J. (2001) Apoptosis-associated release of Smac/DIABLO from mitochondria requires active caspases and is blocked by Bcl-2. *EMBO J.* **20**, 6627–6636.

61. Vucic, D., Deshayes, K., Ackerly, H., Pisabarro, M. T., Kadkhodayan, S., Fairbrother, W. J., and Dixit, V. M. (2002) SMAC negatively regulates the anti-apoptotic activity of ML-IAP. *J. Biol. Chem.* **277**, 12275–12279.

62. Liu, Z., Sun, C., Olejniczak, E. T., Meadows, R. P., Betz, S. F., Oost, T., et al. (2000) Structural basis for binding of Smac/DIABLO to the XIAP BIR3 domain. *Nature* **408**, 1004–1008.

63. Srinivasula, S. M., Datta, P., Fan, X. K., Fernandes-Alnemri, T., Huang, Z., and Alnemri, E. S. (2000) Molecular determinants of the caspase-promoting activity of Smac/DIABLO and its role in the death receptor pathway. *J. Biol. Chem.* **275**, 36152–36157.

64. Wu, G., Chai, J., Suber, T. L., Wu, J. W., Du, C., Wang, X., and Shi, Y. (2000) Strucutral basis of IAP recognition by Smac/DIABLO. *Nature* **408**, 1008–1012.

65. Chai, J., Du, C., Wu, J. W., Kyin, S., Wang, X., and Shi, Y. (2000) Structural and biochemical basis of apoptotic activation by Smac/DIABLO. *Nature* **406**, 855–862.

66. Ekert, P. G., Silke, J., Hawkins, C. J., Verhagen, A. M., and Vaux, D. L. (2001) DIABLO promotes apoptosis by removing MIHA/XIAP from processed caspase 9. *J. Cell Biol.* **152**, 483–490.

67. Song, Z., Guan, B., Bergman, A., Nicholson, D. W., Thornberry, N. A., Peterson, E. P., and Steller, H. (2000) Biochemical and genetic interactions between Drosophila caspases and the proapoptotic genes rpr, hid, and grim. *Mol. Cell. Biol.* **20**, 2907–2914.

68. Wu, J. W., Cocina, A. E., Chai, J., Hay, B. A., and Shi, Y. (2001) Structural analysis of a functional DIAP1 fragment bound to grim and hid peptides. *Mol. Cell* **8**, 95–104.

69. Van Loo, G., van Gurp, M., Depuydt, B., Srinivasula, S. M., Rodriguez, I., Alnemri, E. S., et al. (2002) The serine protease Omi/HtrA2 is released from mitochondria during apoptosis. Omi interacts with caspase-inhibitor XIAP and induces enhanced caspase activity. *Cell Death Differ.* **9**, 20–26.

70. Hegde, R., Srinivasula, S. M., Zhang, Z., Wassell, R., Mukattash, R., Cilenti, L., et al. (2002) Identification of Omi/HtrA2 as a mitochondrial apoptotic serine protease that disrupts inhibitor of apoptosis protein-caspase interaction. *J. Biol. Chem.* **277**, 432–438.

71. Verhagen, A. M., Silke, J., Ekert, P. G., Pakusch, M., Kaufmann, H., Connolly, L. M., et al. (2002) HtrA2 promotes cell death through its serine protease activity and its ability to antagonize inhibitor of apoptosis proteins. *J. Biol. Chem.* **277**, 445–454.

72. Martins, L. M., Iaccarino, I., Tenev, T., Gschmeissner, S., Totty, N. F., Lemoine, N. R., et al. (2002) The serine protease Omi/HtrA2 regulates apoptosis by binding XIAP through a reaper-like motif. *J. Biol. Chem.* **277**, 439–444.

73. Suzuki, Y., Imai, Y., Nakayama, H., Takahashi, K., Takio, K., and Takahashi, R. (2001) A serine protease, HtrA2, is released from the mitochondria and interacts with XIAP, inducing cell death. *Mol. Cell.* **8**, 613–621.

74. Savopoulos, J. W., Carter, P. S., Turconi, S., Pettman, G. R., Karran, E. H., Gray, C. W., et al. (2000) Expression, purification, and functional analysis of the human serine protease HtrA2. *Protein Expr. Purif.* **19**, 227–234.

75. McCarrey, J. R. and Thomas, K. (1987) Human testis-specific PGK gene lacks introns and possesses characteristics of a processed gene. *Nature* **326**, 501–505.

76. Holcik, M., Lefebvre, C., Yeh, C., Chow, T., and Korneluk, R. G. (1999) A new internal-ribosome-entry-site motif potentiates XIAP-mediated cytoprotection. *Nat. Cell Biol.* **1**, 190–192.

77. Holcik, M., Sonenberg, N., and Korneluk, R. G. (2000) Internal ribosome initiation of translation and the control of cell death. *Trends Genet.* **16**, 469–473.

78. Shu, H. B., Takeuchi, M., and Goeddel, D. V. (1996) The tumor necrosis factor receptor 2 signal transducers TRAF2 and c-IAP1 are components of the tumor necrosis factor receptor 1 signaling complex. *Proc. Natl. Acad. Sci. USA* **93**, 13973–13978.

79. Lee, R. and Collins, T. (2001) Nuclear factor-kappaB and cell survival: IAPs call for support. *Circ. Res.* **88**, 262–264.

80. Chu, Z. L., McKinsey, T. A., Liu, L., Gentry, J. J., Malim, M. H., and Ballard, D. W. (1997) Suppression of tumor

necrosis factor-induced cell death by inhibitor of apoptosis c-IAP2 is under NF-kappaB control. *Proc. Natl. Acad. Sci. USA* **94,** 10057–10062.

81. Hofer-Warbinek, R., Schmid, J. A., Stehlik, C., Binder, B. R., Lipp, J., and de Martin, R. (2000) Activation of NF-kappa B by XIAP, the X chromosome-linked inhibitor of apoptosis, in endothelial cells involves TAK1. *J. Biol. Chem.* **275,** 22064–22068.

82. Wright, M. E., Han, D. K., and Hockenbery, D. M. (2000) Caspase-3 and inhibitor of apoptosis protein(s) interactions in Saccharomyces cerevisiae and mammalian cells. *FEBS Lett.* **481,** 13–18.

83. Mattson, M. P. (2000) Apoptosis in neurodegenerative disorders. *Nat. Rev. Mol. Cell Biol.* **1,** 120–129.

84. Martin, L. J. (2001) Neuronal cell death in nervous system development, diseases, and injury. *Int. J. Mol. Med.* **7,** 455–478.

85. Mattson, M. P., Duan, W., Pedersen, W. A., and Culmsee, C. (2001) Neurodegenerative disorders and ischemic brain diseases. *Apoptosis* **6,** 69–81.

86. Xu, D. G., Crocker, S. J., Doucet, J. P., St. Jean, M., Tamai, K., Hakim, A. M., et al. (1997) Elevation of neuronal expression of NAIP reduces ischemic damage in the rat hippocampus. *Nat. Med.* **3,** 997–1004.

87. Xu, D., Bureau, Y., McIntyre, D. C., Nicholson, D. W., Liston, P., Zhu, Y., et al. (1999) Attenuation of ischemia-induced cellular and behavioral deficits by X chromosome-linked inhibitor of apoptosis protein overexpression in the rat hippocampus. *J. Neurosci.* **19,** 5026–5033.

88. Perrelet, D., Ferri, A., MacKenzie, A. E., Smith, G. M., Korneluk, R. G., Liston, P., et al. (2000) IAP family proteins delay motoneuron cell death in vivo. *Eur. J. Neurosci.* **12,** 2059–2067.

89. Perrelet, D., Ferri, A., Liston, P., Muzzin, P., Korneluk, R. G., and Kato, A. C. (2002) IAPs are essential for GDNF-mediated neuroprotective effects in injured motor neurons in vivo. *Nat. Cell Biol.* **4,** 175–179.

90. Kugler, S., Straten, G., Kreppel, F., Isenmann, S., Liston, P., and Bahr, M. (2000) The X-linked inhibitor of apoptosis (XIAP) prevents cell death in axotomized CNS neurons in vivo. *Cell Death Differ.* **7,** 815–824.

91. Crocker, S. J., Wigle, N., Liston, P., Thompson, C. S., Lee, C. J., Xu, D., et al. (2001) NAIP protects the nigrostriatal dopamine pathway in an intrastriatal 6-OHDA rat model of Parkinson's disease. *Eur. J. Neurosci.* **14,** 391–400.

92. Evan, G. I. and Vousden, K. H. (2001) Proliferation, cell cycle and apoptosis in cancer. *Nature* **411,** 342–348.

93. Hanahan, D. and Weinberg, R. A. (2000) The hallmarks of cancer. *Cell* **100,** 57–70.

94. Johnstone, R. W., Ruefli, A. A., and Lowe, S. W. (2002) Apoptosis. A link between cancer genetics and chemotherapy. *Cell* **108,** 153–164.

95. Tamm, I., Kornblau, S. M., Segall, H., Krajewski, S., Welsh, K., Kitada, S., et al. (2000) Expression and prognostic significance of IAP-family genes in human cancers and myeloid leukemias. *Clin. Cancer Res.* **6,** 1796–1803.

96. Li, J., Feng, Q., Kim, J. M., Schneiderman, D., Liston, P., Li, M., et al. (2001) Human ovarian cancer and cisplatin resistance: possible role of inhibitor of apoptosis proteins. *Endocrinology* **142,** 370–380.

97. Mesri, M., Wall, N. R., Li, J., Kim, R. W., and Altieri, D. C. (2001) Cancer gene therapy using a survivin mutant adenovirus. *J. Clin. Invest.* **108,** 981–990.

98. Sharp, J. D., Hausladen, D. A., Maher, M. G., Wheeler, M. A., Altieri, D. C., and Weiss, R. M. (2002) Bladder cancer detection with urinary survivin, an inhibitor of apoptosis. *Front. Biosci.* **7,** E36–E41.

99. Ferreira, C. G., van der Valk, P., Span, S. W., Ludwig, I., Smit, E. F., Kruyt, F. A., et al. (2001) Expression of X-linked inhibitor of apoptosis as a novel prognostic marker in radically resected non-small cell lung cancer patients. *Clin. Cancer Res.* **7,** 2468–2474.

100. Ferreira, C. G., van der Valk, P., Span, S. W., Jonker, J. M., Postmus, P. E., Kruyt, F. A., and Giaccone, G. (2001) Assessment of IAP (inhibitor of apoptosis) proteins as predictors of response to chemotherapy in advanced non-small-cell lung cancer patients. *Ann. Oncol.* **12,** 799–805.

101. Liu, S. S., Tsang, B. K., Cheung, A. N., Xue, W. C., Cheng, D. K., Ng, T. Y., et al. (2001) Anti-apoptotic proteins, apoptotic and proliferative parameters and their prognostic significance in cervical carcinoma. *Eur. J. Cancer* **37,** 1104–1110.

102. Imoto, I., Yang, Z. Q., Pimkhaokham, A., Tsuda, H., Shimada, Y., Imamura, M., et al. (2001) Identification of cIAP1 as a candidate target gene within an amplicon at 11q22 in esophageal squamous cell carcinomas. *Cancer Res.* **61,** 6629–6634.

103. Uren, A. G., O'Rourke, K., Aravind, L. A., Pisabarro, M. T., Seshagiri, S., Koonin, E. V., and Dixit, V. M. (2000) Identification of paracaspases and metacaspases: two ancient families of caspase-like proteins, one of which plays a key role in MALT lymphoma. *Mol. Cell* **6,** 961–967.

104. Vega, F. and Madeiros, L. J. (2001) Marginal-zone B-cell lymphoma of extranodal mucosa-associated lymphoid tissue type: molecular genetics provides new insights into pathogenesis. *Adv. Anat. Pathol.* **8,** 313–326.

105. Yoneda, T., Imaizumi, K., Maeda, M., Yui, D., Manabe, T., Katayama, T., et al. (2000) Regulatory mechanisms of TRAF2–mediated signal transduction by Bcl10, a MALT lymphoma-associated protein. *J. Biol. Chem.* **275,** 11114–11120.

106. Yui, D., Yoneda, T., Oono, K., Katayama, T., Imaizumi, K., and Tohyama, M. (2001) Interchangeable binding of Bcl10 to TRAF2 and cIAPs regulates apoptosis signaling. *Oncogene* **20,** 4317–4323.

107. Baens, M., Maes, B., Steyls, A., Geboes, K., Marynen, P., and De Wolf-Peeters, C. (2000) The product of the t(11;18), an API2–MLT fusion, marks nearly half of gastric MALT type lymphomas without large cell proliferation. *Am. J. Pathol.* **156,** 1433–1439.

108. Liu, H., Ruskon-Fourmestraux, A., Lavergne-Slove, A., Ye, H., Molina, T., Bouhnik, Y., et al. (2001) Resistance of t(11;18) positive gastric mucosa-associated lymphoid tissue lymphoma to Helicobacter pylori eradication therapy. *Lancet* **357,** 39–40.

109. Erl, W., Hansson, G. K., de Martin, R., Draude, G., Weber, K. S., and Weber, C. (1999) Nuclear factor-kappa B regulates induction of apoptosis and inhibitor of apoptosis protein-1 expression in vascular smooth muscle cells. *Circ. Res.* **84,** 668–677.
110. Hong, S. Y., Yoon, W. H., Park, J. H., Kang, S. G., Ahn, J. H., and Lee, T. H. (2000) Involvement of two NF-κB binding elements in tumor necrosis factor alpha-, CD40-, and epstein-barr virus latent membrane protein 1-mediated induction of the cellular inhibitor of apoptosis protein 2 gene. *J. Biol. Chem.* **275,** 18022–18028.
111. Sasaki, H., Sheng, Y., Kosuji, F., and Tsang, B. K. (2000). Down-regulation of X-linked inhibitor of apoptosis protein induces apoptosis in chemoresistant human ovarian cancer cells. *Cancer Res.* **60,** 5659–5666.
112. Holcik, M., Yeh, C., Korneluk, R. G., and Chow, T. (2000) Translational upregulation of X-linked inhibitor of apoptosis (XIAP) increases resistance to radiation induced cell death. *Oncogene* **19,** 4174–4177.
113. Li, F., Ackermann, E. J., Bennett, C. F., Rothermel, A. L., Plescia, J., Tognin, S., et al. (1999) Pleiotropic cell-division defects and apoptosis induced by interference with survivin function. *Nat. Cell Biol.* **1,** 461–466.
114. Chen, J., Wu, W., Tahir, S. K., Kroeger, P. E., Rosenberg, S. H., Cowsert, L. M., et al. (2000) Down-regulation of survivin by antisense oligonucleotides increases apoptosis, inhibits cytokinesis and anchorage-independent growth. *Neoplasia* **2,** 235–241.
115. Grossman, D., Kim, P. J., Schechner, J. S., and Altieri, D. C. (2001) Inhibition of melanoma tumor growth in vivo by survivin targeting. *Proc. Natl. Acad. Sci. USA* **98,** 635–640.
116. Olie, R. A., Simoes-Wust, A. P., Baumann, B., Leech, S. H., Fabbro, D., Stahel, R. A., and Zangemeister-Wittke, U. (2000) A novel antisense oligonucleotide targeting survivin expression induces apoptosis and sensitizes lung cancer cells to chemotherapy. *Cancer Res.* **60,** 2805–2809.
117. Kanwar, J. R., Shen, W.-P., Kanwar, R. K., Berg, R. W., and Krissansen, G. W. (2001) Effects of Survivin antagonists on growth of established tumors and B7-1 immunogene therapy. *J. Natl. Cancer Inst.* **93,** 1541–1552.
118. Shankar, S. L., Mani, S., O'Guin, K. N., Kandimalla, E. R., Agrawal, S., and Shafit-Zagardo, B. (2001) Survivin inhibition induces human neural tumor cell death through caspase-independent and -dependent pathways. *J. Neurochem.* **79,** 426–436.
119. Vaughan, A. T. M., Betti, C. J., and Villalobos, M. J. (2002) Surviving apoptosis. *Apoptosis* **7,** 173–177.

4

Structural Biology of Programmed Cell Death

Yigong Shi

INTRODUCTION

Structural biology is an important and integral component of the modern experimental biology. Insights revealed by X-ray crystallography, nuclear magnetic resonance (NMR), electron microscopy, and other biophysical methods have fundamentally changed our way of thinking and tremendously improved our understanding of the biological system. Similar to other research disciplines, the concept of apoptosis and many aspects of its mechanisms have been made crystal clear through the last decade of structural and biochemical investigation *(1,2)*.

In mammalian cells, apoptosis can be triggered by a wide spectrum of stimuli, from both intra- and extracellular environments. The intracellular stimuli, such as DNA damage, generally cause the activation of the BH3-only Bcl-2 family proteins, which invariably leads to the release of pro-apoptotic factors from the inter-membrane space of mitochondria into the cytoplasm. One of these factors, cytochrome *c*, directly activates Apaf-1 and, in the presence of dATP or ATP, induces the formation of a large multimeric complex "apoptosome." The apoptosome recruits and mediates the auto-activation of the initiator caspase, caspase-9, which goes on to activate caspase-3 and caspase-7, triggering a cascade of caspase cleavage and activation. The active caspases are subject to inhibition by the inhibitor of apoptosis (IAP) family of proteins. Another mitochondria-derived protein, Smac/DIABLO, physically interacts with multiple IAPs and removes IAP-mediated caspase inhibition during apoptosis. Thus mitochondria play an indispensable role in the intrinsic form of apoptosis (*see* Chapter 6).

The extracellular stimuli, such as withdrawal of growth factors, directly activate the death receptors through ligand-induced trimerization and assembly of a large death-inducing signaling complex (DISC) at the plasma membrane. Although the constituents of DISC have not been fully identified, one adapter protein, the Fas-associated death domain or FADD, appears to be the obligate component, which recruits and mediates the auto-activation of the initiator caspase, procaspase-8. The active caspase-8 cleaves and activates caspase-3 and -7. Thus both extrinsic and intrinsic cell death results in the activation of caspase-3 and -7.

One physiological target of the active caspase-8 is Bid, a BH3-only protein, which lacks a transmembrane region. After cleavage, the C-terminal fragment of Bid (truncated Bid or tBid) translocates to the outer mitochondrial membrane and induces the release of pro-apoptotic factors. Thus Bid mediates the crosswalk between the extrinsic and intrinsic forms of cell death.

From: *Essentials of Apoptosis: A Guide for Basic and Clinical Research*
Edited by: X-M. Yin and Z. Dong © Humana Press Inc., Totowa, NJ

Structural information is now available on the death ligands and receptors, all three subfamilies of the Bcl-2 proteins, both initiator and effector caspases, IAPs, Smac/DIABLO, several classes of signaling motifs, and the apoptosome. These structures and associated biochemical studies have revealed significant insights into the molecular mechanisms of the initiation, execution, and regulation of programmed cell death.

DEATH RECEPTORS

Death receptors, located on the cell surface and characterized by multiple cysteine-rich extracellular domains, belong to the tumor necrosis factor (TNF) family of proteins *(3)* (*see* Chapter 5). They transmit apoptotic signals initiated by specific death ligands. The best-characterized death receptors include TNFR1 (also p55 or CD120a), Fas (also Apo1 or CD95), and DR3 (also Apo3 or Wsl1). Each of the death receptors contains a single death domain in the intracellular compartment, which is responsible for recruiting adapter proteins, such as FADD and TRADD, through homotypic interactions.

The activated death ligands are homo-trimeric and induces trimerization of the death receptors upon binding. The associated adapter proteins further recruit other effector proteins. For example, the death effector domain of FADD interacts with the prodomain of procaspase-8, thus bringing three procaspase-8 molecules into close proximity of one another and presumably facilitating their auto-activation.

Structures of the death ligands TNF-α and TNF-β have been determined in their trimeric forms, which reveal a highly similar fold *(4–6)*. Each monomer contains a β-sandwich with a canonical jellyroll topology (Fig. 1A). Three monomers intimately associate to form a bell-shaped homo-trimer (Fig. 1A). The extensive trimeric interface between adjacent monomers involves a mixture of hydrophobic contacts and hydrogen bond interactions. The co-crystal structure of the soluble extracellular domain of TNFR1 in complex with TNF-β reveals the specific recognition of a death ligand and the activated state of the death receptor *(7)*. In this complex, the four cysteine-rich domains (CRDs) within one TNFR1 molecule stack up vertically over a distance of 80 Å to form an elongated rod-like structure (Fig. 1B). Each TNFR1 rod binds to two adjacent TNF-β monomers at their interface and the three TNFR1 molecules do not directly interact with one another. Protruding loops from CRD2 (named the "50s loop") and CRD3 (named the "90s loop") bind to two distinct regions of TNF-β (Fig. 1B). This structure allows the construction of a model, in which the cytoplasmic regions of the receptors are proposed to cluster together.

Although TNFR1 contains four CRDs, only the second and third directly bind to the ligand. Another death receptor DR5 contains only two CRDs. The structure of DR5 in a complex with another death ligand TRAIL reveals a similar overall binding topology but significant differences that determine the specificity of this interaction *(8,9)*. The two CRDs of DR5 correspond to CRD2 and CRD3 of TNFR1. Although CRD1 of DR5 interacts with TRAIL in a similar fashion as CRD2 of TNFR1, CRD2 of DR5 binds to TRAIL in a different conformation compared to CRD3 of TNFR1. These structural differences underlie distinct signaling specificity by different death ligands and may have important ramifications for the design of specific therapeutic agents.

In addition to the ligand-bound activated death receptors, the structure of the isolated TNFR1 was also reported *(10,11)*. These structures reveal that the free receptors associate into dimers of two distinct types *(10)*. In one case, the two receptors are arranged in an anti-parallel fashion, which would result in the separation of their cytoplasmic domains by a distance of over 100 Å (Fig. 1C). Under this circumstance these death receptors would prevent the intracellular adapter proteins from forming a productive signaling complex. Thus the structure of this dimeric TNFR1 may represent the inactive form. In the other case, the two receptors are placed parallel to each other, with their TNF-binding surfaces fully exposed (Fig. 1D). Each of the two receptors is capable of forming a trimeric assembly upon ligand binding; thus this arrangement would result in clustering of TNF/TNFR1 trimers, a scenario that may enhance signaling efficiency.

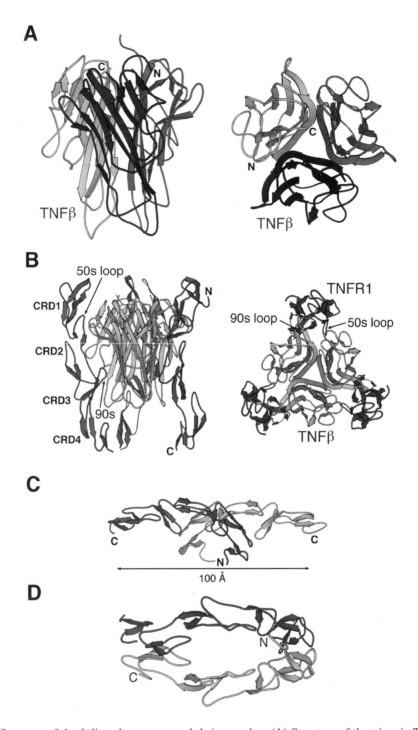

Fig. 1. Structure of death ligand, receptor, and their complex. **(A)** Structure of the trimeric TNF-β. Two perpendicular views are shown. **(B)** Structure of a complex between the soluble extra-cellular domain of TNFR1 and TNF-β. The four CRDs as well as the two important loops are labeled. **(C)** The suggested inactive form of a dimeric TNFR1 (extracellular domain). This configuration presumably prevents the formation of a productive intracellular signaling complex. **(D)** The suggested active form of a dimeric TNFR1 (extracellular domain). All figures were prepared using MOLSCRIPT *(71)*.

BCL-2 FAMILY OF PROTEINS

The Bcl-2 family of proteins controls the mitochondria-initiated intrinsic apoptosis and regulates the death receptor-initiated extrinsic cell death. More than two dozen Bcl-2 family proteins have been identified in multicellular organisms examined to date *(12)* (*see* Chapter 2). On the basis of function and sequence similarity, the diverse Bcl-2 members can be grouped into three subfamilies. The anti-apoptosis subfamily, represented by Bcl-2/Bcl-x_L in mammals and CED-9 in worms, inhibits programmed cell death by distinct mechanisms. For example, Bcl-2/Bcl-x_L functions by preventing the release of mitochondrial proteins, whereas CED-9 binds CED-4 and prevents CED-4-mediated activation of CED-3. The pro-apoptosis proteins constitute two subfamilies, represented by Bax/Bak and Bid/Bim. Members in the Bcl-2/Bcl-x_L subfamily contain all four conserved Bcl-2 homology domains (BH4, BH3, BH1, and BH2), whereas the Bax/Bak subfamily lack the BH4 domain and the Bid/Bim subfamily only contain the BH3 domain. Most members of the Bcl-2 family contain a single membrane-spanning region at their C-termini. Members of the opposing subfamilies as well as between the two pro-apoptotic subfamilies can dimerize, mediated by the amphipathic BH3 helix.

The first structure of the Bcl-2 family of proteins was determined on Bcl-x_L by both X-ray crystallography and NMR spectroscopy *(13)*. This structure reveals two centrally located hydrophobic α helices (α5 and α6), packed by five amphipathic helices on both sides (Fig. 2A). Interestingly, the Bcl-x_L structure closely resembles the pore-forming domains of bacterial toxins such as diphtheria toxin. This similarity raised an interesting hypothesis that the Bcl-2 family of proteins may form pores at the outer mitochondrial membrane to regulate ion exchange. Indeed, this conjecture has been proven for nearly all members of the Bcl-2 family examined in vitro. However, it is unclear whether the pH-dependent ion-conducting property of Bcl-2 proteins occurs in vivo and, if so, how it contributes to the regulation of apoptosis.

Because most members of the Bcl-2 family associate with lipid membranes using their C-terminal transmembrane region, it is important to examine the structure of membrane-associated Bcl-2 family proteins. Unfortunately, no such structure is yet available due to technical difficulty. Towards this ultimate goal, Bcl-x_L was characterized in detergent micelles and, compared to the aqueous solution, exhibited significant structural difference *(14)*.

Bcl-x_L or Bcl-2 interacts with the BH3-only subfamily of Bcl-2 proteins such as Bad, Bim, and Bid. The recognition mode was revealed by the solution structure of Bcl-x_L bound to a BH3 peptide from Bak *(15)* (Fig. 2B). The BH3 peptide forms an amphipathic α helix and interacts with a deep hydrophobic groove on the surface of Bcl-x_L. The binding of the BH3 domain causes a significant conformational change in Bcl-x_L, including the melting of a short α-helix (α3).

The structures of the other two groups of Bcl-2 family members have also been determined. Bid, containing only the BH3 domain and displaying very weak sequence homology with Bcl-x_L, exhibits a conserved structure with that of Bcl-x_L *(16,17)*. The minor differences include the length and relative orientation of several helices as well as an extra α helix between helices α1 and α2 (Fig. 2C). On the basis of the structure, a model was proposed to explain how the caspase-8-mediated cleavage of Bid (after residue 59) improves its pro-apoptotic function. In this mechanism, removal of the N-terminal 59 amino acids leads to the exposure of the BH3 domain, which mediates binding to the other two subfamilies of Bcl-2 proteins.

Similar to Bid, the structure of the full-length Bax in aqueous solution closely resembles that of Bcl-x_L *(18)*. Intriguingly, the C-terminal membrane-spanning region folds back to bind a hydrophobic groove that normally accommodates the BH3 domain of another Bcl-2 protein (Fig. 2D). Thus this structure may represent the inactive or closed form of the Bax/Bak group and suggests a regulatory role for their C-termini in the absence of apoptotic stimuli.

Bax and Bak exist as monomers in aqueous solution but can form homo-oligomers in the presence of detergents. These large homo-oligomers are thought to form channels with a pore size large enough to allow passage of proteins such as cytochrome *c*, though direct biochemical and structural evidence

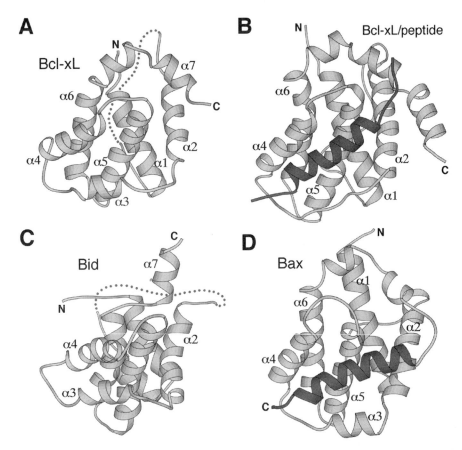

Fig. 2. Structure of the Bcl-2 family proteins. (**A**) Structure of Bcl-x_L. The flexible loop linking helices $\alpha 1$ and $\alpha 2$ are represented by a dotted line. (**B**) Structure of Bcl-x_L bound to a BH3 peptide from Bad. This Bad peptide exists as an amphipathic helix, with the hydrophobic side binding to Bcl-x_L. (**C**) Structure of the uncleaved form of Bid. Cleavage after Asp59 results in the activation of Bid, presumably due to exposure of the BH3 helix. (**D**) Structure of the full-length Bax. Note that the C-terminal amphipathic helix folds back to bind a hydrophobic surface groove, resembling the Bcl-x_L-bound Bad peptide.

is lacking. Despite structural information on all three subfamilies of Bcl-2 members, how these proteins regulate apoptosis remains largely unknown. Biophysical and structural characterization of the Bcl-2 protein complexes under membrane-like conditions will likely reveal some surprising insights, although it is also possible that other important regulators of Bcl-2 proteins are yet to be identified.

CASPASES: EXECUTIONERS OF APOPTOSIS

Caspases are a family of highly conserved cysteine proteases that cleave after an aspartate in their substrates (*19*). The critical involvement of a caspase in apoptosis was first documented in 1993 (*20*), in which CED3 was found to be indispensable for the programmed cell death in the nematode *Caenorhabditis elegans*. Since then, compelling evidence has demonstrated that the mechanism of apoptosis is evolutionarily conserved, executed by caspases from worms to mammals. At least 14 distinct mammalian caspases have been identified (*21*) (*see* Chapter 1).

Caspases involved in apoptosis are divided into two groups: the initiator caspases, which include caspases-1, -2, -8, -9, and -10; and the effector caspases, which include caspases-3, -6, and -7. An

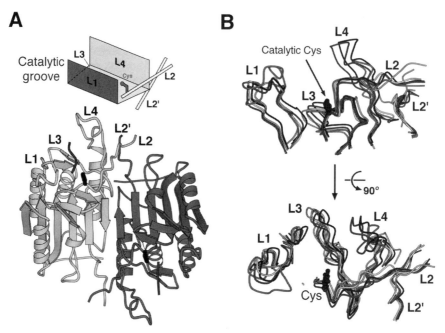

Fig. 3. Structural features of caspases. (**A**) A representative structure of the inhibitor-bound caspase-3 (PDB code 1DD1). The bound peptide inhibitor is shown in black. The four surface loops that constitute the catalytic groove of one hetero-dimer are labeled. The apostrophe denotes the other hetero-dimer. Note that L2' stabilizes the active site of the adjacent hetero-dimer. The substrate-binding groove is schematically shown above. (**B**) The active-site conformation of all known caspases is conserved. Of the four loops, L1 and L3 are relatively constant while L2 and L4 exhibit greater variability. The catalytic residue Cys is shown in black. Two perpendicular views are shown.

initiator caspase invariably contains an extended N-terminal prodomain (>90 amino acids) important for its function, whereas an effector caspase contains only 20–30 residues in its prodomain sequence. All caspases are synthesized in cells as catalytically inactive zymogens and must undergo proteolytic activation. The activation of an effector caspase, such as caspases-3 or -7, is performed by an initiator caspase, such as caspase-9, through internal cleavages to separate the large and small subunits. The initiator caspases, however, are auto-activated under apoptotic conditions.

The first caspase structure was determined on caspase-1 (or ICE, interleukin-1β-converting enzyme) bound with a covalent peptide inhibitor *(22,23)*. Structural information is now available on caspase-3 *(24,25)*, caspase-7 *(26)*, caspase-8 *(27,28)*, and more recently, caspase-9 *(29)*. In each case, caspase is bound to a synthetic peptide inhibitor (Fig. 3A). These structures reveal that the functional caspase unit is a homo-dimer, with each monomer comprising a large (~20 kDa) and a small (~10 kDa) subunit. Homo-dimerization is mediated by hydrophobic interactions, with six anti-parallel β-strands from each monomer forming a single contiguous 12-stranded β-sheet (Fig. 3A). Five α helices and five short ß strands are located on either side of the central ß-sheet, giving rise to a globular fold. The active sites, highly conserved among all caspases and located at two opposite ends of the β-sheet, are formed by four protruding loops (L1, L2, L3, and L4) from the scaffold.

Caspases recognize at least four contiguous amino acids, named P4–P3–P2–P1, in their substrates, and cleave after the C-terminal residue (P1), usually an Asp. The binding sites for P4–P3–P2–P1 are named S4–S3–S2–S1, respectively, in caspases. These sites are located in the catalytic groove. The L1 and L4 loops constitute two sides of the groove (Fig. 3). Loop L3 and the following β-hairpin, collectively referred to as L3, is located at the base of the groove. Loop L2, which harbors the catalytic residue Cys, is positioned at one end of the groove with Cys poised for binding and catalysis. These four loops, of which L1 and L3 exhibit conserved length as well as composition, determine the sequence specificity of the substrates.

The S1 and S3 sites are nearly identical among all caspases. The P1 residue (Asp) is coordinated by three invariant residues at the S1 site, an Arg from the L1 loop, a Gln at the beginning of the L2 loop, and an Arg at the end of the L3 loop. The Arg residue on the L3 loop also coordinates the P3 residue (Glu). The S2 and S4 sites are coordinated mainly by the L3 and L4 loops. Because the sequence of L4 is most divergent among caspases, the P2 and P4 residues exhibit greater sequence variation. For example, the L4 loop in caspases-1, -8, or -9 is considerably shorter than that in caspase-3 or -7, resulting in a shallower substrate-binding groove. This observation is consistent with a bulky hydrophobic residue as the preferred P4 residue for caspases-1, -8, or -9.

The conformational similarity at the active site is extended to surrounding structural elements. In particular, loops L4 and L2 from one catalytic subunit are stabilized by the N-terminus (loop L2') of the small subunit of the other catalytic subunit, forming the so-called "loop-bundle" *(30)*.

Most structural information is derived from the inhibitor-bound caspases, which share the same topology at the active site. These observations give the impression that the substrate-binding grooves of caspases are pre-formed. However, in the structure of the free caspase-7 *(30)*, these loops are flexible and quite different from those in the inhibitor-bound caspase-7 (Fig. 4), suggesting that the inhibitor-bound state is transient and trapped by the covalent peptide inhibitors. Thus, substrate binding and catalysis may be a process of induced-fit, accompanied by some large conformation changes, such as the back-and-forth flipping of the critical L2' loop.

Why are procaspase zymogens (except procaspase-9) catalytically inactive? The answer was partially provided by the crystal structure of procaspase-7 *(30,31)*, which reveals significant conformational changes in the four active site loops (Fig. 4). Except L1, all three other loops move away from their productive positions, unraveling the substrate-binding groove. Most notably, the loop-bundle seen in the inhibitor-bound caspases is missing in the procaspase-7 zymogen as the L2' loop is flipped by 180 degrees, existing in a "closed" conformation. This closed conformation is locked by the unprocessed nature of the procaspase-7 zymogen.

The ability of L2' to move freely in response to inhibitor/substrate binding is a decisive feature for the active caspase-7. This ability is acquired through activation cleavage after Asp198 in procaspase-7. Because L2' is at the N-terminus of the small subunit, inverting the order of primary sequences of the large and small subunits could free L2' and hence constitutively activate caspases. Indeed, this prediction was confirmed for caspases-3 and -6 *(32)* as well as for the *Drosophila* caspase drICE *(33)*.

In contrast to most other caspases, procaspase-9 exhibits a basal level of activity prior to proteolytic activation *(34)*. The surprising feature may be explained by the fact that procaspase-9 contains an expanded L2 loop, which could allow enough conformational flexibility such that procaspase-9 does not need an inter-domain cleavage to have the L2' loop move to its productive conformation.

INHIBITORS OF APOPTOSIS (IAP)

The inhibitor of apoptosis (IAP) family of proteins, originally identified in the genome of baculovirus, suppress apoptosis by interacting with and inhibiting the enzymatic activity of mature caspases *(35)* (*see* Chapter 3). At least eight distinct mammalian IAPs including XIAP, c-IAP1, c-IAP2, and ML-IAP/Livin, have been identified, and they all have anti-apoptotic activity in cell culture. A structural feature common to all IAPs is the presence of at least one BIR (baculoviral IAP repeat) domain, characterized by a conserved zinc-coordinating Cys/His motif ($CX_2CX_{16}HX_6C$). Some IAPs, such as XIAP and ML-IAP/Livin, also contain a C-terminal RING finger, a C_3HC_4-type

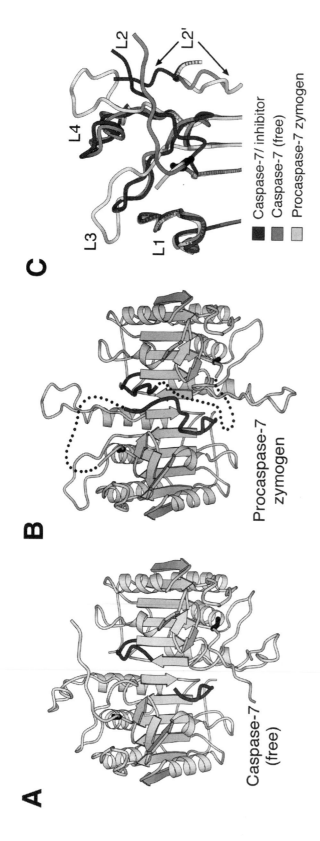

Fig. 4. Mechanisms of procaspase-7 activation and substrate binding. (**A**) Structure of an active and free caspase-7 (PDB code 1K88). The active-site loops are flexible. Despite an inter-domain cleavage, the L2' loop still exists in the closed conformation, indicating an induced-fit mechanism for binding to inhibitors/ substrates. (**B**) Structure of a procaspase-7 zymogen (PDB code 1K86). Compared to that of the inhibitor-bound caspase-7, the conformation of the active site loops does not support substrate-binding or catalysis. The L2' loop, locked in a closed conformation by covalent linkage, is occluded from adopting its productive and open conformation. (**C**) Comparison of the conformation of the active site loops. Compared to the procaspase-7 zymogen or the free caspase-7, the L2' loop is flipped 180° in the inhibitor-bound caspase-7, the L2' loop to stabilize loops L2 and L4.

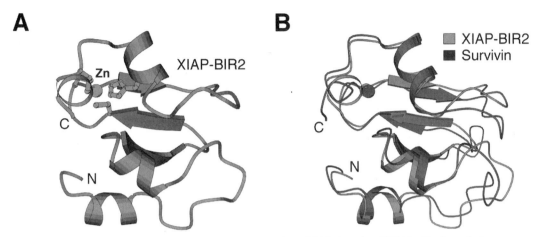

Fig. 5. Structure of the BIR domains. (**A**) Structure of the BIR2 domain of XIAP. The bound zinc atom as well as the four conserved Cys/His residues are labeled. (**B**) Superposition of the structure of XIAP-BIR2 with that of human survivin.

zinc-binding module. Most mammalian IAPs have more than one BIR domain, with the different BIR domains exhibiting distinct functions. For example, in XIAP, c-IAP1, and c-IAP2, the third BIR domain (BIR3) potently inhibits the activity of processed caspase-9, whereas the linker region between BIR1 and BIR2 selectively targets active caspase-3. The RING fingers were found to exhibit ubiquitin ligase (E3) activity, which may regulate self-destruction or degradation of active caspases through the 26S proteasome pathway.

The structures of various BIR domains, determined by both NMR and X-ray crystallography *(36–38)*, reveal a highly conserved topology, with a three-stranded anti-parallel β sheet and four α helices (Fig. 5A). The three cysteine and one histidine residues, invariant in all BIRs, coordinate a zinc atom. Although the BIR2 and BIR3 domains of XIAP share an identical fold, structure-based mutational analysis revealed that different regions are involved in the interaction with and the inhibition of caspases-3 and -9 *(36,37)*. Several amino acids in the linker sequence preceding BIR2 were found to be essential in targeting caspase-3, while residues on the surface of XIAP-BIR3 inhibited caspase-9.

The smallest IAP is survivin, with only one BIR domain and a C-terminal acidic stretch. In contrast to the relatively stable expression levels of other IAPs, expression of survivin oscillates with cell cycle and peaks at the G_2/M phase *(39)*. Recombinant survivin does not inhibit caspase activity in vitro and appears to play an important role in mitosis. Although the structures of survivin reveal that the BIR domain adopts the canonical fold (Fig. 5B), two contrasting modes of dimerization were proposed *(40–42)*, each with supporting evidence.

CASPASE-IAP COMPLEX

The IAP-bound structures were determined for two highly conserved effector caspases, caspases-3 and -7 *(43–45)* (Fig. 6A,B). In the structures, the linker peptide N-terminal to XIAP-BIR2 forms highly similar interactions with both caspases-3 and -7 (Fig. 6C). Compared to the covalent peptide inhibitors, the linker segment of XIAP occupies the active site of caspases, resulting in a blockade of substrate entry. Four residues from the XIAP linker peptide, Gly144–Val146–Val147–Asp148, occupy the corresponding positions for the P1–P2–P3–P4 residues of the substrates, respectively (Fig. 6D). The P1 position is occupied by the N-terminal Gly144 of these four residues. Thus this orientation is the reverse of that observed for the tetrapeptide caspase inhibitors, in which the P1 position is occupied by the C-terminal Asp. Interestingly, despite a reversal of relative orientation, a

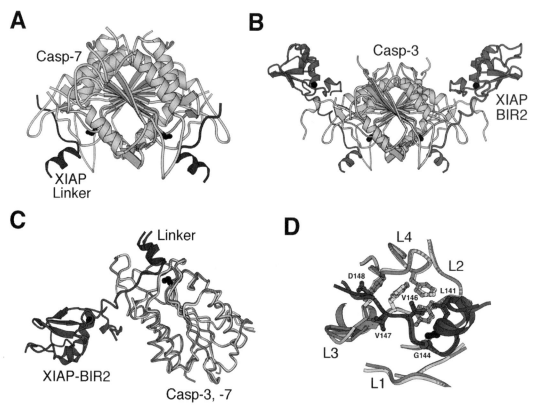

Fig. 6. Mechanisms of IAP-mediated inhibition of effector caspases. **(A)** Structure of caspase-7 bound with an XIAP linker peptide preceding the BIR2 domain. **(B)** Structure of caspase-3 bound with an XIAP fragment including BIR2 and its preceding linker peptide. **(C)** Superposition of the structures of caspases-3 and -7 together with their bound XIAP fragments. The interactions primarily occur between a linker segment N-terminal to the BIR2 domain of XIAP and the active site of caspases-3 or -7. **(D)** Close-up view of the active sites of caspases-3 and -7 bound to their respective XIAP fragments. Two hydrophobic residues of XIAP, Leu141 and Val146, make multiple van der Waals interactions with a conserved hydrophobic pocket on caspases-3 or -7. Asp148 of XIAP, occupying the S4 pocket, hydrogen bonds to neighboring residues in caspases-3 or -7.

subset of interactions between caspases-7 or -3 and XIAP closely resemble those between caspases-7 or -3 and its tetrapeptide inhibitor DEVD-CHO. Asp148 of XIAP binds the S4 pocket in the same manner as the P4 residue of the covalent peptide inhibitors. In addition, Val146 makes a similar set of van der Waals contacts to surrounding caspase residues as does the P2 residue.

Interestingly, although the linker sequence between BIR1 and BIR2 of XIAP plays a dominant role in inhibiting caspases-3 and -7, this fragment in isolation is insufficient. However, an engineered protein with the linker peptide fused either N- or C-terminal to BIR1 was able to bind and inhibit caspase-3 while neither BIR1 nor BIR2 in isolation exhibited any effect *(36)*. These observations suggest that the linker peptide needs to be presented in a "productive" conformation by surrounding BIR domains while their identities do not matter. In support of this hypothesis, the linker peptide fused to glutathione S-transferase (GST) was able to inhibit caspases-3 and -7 *(43,44)*. Nevertheless, the BIR domains also contribute to the inhibition of caspases, as XIAP exhibits about 20-fold higher potency than the GST-linker peptide fusion. Consistent with this observation, XIAP-BIR2 also makes direct contacts to caspase-3 in the crystal structure *(45)* (Fig. 6B).

Although the mode of IAP-mediated inhibition of the effector caspases is well-characterized, how IAPs inhibit the initiator caspase, caspase-9, remain unclear. Nevertheless, biochemical investigation has revealed an encouraging hint. Although both BIR2 and BIR3 of XIAP can inhibit caspase-9, BIR3 displays tighter binding affinity and higher potency. Mutation of Trp310 or Glu314 in BIR3 completely abrogated XIAP-mediated inhibition of caspase-9 while amino acids outside of BIR3 are unnecessary for this inhibition *(37)*. In the structure, these two residues are located close to each other, suggesting that this region is involved in binding and inhibiting caspase-9. Indeed, BIR3 of XIAP binds to the N-terminus of the small subunit of caspase-9 that becomes exposed after proteolytic processing *(46)*. This binding presumably brings BIR3 close to the active site of caspase-9, which may block substrate entry and subsequent catalysis. Supporting this hypothesis, when the N-terminal four residues of the caspase-3 small subunit were replaced with those from caspase-9, the resulting protein could be inhibited by BIR3. Because BIR3 binds the N-terminus of a flexible 20-residue loop in caspase-9, exactly how BIR3 prevents substrate entry remains to be investigated by structural approaches.

CASPASE-P35 COMPLEX

IAPs are not the only natural inhibitors to caspases. In contrast to XIAP, which only affects caspases-3, -7, and -9, the baculoviral p35 protein is a pan-caspase inhibitor, and it potently targets most caspases both in vivo and in vitro *(47)*. Caspase inhibition by p35 correlates with the cleavage of its reactive site loop after Asp87, which leads to the translocation of the N-terminus of p35 into the active site of caspases *(48,49)*. The crystal structure of caspase-8 in complex with p35 reveals that the catalytic residue Cys360 of caspase-8 is covalently linked to the Asp87 of p35 through a thioester bond *(50)* (Fig. 7). Although a thioester bond is generally susceptible to hydrolysis, this bond is protected by the neighboring N-terminus of p35, which occludes access by water molecules.

Compared with the structure of the uncleaved p35 *(51)*, the cleavage after Asp87 of the reactive site loop produces a dramatic conformational switch *(50,52)*. The C-terminal end of the reactive site loop flips 180 degrees and folds back to form a ß strand as part of an anti-parallel ß-sheet. The vacated space is in turn occupied by the N-terminus of p35, which springs from a closed conformation into an open form, moving over a distance of 20 Å (Fig. 7C). Much of this conformational change occurs in the absence of binding to caspase-8.

Another protein, the serpin CrmA derived from the cowpox virus, can also inhibit several caspases, likely through similar covalent modification. This unique covalent mechanism adds to the complexity of caspase inhibition by natural proteins. It is important to note, however, that the equivalent of p35 in mammalian genome has not been identified.

SMAC/DIABLO

In normal surviving cells that have not received an apoptotic stimulus, aberrant activation of caspases can be inhibited by IAPs. In cells signaled to undergo apoptosis, however, this inhibitory effect must be suppressed. This process is mediated by a mitochondrial protein named Smac (second mitochondria-derived activator of caspases) *(53)* or DIABLO (direct IAP binding protein with low pI) *(54)*. Smac, synthesized in the cytoplasm, is targeted to the inter-membrane space of mitochondria. Upon apoptotic stimuli, Smac is released from mitochondria into the cytosol, together with cytochrome *c*. Whereas cytochrome *c* directly activates Apaf-1 and caspase-9, Smac interacts with multiple IAPs and relieves their inhibitory effect on both initiator and effector caspases.

Structural analysis reveals that Smac exists as an elongated dimer in solution, spanning over 130 Å in length *(55)* (Fig. 8A). The wild-type Smac protein binds to both the BIR2 and the BIR3 domains of XIAP but not BIR1. In contrast, the monomeric Smac mutants retain strong interaction with BIR3 but can no longer form a stable complex with BIR2. Because the linker sequence immediately preceding BIR2 is involved in binding and inhibiting caspase-3, Smac monomers cannot relieve the IAP-

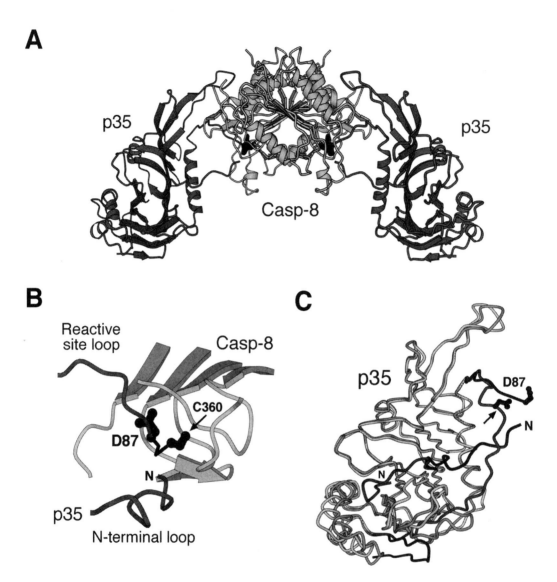

Fig. 7. Mechanisms of p35-mediated pan-caspase inhibition. (**A**) An overall view of caspase-8 covalently bound to its inhibitor p35. (**B**) A close-up view of the covalent inhibition of caspase-8 by p35. The thioester intermediate is shown between Asp87 of p35 and Cys360 (active site residue). The N-terminus of p35 restricts solvent access to this intermediate. (**C**) Superposition of the structure of uncleaved p35 with that bound to caspase-8. The arrow indicates the position of proteolytic cleavage.

mediated inhibition of caspase-3. Despite maintenance of interactions with BIR3, the monomeric Smac mutants also exhibit compromised activity in relieving IAP-mediated inhibition of caspase-9.

The N-terminal mitochondria-targeting sequence of Smac is proteolytically removed upon import. The freshly exposed N-terminal residues play an indispensable role for Smac function; a 7-residue peptide including these residues can remove the IAP-mediated inhibition of caspase-9. Strikingly, a missense mutation of the N-terminal residue Ala to Met in Smac leads to a complete loss of interactions with XIAP and the concomitant loss of Smac function *(55)*.

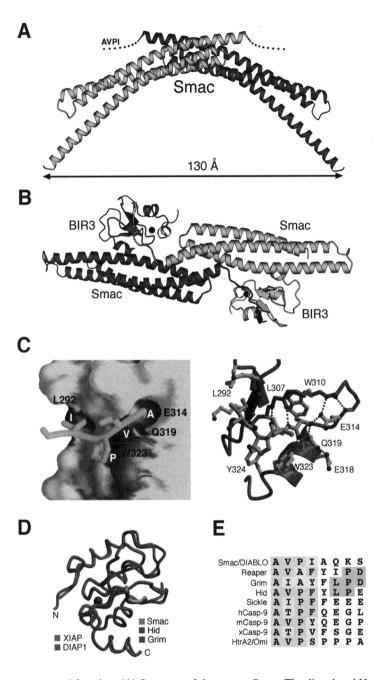

Fig. 8. Smac structure and function. (**A**) Structure of the mature Smac. The disordered N-terminal residues are shown as dotted lines. (**B**) Structure of a monomeric Smac bound to the BIR3 domain of XIAP. (**C**) A close-up view of the Smac N-terminal tetrapeptide bound to the BIR3 surface groove. The BIR3 domain is shown either by degree of hydrophobicity (left panel) or in ribbon diagram (right panel) to highlight the interactions. The amino and carbonyl groups of the N-terminal Ala make several hydrogen bonds to conserved residues in XIAP. (**D**) A conserved IAP-binding mode from mammals to fruit flies. The structure of DIAP1-BIR2 is super-imposed with that of the XIAP-BIR3 domain, with their corresponding bound peptides Hid, Grim, and Smac/DIABLO. (**E**) A conserved motif of IAP-binding tetrapeptides. The tetrapeptide motif has the consensus sequence A-(V/T/I)-(P/A)-(F/Y/I/V/S). The *Drosophila* proteins have an additional binding component (conserved 6th-8th residues).

In fruit flies, the anti-death function of the *Drosophila* IAP, DIAP1, is removed by four pro-apoptotic proteins, Reaper, Grim, Hid, and Sickle, which physically interact with the BIR2 domain of DIAP1 and remove its inhibitory effect on *Drosophila* caspases *(56)*. Thus Reaper, Grim, Hid, and Sickle represent the functional homologs of the mammalian protein Smac/DIABLO. Although generally regarded as having similar functions in *Drosophila* development, Hid, Grim, Reaper, and Sickle only share homology in the N-terminal four residues in their primary sequences. Interestingly, these four residues are very similar to the N-terminal residues of mature Smac.

SMAC-IAP COMPLEX

The molecular explanation for the indispensable role of the Smac N-terminal sequences is provided by structures of XIAP-BIR3 bound to either a monomeric Smac protein *(57)* (Fig. 8B) or a nine-residue Smac peptide *(58)*. The Smac N-terminal tetrapeptide (Ala-Val-Pro-Ile) recognizes an acidic surface groove on BIR3, with the first residue Ala binding a hydrophobic pocket and making hydrogen bonds to neighboring XIAP residues (Fig. 8C). The next three residues also interact with surrounding hydrophobic residues in BIR3. To accommodate these interactions, the N-terminus of Smac must be free, thus explaining why only the mature Smac can bind to IAPs. Modeling studies indicate that replacement of Ala by a bulkier residue will cause steric hindrance, whereas Gly substitution may result in an entropic penalty as well as loss of binding in the hydrophobic pocket. This analysis explains why mutation of Ala to Met or Gly abrogated interactions with the BIR domains.

The Smac-binding surface groove on XIAP-BIR3 comprises highly conserved residues among the BIR3 domains of c-IAP1 and c-IAP2 and the BIR2 domain of DIAP1, suggesting a conserved binding mode. Indeed, the crystal structures of DIAP1-BIR2 by itself and in complex with the N-terminal peptides from Grim, Hid, and Sickle reveal that the binding of these N-terminal tetrapeptides precisely match that of the Smac-XIAP interactions *(59)* (Fig. 8D). For Grim and Hid, the next three conserved residues also contribute to DIAP1-binding through hydrophobic interactions.

Thus the tetrapeptides in the N-termini of the mammalian Smac/DIABLO and the Drosophila Reaper, Grim, Hid, and Sickle define an evolutionarily conserved family of IAP-binding motif (Fig. 8E). Interestingly, caspase-9 also contains such a tetrapeptide motif (Ala-Thr-Pro-Phe) in the N-terminus of the small subunit. Subsequent experiments confirmed that this sequence is indeed primarily responsible for the interactions between the processed caspase-9 and XIAP *(46)*. In the absence of proteolytic processing, procaspase-9 is unable to interact with IAPs. Proteolytic processing of procaspase-9 at Asp315 leads to the exposure of an internal tetrapeptide motif, which recruits IAPs to inhibit caspase-9. The mature Smac binds IAPs, again using a similar N-terminal tetrapeptide. The conserved IAP-binding motif in Smac competes with that in caspase-9 for the binding to IAPs and thus releases caspase-9 from the inhibition by IAPs.

During apoptosis, caspase-9 can be further cleaved after Asp330 by downstream caspases such as caspase-3. This positive feedback not only permanently removes XIAP-mediated caspase-9 inhibition but also releases a 15-residue peptide that is able to relieve IAP-mediated inhibition of other caspases. This mechanism ensures that less than stoichiometric amount of Smac can remove the IAP-mediated caspase-9 inhibition as transient activation of caspase-9 may lead to the activation of caspase-3 and ensuing positive feedback.

Although the Smac tetrapeptide in isolation can remove IAP-mediated caspase-9 inhibition, it plays a less direct role in the removal of IAP-mediated inhibition of effector caspases. The binding site for this tetrapeptide motif maps to the surface of BIR2 or BIR3 whereas the fragment responsible for inhibiting caspases-3 or -7 is located between BIR1 and BIR2 of XIAP. Although a conclusive mechanism remains elusive, modeling studies of a Smac/BIR2/caspase-3 complex suggest that steric clashes preclude XIAP-BIR2 from simultaneously binding to caspase-3 and Smac *(43)*. In this model, binding to the BIR2 domain requires not only the N-terminal tetrapeptide of Smac but also an extensive surface available only in the wild-type dimeric Smac protein. This model is consistent with the observation that monomeric Smac mutants only weakly interacted with BIR2 and were unable to remove the IAP-mediated caspase-3 inhibition.

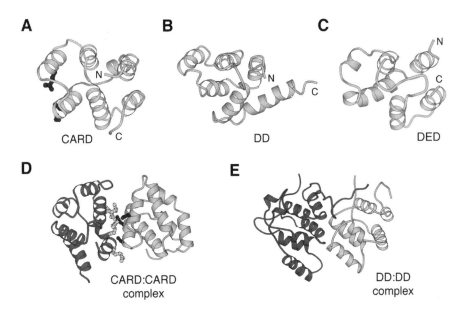

Fig. 9. Structure of signaling modules. **(A)** Structure of Apaf-1 CARD. The acidic residues important for caspase-9 binding are shown in black. **(B)** Structure of the death domain (DD) of Fas. **(C)** Structure of the death effector domain (DED) of FADD. **(D)** Structure of a hetero-dimer between the CARD domains of Apaf-1 and caspase-9. Critical interface residues are shown. **(E)** Structure of a hetero-dimer between the death domains of Pelle and Tube proteins in *Drosophila*.

SIGNALING MODULES IN APOPTOSIS

During apoptosis, caspase recruitment and activation require three highly conserved families of signaling modules, the caspase recruitment domain (CARD), the death domain (DD), and the death effector domain (DED) *(1)*. These motifs share the same structural topology with different surface features, which give rise to specific homotypic recognition.

Structures of representative members of these signaling motifs have been determined by both NMR and X-ray crystallography *(60–62)*. In each case, the topology consists of six anti-parallel helices, with minor variation in inter-helical packing (Fig. 9). Mutational analyses on these domains revealed critical residues important for interactions with other signaling motifs. In two cases, these mutagenesis results were further confirmed with crystal structures of complexes, each involving two similar domains.

The recognition of procaspase-9 by Apaf-1, primarily through a CARD-CARD interaction, is essential to the formation of the apoptosome holoenzyme and subsequent activation of caspases. The positively charged surface of procaspase-9 CARD formed by the helices H1a/H1b and H4 is recognized by Apaf-1 CARD through a negatively charged surface formed by the helices H2 and H3 *(63)* (Fig. 9D). In *Drosophila*, recruitment of the Ser/Thr kinase Pelle to the plasma membrane by the adapter protein Tube is important for embryogenesis. The structure of a death domain complex between Pelle and Tube reveals an interesting addition to the diverse recognition mechanisms by these simple motifs *(64)* (Fig. 9E). Despite these advances, it is structurally unclear how these signaling motifs function in the context of a large signaling complex such as the DISC or the apoptosome.

APOPTOSOME AND THE ACTIVATION OF INITIATOR CASPASES

The initiator caspases invariably contain one of two protein-protein interaction motifs, the CARD or the DED. These motifs directly interact with similar motifs present on oligomerized adapter proteins, thus bringing multiple initiator caspase molecules into close proximity of one another and presumably facilitating their auto-activation. This hypothesis is summarized as the induced-proximity model *(65)*. For example, procaspase-8 contains two copies of the DED, which interact with the DED of FADD. Through association with FADD, procaspase-8 is brought into the DISC, resulting in its auto-activation. Another well-characterized paradigm involves procaspase-9 activation. During apoptosis, cytochrome *c* is released from mitochondria into the cytoplasm, where it assembles Apaf-1 into an apoptosome in the presence of dATP or ATP *(2)*. The primary function of the apoptosome is to recruit procaspase-9 and facilitate its auto-activation.

The induced proximity model is consistent with a number of observations. For example, bacterially expressed caspases can be processed to their mature forms, presumably due to high local concentrations. In mammalian cells, forced oligomerization of procaspase-8 led to its activation and subsequent apoptosis *(66,67)*. Similar results were obtained for the mammalian caspase-9 and the *C. elegans* CED3 *(68)*.

Although induced proximity undoubtedly leads to caspase activation, it remains unclear whether this is indeed how the initiator caspases are activated under physiological conditions. The strongest supporting evidence for the induced proximity hypothesis also serves as a cautionary reminder. For example, effector caspases can also be auto-activated through induced proximity; yet they are activated in vivo by the initiator caspases. In addition, unprocessed procaspase-9 is nearly as active as the mature caspase-9 and the primary function of the apoptosome is to allosterically enhance caspase-9 activity rather than to facilitate its auto-activation. Furthermore, forced oligomerization of the initiator caspases may not recapitulate the physiological context, in terms of protein expression levels, and more importantly, in terms of the specific protein-protein interactions that are required for the precise positioning and activation of the initiator caspases.

Surprisingly, the isolated caspase-9 is only marginally active in the absence of the apoptosome, prompting the concept of a holo-enzyme *(69)*. Through association with the apoptosome, the catalytic activity of caspase-9 is enhanced about three orders of magnitude. Intriguingly, the unprocessed caspase-9 can be similarly maintained in this "hyperactive" state, demonstrating that the proteolytic processing is unnecessary for the activation of procaspase-9.

Although most other caspases exist exclusively as a homo-dimer in solution, caspase-9 was found to exist mostly as a monomer at micromolar concentrations *(29)*. Fractions corresponding to dimers and monomers from gel filtration were separately analyzed for their catalytic activity, and only the dimer fractions were found to be active. Biochemical as well as structural analyses revealed that dimerization resulted in the formation of only one functional active site. Based on these observations, it was proposed that dimer formation may drive the activation of caspase-9 *(29)*.

The three-dimensional structure of the ~1.4 MDa apoptosome at 27 Å resolution reveals a wheel-shaped heptameric complex, with the CARD domains located at the central hub and the WD40 repeats at the extended spokes *(70)* (Fig. 10A). Docking of caspase-9 to this apoptosome resulted in a dome-shaped structure in the center; however, the bulk of the caspase-9 was not visible in these EM studies (Fig. 10B). The central domes were thought to be complexes involving the CARD domains from Apaf-1 and caspase-9.

Based on structural information on caspase-9 and the apoptosome, a model was proposed to explain the activation of procaspase-9 *(70)*. In this model, a heptameric apoptosome binds seven monomers of inactive caspase-9. The high local concentrations of caspase-9 within this apoptosome drive the efficient recruitment of additional inactive caspase-9 monomers, which become activated upon binding. This interesting model has a strong assumption that caspase-9 activity in the active dimers is identical to that in the apoptosome holoenzyme, which remains to be experimentally tested.

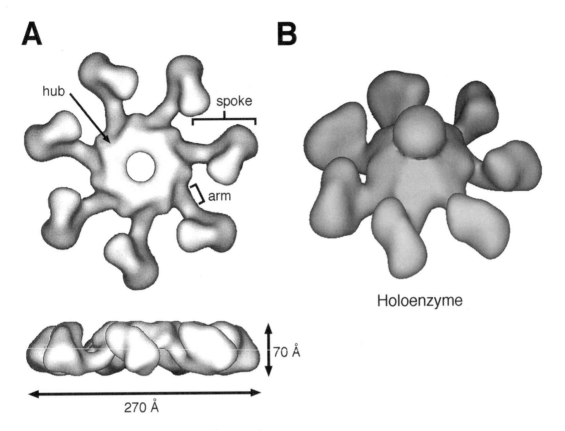

Fig. 10. Structure of the apoptosome. **(A)** Structure of the apoptosome at 27 Å resolution by electron micro-scopy. The CARD domain of Apaf-1 was interpreted to reside in the central hub of this wheel-shaped structure. **(B)** Structure of the caspase-9-bound apoptosome. Caspase-9 binding induces a dome-shaped structure in the center of the apoptosome.

CONCLUSION

The rapid progress in the characterization of apoptosis by structural biology has significantly enhanced our understanding of the underlining mechanisms. However, many daunting tasks remain. For example, we still know very little about the activation mechanisms of the initiator caspases. In this respect, future effort should be directed at solving structures of higher-order protein complexes and performing associated biochemical and biophysical analysis. Furthermore, the apoptotic mechanisms will likely become more complex with the discovery and characterization of additional players and pathways, which will present structural biologists exciting new challenges for years to come.

REFERENCES

1. Fesik, S. W. (2000) Insights into programmed cell death through structural biology. *Cell* **103**, 273–282.
2. Shi, Y. (2001) A structural view of mitochondria-mediated apoptosis. *Nature Struct. Biol.* **8**, 394–401.
3. Ashkenazi, A. and Dixit, V. M. (1998) Death receptors: signaling and modulation. *Science* **281**, 1305–1308.
4. Jones, E. Y., Stuart, D. I., and Walker, N. P. (1989) Structure of tumour necrosis factor. *Nature* **338**, 225–228.
5. Eck, M. J. and Sprang, S. R. (1989) The structure of tumor necrosis factor alpha at 2.6 Å resolution: implications for receptor binding. *J. Biol. Chem.* **264**, 17595–17605.
6. Eck, M. J., Ultsch, M., Rinderknecht, E., de Vos, A. M., and Sprang, S. R. (1992) The structure of human lymphotoxin (tumor necrosis factor beta) at 1.9 Å resolution. *J. Biol. Chem.* **267**, 2119–2122.

7. Banner, D. W., D'Acry, A., Janes, W., Gentz, R., Schoenfeld, H.-J., Broger, C., et al. (1993) Crystal structure of the soluble human 55 kd TNF receptor-human TNF-ß complex: implications for TNF receptor activation. *Cell* **73,** 431–445.

8. Hymowitz, S. G., Christinger, H. W., Fuh, G., Ultsch, M., O'Connell, M., Kelley, R. F., et al. (1999) Triggering cell death: the crystal structure of Apo2L/TRAIL in a complex with death receptor 5. *Mol. Cell* **4,** 563–571.

9. Mongkolsapaya, J., Grimes, J. M., Chen, N., Xu, X.-N., Stuart, D. I., Jones, E. Y., and Screaton, G. R. (1999) Structure of the TRAIL-DR5 complex reveals mechanisms conferring specificity in apoptotic initiation. *Nature Struct. Biol.* **6,** 1048–1053.

10. Naismith, J. H., Devine, T. Q., Brandhuber, B. J., and Sprang, S. R. (1995) Crystallographic evidence for dimerization of unliganded tumor necrosis factor receptor. *J. Biol. Chem.* **270,** 13303–13307.

11. Naismith, J. H., Devine, T. Q., Kohno, T., and Sprang, S. R. (1996) Structures of the extracellular domain of the type I tumor necrosis factor receptor. *Structure* **4,** 1251–1262.

12. Adams, J. and Cory, S. (1998) The Bcl-2 protein family: arbiters of cell survival. *Science* **281,** 1322–1326.

13. Muchmore, S. W., Sattler, M., Liang, H., Meadows, R. P., Harlan, J. E., Yoon, H. S., et al. (1996) X-ray and NMR structure of human Bcl-xL, an inhibitor of programmed cell death. *Nature* **381,** 335–341.

14. Losonczi, J. A., Olejniczak, E. T., Betz, S. F., Harlan, J. E., Mack, J., and Fesik, S. W. (2000) NMR studies of the anti-apoptotic protein Bcl-x_L in micelles. *Biochemistry* **39,** 11024–11033.

15. Sattler, M., Yoon, H. S., Nettesheim, D., Meadows, R. P., Harlan, J. E., Eberstadt, M., et al. (1997) Structure of Bcl-xL-Bak peptide complex: recognition between regulators of apoptosis. *Science* **275,** 983–986.

16. Chou, J. J., Li, H., Salvesen, G. S., Yuan, J., and Wagner, G. (1999) Solution structure of BID, an intracellular amplifier of apoptotic signaling. *Cell* **96,** 615–624.

17. McDonnell, J. M., Fushman, D., Milliman, C. L., Korsmeyer, S. J., and Cowburn, D. (1999) Solution structure of the proapoptotic molecule BID: a structural basis for apoptotic agonists and antagonists. *Cell* **96,** 625–634.

18. Suzuki, M., Youle, R. J., and Tjandra, N. (2000) Structure of Bax: coregulation of dimer formation and intracellular localization. *Cell* **103,** 645–654.

19. Thornberry, N. A. and Lazebnik, Y. (1998) Caspases: enemies within. *Science* **281,** 1312–1316.

20. Yuan, J., Shaham, S., Ledoux, S., Ellis, H. M., and Horvitz, H. R. (1993) The *C. elegans* cell death gene Ced-3 encodes a protein similar to mammalian interleukin-1 beta-converting enzyme. *Cell* **75(4),** 641–652.

21. Shi, Y. (2002) Mechanisms of caspase inhibition and activation during apoptosis. *Mol. Cell* **9,** 459–470.

22. Walker, N. P., Talanian, R. V., Brady, K. D., Dang, L. C., Bump, N. J., Ferenz, C. R., et al. (1994) Crystal structure of the cysteine protease interleukin-1β-converting enzyme: a (p20/p10)2 homodimer. *Cell* **78,** 343–352.

23. Wilson, K. P., Black, J.-A., Thomson, J. A., Kim, E. E., Griffith, J. P., Navia, M. A., et al. (1994) Structure and mechanism of interleukin-1β converting enzyme. *Nature* **370,** 270–275.

24. Rotonda, J., Nicholson, D. W., Fazil, K. M., Gallant, M., Gareau, Y., Labelle, M., et al. (1996) The three-dimensional structure of apopain/CPP32, a key mediator of apoptosis. *Nature Struct. Biol.* **3,** 619–625.

25. Mittl, P. R., Di Marco, S., Krebs, J. F., Bai, X., Karanewsky, D. S., Priestle, J. P., et al. (1997) Structure of recombinant human CPP32 in complex with the tetrapeptide acetyl-Asp-Val-Ala-Asp fluoromethyl ketone. *J. Biol. Chem.* **272,** 6539–6547.

26. Wei, Y., Fox, T., Chambers, S. P., Sintchak, J.-A., Coll, J. T., Golec, J. M. C., et al. (2000) The structures of caspases-1, -3, -7 and -8 reveal the basis for substrate and inhibitor selectivity. *Chem. Biol.* **7,** 423–432.

27. Blanchard, H., Kodandapani, L., Mittl, P. R. E., Di Marco, S., Krebs, J. K., Wu, J. C., et al. (1999) The three dimensional structure of caspase-8: an initiator enzyme in apoptosis. *Structure* **7,** 1125–1133.

28. Watt, W., Koeplinger, K. A., Mildner, A. M., Heinrikson, R. L., Tomasselli, A. G., and Watenpaugh, K. D. (1999) The atomic-resolution structure of human caspase-8, a key activator of apoptosis. *Structure* **7,** 1135–1143.

29. Renatus, M., Stennicke, H. R., Scott, F. L., Liddington, R. C., and Salvesen, G. S. (2001) Dimer formation drives the activation of the cell death protease caspase 9. *Proc. Natl. Acad. Sci. USA* **98,** 14250–14255.

30. Chai, J., Shiozaki, E., Srinivasula, S. M., Alnemri, E. S., and Shi, Y. (2001) Crystal structure of a procaspase-7 zymogen: mechanisms of activation and substrate binding. *Cell* **107,** 399–407.

31. Riedl, S. J., Fuentes-Prior, P., Renatus, M., Kairies, N., Krapp, S., Huber, R., et al. (2001) Structural basis for the activation of human procaspase-7. *Proc. Natl. Acad. Sci. USA* **98,** 14790–14795.

32. Srinivasula, S. M., Ahmad, M., MacFarlane, M., Luo, Z., Huang, Z., Fernandes-Alnemri, T., and Alnemri, E. S. (1998) Generation of constitutively active recombinant caspase-3 and -6 by rearrangement of their subunits. *J. Biol. Chem.* **273,** 10107–10111.

33. Wang, S., Hawkins, C., Yoo, S., Muller, H.-A., and Hay, B. (1999) The Drosophila caspase inhibitor DIAP1 is essential for cell survival and is negatively regulated by HID. *Cell* **98,** 453–463.

34. Stennicke, H. R., Deveraux, Q. L., Humke, E. W., Reed, J. C., Dixit, V. M., and Salvesen, G. S. (1999) Caspase-9 can be activated without proteolytic processing. *J. Biol. Chem.* **274,** 8359–8362.

35. Deveraux, Q. L. and Reed, J. C. (1999) IAP family proteins: suppressors of apoptosis. *Genes Dev.* **13,** 239–252.

36. Sun, C., Cai, M., Gunasekera, A. H., Meadows, R. P., Wang, H., Chen, J., et al. (1999) NMR structure and mutagenesis of the inhibitor-of-apoptosis protein XIAP. *Nature* **401,** 818–822.

37. Sun, C., Cai, M., Meadows, R. P., Xu, N., Gunasekera, A. H., Herrmann, J., et al. (2000) NMR structure and mutagenesis of the third BIR domain of the inhibitor of apoptosis protein XIAP. *J. Biol. Chem.* **275,** 33777–33781.

38. Hinds, M. G., Norton, R. S., Vaux, D. L., and Day, C. L. (1999) Solution structure of a baculoviral inhibitor of apoptosis (IAP) repeat. *Nature Struct. Biol.* **6(7),** 648–651.

39. Li, F., Ambrosini, G., Chu, E. Y., Plescia, J., Tognin, S., Marchisio, P. C., and Altieri, D. C. (1998) Control of apoptosis and mitotic spindle checkpoint by survivin. *Nature* **396,** 580–584.

40. Muchmore, S. W., Chen, J., Jakob, C., Zakula, D., Matayoshi, E. D., Wu, W., et al. (2000) Crystal structure and mutagenic analysis of the inhibitor-of-apoptosis protein survivin. *Mol. Cell* **6,** 173–182.

41. Verdecia, M. A., Huang, H.-K., Dutil, E., Kaiser, D. A., Hunter, T., and Noel, J. P. (2000) Structure of the human anti-apoptotic protein survivin reveals a dimeric arrangement. *Nat. Struc. Biol.* **7,** 602–608.

42. Chantalat, L., Skoufias, D. A., Kleman, J.-P., Jung, B., Dideberg, O., and Margolis, R. L. (2000) Crystal structure of human survivin reveals a bow-tie shaped dimer with two unusual alpha helical extensions. *Mol. Cell* **6,** 183–189.

43. Chai, J., Shiozaki, E., Srinivasula, S. M., Wu, Q., Datta, P., Alnemri, E. S., and Shi, Y. (2001) Structural basis of caspase-7 inhibiton by XIAP. *Cell* **104,** 769–780.

44. Huang, Y., Park, Y. C., Rich, R. L., Segal, D., Myszka, D. G., and Wu, H. (2001) Structural basis of caspase inhibition by XIAP: differential roles of the linker versus the BIR domain. *Cell* **104,** 781–790.

45. Riedl, S. J., Renatus, M., Schwarzenbacher, R., Zhou, Q., Sun, C., Fesik, S. W., et al. (2001) Structural basis for the inhibition of caspase-3 by XIAP. *Cell* **104,** 791–800.

46. Srinivasula, S. M., Saleh, A., Hedge, R., Datta, P., Shiozaki, E., Chai, J., et al. (2001) A conserved XIAP-interaction motif in caspase-9 and Smac/DIABLO mediates opposing effects on caspase activity and apoptosis. *Nature* **409,** 112–116.

47. Miller, L. K. (1999) An exegesis of IAPs: salvation and surprises from BIR motifs. *Trends Cell Biol.* **9,** 323–328.

48. Bump, N. J., Hackett, M., Hugunin, M., Seshagiri, S., Brady, K., Chen, P., et al. (1995) Inhibition of ICE family proteases by baculovirus antiapoptotic protein p35. *Science* **269,** 1885–1888.

49. Zhou, Q., Krebs, J. F., Snipas, S. J., Price, A., Alnemri, E. S., Tomaselli, K. J., and Salvesen, G. S. (1998) Interaction of the baculovirus anti-apoptotic protein p35 with caspases. Specificity, kinetics, and characterization of the caspase/p35 complex. *Biochemistry* **37,** 10757–10765.

50. Xu, G., Cirilli, M., Huang, Y., Rich, R. L., Myszka, D. G., and Wu, H. (2001) Covalent inhibition revealed by the crystal structure of the caspase-8/p35 complex. *Nature,* **410,** 494–497.

51. Fisher, A. J., Cruz, W. D., Zoog, S. J., Schneider, C. L., and Friesen, P. D. (1999) Crystal structure of baculovirus p35: role of a novel reactive site loop in apoptotic caspase inhibition. *EMBO J.* **18,** 2031–2040.

52. Dela Cruz, W. P., Friesen, P. D., and Fisher, A. J. (2001) Crystal structure of baculovirus p35 reveals a novel conformational change in the reactive site loop after caspase cleavage. *J. Biol. Chem.* **276,** 32933–32939.

53. Du, C., Fang, M., Li, Y., and Wang, X. (2000) Smac, a mitochondrial protein that promotes cytochrome *c*-dependent caspase activation during apoptosis. *Cell* **102,** 33–42.

54. Verhagen, A. M., Ekert, P. G., Pakusch, M., Silke, J., Connolly, L. M., Reid, G. E., et al. (2000) Identification of DIABLO, a mammalian protein that promotes apoptosis by binding to and antagonizing IAP proteins. *Cell* **102,** 43–53.

55. Chai, J., Du, C., Wu, J.-W., Kyin, S., Wang, X., and Shi, Y. (2000) Structural and biochemical basis of apoptotic activation by Smac/DIABLO. *Nature* **406,** 855–862.

56. Hay, B. A. (2000) Understanding IAP function and regulation: a view from *Drosophila. Cell Death Differ.* **7(11),** 1045–1056.

57. Wu, G., Chai, J., Suber, T. L., Wu, J.-W., Du, C., Wang, X., and Shi, Y. (2000) Structural basis of IAP recognition by Smac/DIABLO. *Nature* **408,** 1008–1012.

58. Liu, Z., Sun, C., Olejniczak, E. T., Meadows, R. P., Betz, S. F., Oost, T., et al. (2000) Structural basis for binding of Smac/DIABLO to the XIAP BIR3 domain. *Nature* **408,** 1004–1008.

59. Wu, J.-W., Cocina, A. E., Chai, J., Hay, B. A., and Shi, Y. (2001) Structural analysis of a functional DIAP1 fragment bound to grim and hid peptides. *Mol. Cell* **8,** 95–104.

60. Huang, B., Eberstadt, M., Olejniczak, E. T., Meadows, R. P., and Fesik, S. W. (1996) NMR structure and mutagenesis of the Fas (APO-1/CD95) death domain. *Nature* **384,** 638–641.

61. Eberstadt, M., Huang, B., Chen, Z., Meadows, R. P., Ng, S.-C., Zheng, L., et al. (1998) NMR structure and mutagenesis of the FADD (Mort1) death-effector domain. *Nature* **392,** 941–945.

62. Chou, J. J., Matsuo, H., Duan, H., and Wagner, G. (1998) Solution structure of the RAIDD CARD and model for CARD/CARD interaction in caspase-2 and caspase-9 recruitment. *Cell* **94,** 171–180.

63. Qin, H., Srinivasula, S. M., Wu, G., Fernandes-Alnemri, T., Alnemri, E. S., and Shi, Y. (1999) Structural basis of procaspase-9 recruitment by the apoptotic protease-activating factor 1. *Nature* **399,** 547–555.

64. Xiao, T., Towb, P., Wasserman, S. A., and Sprang, S. R. (1999) Three-dimensional structure of a complex between the death domains of Pelle and Tube. *Cell* **99,** 545–555.

65. Salvesen, G. S. and Dixit, V. M. (1999) Caspase activation: The induced-proximity model. *Proc. Natl. Acad. Sci. USA* **96,** 10964–10967.

66. Muzio, M., Stockwell, B. R., Stennicke, H. R., Salvesen, G. S., and Dixit, V. M. (1998) An induced proximity model for caspase-8 activation. *J. Biol. Chem.* **273(5),** 2926–2930.

67. Yang, X., Chang, H. Y., and Baltimore, D. (1998) Autoproteolytic activation of pro-caspases by oligomerization. *Mol. Cell* **1,** 319–325.

68. Yang, X., Chang, H. Y., and Baltimore, D. (1998) Essential role of CED-4 oligomerization in CED-3 activation and apoptosis. *Science* **281,** 1355–1357.

69. Rodriguez, J. and Lazebnik, Y. (1999) Caspase-9 and Apaf-1 form an active holoenzyme. *Genes Dev.* **13(24),** 3179–3184.

70. Acehan, D., Jiang, X., Morgan, D. G., Heuser, J. E., Wang, X., and Akey, C. W. (2002) Three-dimensional structure of the apoptosome: implications for assembly, procaspase-9 binding and activation. *Mol. Cell,* **9,** 423–432.

71. Kraulis, P. J. (1991) Molscript: a program to produce both detailed and schematic plots of protein structures. *J. Appl. Crystallogr.* **24,** 946–950.

5

The Death Receptor Family and the Extrinsic Pathway

Maria Eugenia Guicciardi and Gregory J. Gores

INTRODUCTION

Death receptors are cell-surface cytokine receptors belonging to the tumor necrosis factor/nerve growth factor (TNF/NGF) receptor superfamily that trigger apoptosis after binding a group of structurally related ligands or specific antibodies *(1–3)*. The members of this family are type-I transmembrane proteins with a C-terminal intracellular tail, a membrane-spanning region, and an extracellular ligand-binding N-terminal domain. They are characterized by a significant homology in a region containing one to five cysteine-rich repeats in their extracellular domains, and in a 60- to 80-amino acid cytoplasmic sequence known as death domain (DD), which typically enables death receptors to initiate the death signal.

Death receptors are activated through an interaction with their natural ligands, a group of complementary cytokines that belongs to the TNF family of proteins. With the exception of the soluble, lymphocyte-derived cytokine Lymphotoxin alpha (LTα), these proteins (known as death ligands) are type-II transmembrane proteins, comprised of an intracellular N-terminal domain, a transmembrane region, and a C-terminal extracellular tail. Death ligands can also be released as soluble cytokines by the cleavage of metalloproteases.

Signal transduction by death receptors is initiated by the oligomerization of the receptor triggered upon juxtaposition of the intracellular domains that follows the engagement of the ligand to the receptor's extracellular domain. This event leads to recruitment of different adapter proteins, which provide the link between the receptor and the cell-death effectors, namely, the so-called initiator caspases (e.g., caspase-8 and caspase-10). Adapter proteins generally have no enzymatic activity of their own, but are able to associate with receptors through homophilic interaction of the receptor's DD and an analogous DD on the adapter itself. Adapter proteins may also contain a death-effector domain (DED) that mediates the recruitment of caspases through the association with a correspondent DED or a caspase-recruitment domain (CARD) in the prodomain of the inactive initiator caspases (e.g., caspase-8). The resulting complex is called the death-inducing signaling complex (DISC). The proximity of several caspase molecules recruited to the receptor results in self-processing and activation of the caspase, likely through a mild proteolytic activity of the procaspase itself. The activated initiator caspase subsequently starts a cascade of caspase activation by processing and activating the so-called effector caspases (e.g., caspase-3, caspase-6, and caspase-7), which are directly or indirectly

From: *Essentials of Apoptosis: A Guide for Basic and Clinical Research*
Edited by: X-M. Yin and Z. Dong © Humana Press Inc., Totowa, NJ

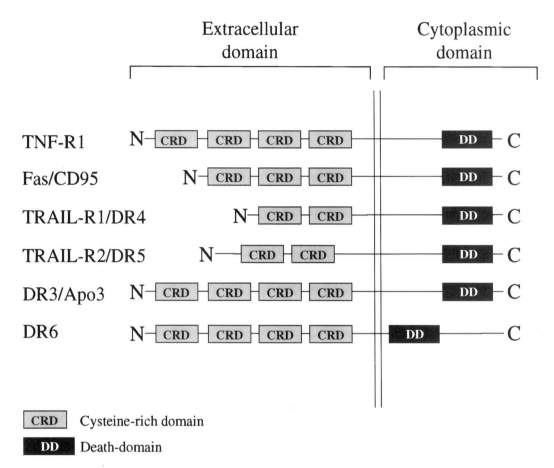

Fig. 1. Structural comparison of the death receptors. Schematic models of the death receptors. The extracellular domains are characterized by the presence of a variable number of cysteine-rich motifs (CRD), whereas the intracellular tails contain the death domain (DD), essential for signaling apoptosis.

responsible for the cleavage and degradation of several crucial cellular proteins, and for the execution of cell death. Alternatively, initiator caspases may cleave other substrates, which induce mitochondrial dysfunction (e.g., cleavage of Bid), and activate the effector caspases through the release of pro-apoptotic mitochondrial factors. After the apoptotic signal is triggered, the DISC is internalized, where it dissociates at low pH.

Currently, six death receptors are known, including the well-characterized death receptors Fas (also called CD95 or APO-1) and TNF-R1 (also called p55 or CD120a), TRAIL-R1 (also called DR4), TRAIL-R2 (also called DR5 or APO-2 or KILLER), death receptor 3 (DR3; also called APO-3 or TRAMP or WSL-1 or LARD), and death receptor 6 (DR6) (Fig. 1).

Since the cloning of the first death receptor about 10 years ago, hundreds of reports have been published providing valuable information on these receptors, yet the understanding of the complex signaling originating from them seems still to be incomplete. The purpose of this chapter is to provide an updated overview of the most important death receptors, and their role in both human and animal physiology and patho-physiology.

FAS (CD95/APO-1) AND FAS LIGAND (FASL/CD95L)

Fas (CD95/APO-1)

Fas (CD95/APO-1) is a glycosylated cell-surface protein, consisting of 325–335 amino acids residues, with a molecular weight of 45–52 kDa. The gene encoding for Fas is located on the long arm of chromosome 10 in humans and on chromosome 19 in the mouse. It is ubiquitously expressed in various tissues, but is particularly abundant in thymus, liver, heart, kidney, and in activated mature lymphocytes or virus-transformed lymphocytes. Although the membrane-bound form is largely predominant, several splice variants have been described that generate soluble forms of the receptor, the function of which is still unclear *(4)*. These soluble forms of the receptor may antagonize Fas-mediated cytotoxicity by binding to and inactivating Fas ligand (FasL), thereby exerting an anti-apototic effect. A tight regulation of Fas-mediated apoptosis is essential for the proper physiology of the cell, and this is achieved through a variety of mechanisms, some of which are operated directly at the level of the receptor. Fas localizes on the plasma membrane as well as in the cytosol, in particular, in the Golgi complex and the trans-Golgi network *(5,6)*. Translocation of Fas-containing vesicles to the cell surface has been observed upon stimulation, providing an effective mechanism to regulate the plasma-membrane density of the death receptor, and avoid its spontaneous activation *(6,7)*. Fas-mediated apoptosis can also be modulated by glycosylation of the receptor *(8)*, as well as at the transcriptional level, by directly regulating Fas expression. A composite binding site for the transcription factor NF-κB has been described at position 295 to 286 of the Fas gene promoter, which regulates activation-dependent Fas expression in lymphocytes *(9)*. A p53-responsive element is also located within the first intron of the Fas gene, and cooperates with three sequences in the promoter to upregulate both Fas receptor and Fas ligand expression during drug-induced apoptosis of leukemic and hepatocellular carcinoma cell lines *(10–12)*. Other mechanisms of regulation of Fas-mediated apoptosis signaling are described later in this chapter.

Fas Ligand (FasL/CD95L)

FasL (CD95L) is a 40 kDa type II transmembrane protein with homotrimeric structure *(12a)*. It is expressed on the cell surface of activated T cells and, together with its receptor, plays an important role in the maintenance of the peripheral T- and B-cell homeostasis, and in the killing of harmful cells, such as virus-infected cells or cancer cells *(13–15)*. FasL can also be proteolytically cleaved by a metalloprotease between Ser126 and Leu127 in its extracellular domain, generating a soluble, trimeric form whose biological activity remains controversial *(16)*. Indeed, soluble FasL with apoptosis-inducing ability has been described. However, serum FasL levels are often high in hepatitis, AIDS, and several types of tumor without apparent consequences *(17)*. Species-related differences in the function of the soluble form are possible, as soluble human FasL is able to induce apoptosis, whereas soluble mouse FasL is not *(18)*. However, recent studies have demonstrated that the apoptotic-inducing capacity of the soluble form is reduced by over 1000-fold compared to the membrane-bound FasL, providing an explanation for the absence of tissue damage in diseases associated with elevated circulating levels of FasL *(16,19,20)*. In lymphoma cells, expression of soluble FasL does not trigger apoptosis, but has been shown to suppress the inflammatory response *(21)*.

Fas Signaling

Engagement of Fas by either agonistic antibodies or FasL leads to the formation of microaggregates of the Fas receptor followed by recruitment of the adaptor molecule FADD (Fas-associated protein with death domain)/MORT-1 (mediator of receptor-induced toxicity) *(22)*. FADD/MORT-1 is a ubiquitously expressed, 28 kDa cytosolic protein with a C-terminal death domain, and a DED at the N-terminus. FADD associates with the receptor through its DD, while its DED is required for self-association and binding procaspase-8. Recruitment and accumulation of procaspase-8 at the DISC

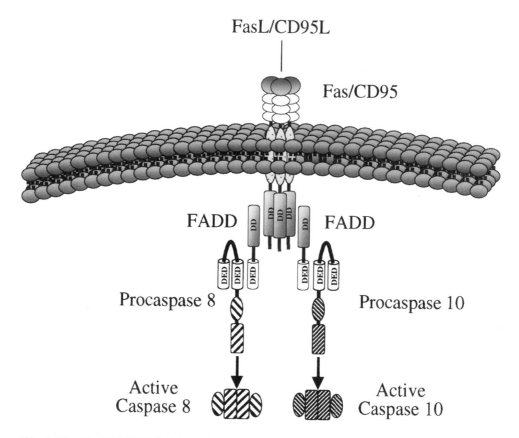

Fig. 2. The Fas/CD95 DISC. Schematic representation of the death-inducing signaling complex (DISC) triggered by engagement of FasL/CD95L to Fas/CD95. The death domain (DD) on the adapter protein FADD interacts with the receptor's death domain, whereas the death effector domain (DED) binds the correspondent death effector domain in the pro-domain of the initiator caspase.

results in spontaneous activation of caspase-8 via autoproteolytic cleavage, and initiation of the apoptosis signal. Procaspase-10 has also been demonstrated to be recruited and activated at the Fas DISC (Fig. 2). Recent studies have shown that procaspases-8 and -10 are processed with similar kinetics, and both can initiate apoptosis independently of each other *(23,24)*.

Several proteins encoded by viral genes are able to prevent apoptosis by inhibiting caspases. One of the most potent inhibitors of caspase-8 is the viral serpin CrmA (cytokine response-modifier A), a toxin produced by the cowpox virus, which blocks apoptosis by binding to the active protease, thus preventing further activation of effector caspases. CrmA is widely used in apoptosis studies to effectively block caspase-8-mediated cell death. It is still controversial whether CrmA also inhibits caspase-10.

The signal downstream of DISC formation differs between cell types. Two classifications, type I and type II, of Fas-mediated apoptosis-signaling pathways have been described *(25)* (Fig. 3). In type I cells, large amounts of caspase-8 are activated at the DISC and closely followed by rapid cleavage of caspase-3. Overexpression of the anti-apoptotic proteins Bcl-2 or Bcl-X_L does not prevent activation of caspase-8 or caspase-3 in these cells, nor does it inhibit apoptosis, suggesting a mitochondria-

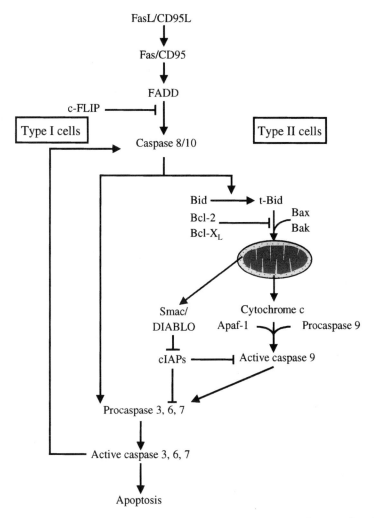

Fig. 3. Fas/CD95-mediated apoptotic pathways. Schematic representation of Fas-mediated apoptotic pathways in type I and type II cells. See text for details.

independent activation of a caspase cascade. In contrast, DISC formation in type II cells is strongly reduced, and activation of caspases, including caspase-8, occurs mainly downstream of mitochondria, because both caspase activation and apoptosis can be blocked by overexpression of Bcl-2 or Bcl-X$_L$. Notably, Fas triggers the activation of mitochondria in both type I and type II cells, and in both cell types, the apoptogenic activity of mitochondria is blocked by overexpression of Bcl-2 or Bcl-X$_L$. However, only in type II cells, and not in type I cells, overexpression of Bcl-X$_L$ or Bcl-2 blocks apoptosis. Therefore, only in type II cells are mitochondria essential for the execution of the apoptotic program, whereas in type I cells, mitochondrial dysfunction likely functions as an amplifier of the apoptotic signal.

Mitochondrial dysfunction during Fas signaling is mediated by caspase-8 cleavage of Bid, a proapoptotic, BH3-only, member of the Bcl-2 family of proteins. The resulting 15 kDa fragment (tBid) translocates to the mitochondria and induces release of apoptogenic factors, such as cytochrome *c*, AIF (apoptosis-inducing factor) *(26)* and the recently identified Smac/Diablo (second mitochondria-

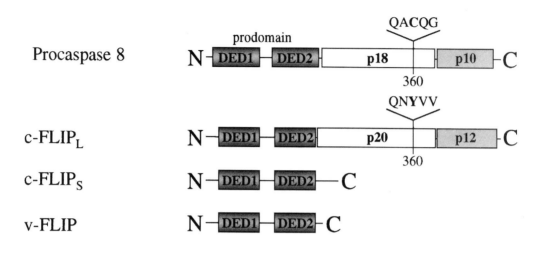

DED1   Death Effector Domain 1
DED2   Death Effector Domain 2

Fig. 4. Structural comparison of caspase-8 and FLIPs. Schematic models of caspase-8, c-FLIP$_{S/L}$, and v-FLIP. Each molecule contains two death effector domains (DED) at the N-terminal. Caspase-8 and c-FLIP$_L$ also show high sequence homology in the C-terminal domain, containing both the large (p18–20) and the small (p10–12) catalytic subunits. The pentapeptide QACQG (or QACRG), highly conserved in all caspases, but lacking in c-FLIP$_L$, is critical for the catalytic activity of the proteins.

derived activator of caspases/direct IAP binding protein with low pI)*(27,28)*. Once in the cytosol, cytochrome *c* binds to the co-factor Apaf-1 (apoptosis-activating factor 1) and to procaspase-9, forming a complex called apoptosome. Through an energy-requiring reaction, procaspase-9 is processed to the mature enzyme and in turn activates caspase-3, starting a caspase cascade downstream the mitochondrium (*see* Chapters 2 and 6).

Endogenous Inhibitors of Fas Signaling

A fine regulation of Fas signaling is required to avoid triggering unnecessary cell death and to ensure proper functioning of the apoptotic machinery. Recently, a family of proteins of viral origin with anti-apoptotic activity was identified and called v-FLIPs (viral FLICE-inhibitory proteins) (Fig. 4). v-FLIPs contain two DEDs that enable them to bind to the Fas DISC, as well as to several other death receptors, and block caspase-8 activation *(29)*. It has not been established yet whether v-FLIP also inhibits caspase-10 activation. A human cellular homolog has also been identified and called c-FLIP (also FLAME-1 or I-FLICE or Casper or CASH or MRIT or CLARP or Usurpin) *(30)*. c-FLIP occurs in a short and a long isoform, as a result of different splice variants. The short form, c-FLIP$_S$, consists only of two DEDs, and structurally resembles v-FLIP. The long form, c-FLIP$_L$ (also called I-FLICE, for inhibitor of FLICE) consists of two DEDs and a caspase-like domain, and closely resembles caspase-8, except that it contains an inactive enzymatic site. In particular, c-FLIP$_L$ lacks the catalytic cysteine embedded in the conserved pentapeptide QACRG or QACQG motif present in all caspases, and other key residues involved in catalysis and substrate binding, therefore showing no cysteine protease activity. Both forms of c-FLIP are recruited and bind to the DISC upon stimulation

Table 1
Experimental Tools to Inhibit Death Receptor-Mediated Apoptosis

Reagents	*Mechanism*
IETD-fmk	Caspase-8 inhibition
CrmA	Caspase-8 inhibition
c-FLIP	DISC disruption
DN FADD	DISC disruption

(31). Whereas c-FLIP$_S$ may competitively inhibit procaspase-8 recruitment to the DISC, c-FLIP$_L$ allows procaspase-8 to be recruited to the DISC and even partially cleaved from a 55 kDa proform to 41 kDa and 43 kDa polypeptides. However, by unclear mechanisms, c-FLIP$_L$ prevents further proteolytic processing of caspase-8 to generate the active 18 kDa (p18) and 10 kDa (p10) subunits *(32)*.

Experimental Tools for Inhibiting Death Receptor-Mediated Apoptosis

From the previous overview, it is clear that several approaches can be used to inhibit Fas-mediated apoptosis, or to implicate death receptor signaling in an apoptotic process (Table 1). The most common approaches include the inhibition of caspase-8 by a pharmacological inhibitor, IETD-fmk, or transfection of the cells with an expression vector encoding CrmA. The assembly of the DISC can also be disrupted by forced overexpression of FLIP, which ultimately prevents caspase-8 activation. Likewise, the overexpression of a dominant negative FADD, which contains the death domain, but lacks the DED and, therefore, cannot bind to caspase-8 or -10, also prevents DISC assembly. All of these tools are frequently used in the laboratory to study the role of death receptors in an apoptotic process.

Physiology and Pathophysiology

A balance between cell death and cell proliferation is required to maintain tissue homeostasis. Not surprisingly, excess or lack of apoptosis always leads to disease pathogenesis. As long as death receptors are appropriately expressed, they represent a powerful tool to execute apoptosis in a controlled manner. However several diseases have been associated with either loss of function or overexpression of death receptors, which results in too little or too much apoptosis, respectively. As far as the Fas/FasL system is concerned, its crucial role in vivo was clarified by the study of two spontaneous recessive mutations in mice, *lpr* (lymphoproliferation) and *gld* (generalized lymphoproliferative disease). These mutations are phenotypically associated with systemic autoimmunity, lymphadenopathy, and splenomegaly with significant accumulation of CD4$^-$ CD8$^-$ T-lymphocytes *(33,34)*. Molecular analyses of *lpr* and *gld* mutations showed that they are loss of function mutations in the Fas and FasL genes, respectively *(35)*. Impairment of the Fas/FasL system causes increased resistance of T-cells to activation-induced apoptosis, and as a result, mature CD4$^-$ CD8$^-$ T-cells can no longer be eliminated, and therefore accumulate in lymph nodes and spleen *(36,37)*. A phenotype similar to that seen in *lpr* mice has also been described in humans affected by autoimmune lymphoproliferative syndrome (ALPS) *(38,39)*. These patients, most of which are children, carry a heterozygous mutation in the Fas gene, which generates a defective protein that either lacks its normal function or works as dominant-negative when expressed with normal Fas. As a consequence, ALPS patients are incapable of effectively downregulating the immune reaction, and develop lymphoadenopathy, splenomegaly, hypergammaglobulinemia, and, in some cases, autoimmune diseases such as hemolytic anemia, thrombocytopenia, and neutropenia, due to the production of antibodies against red blood cells and platelets. Indeed, abnormally prolonged survival of activated lymphocytes may result in persistent humoral immune responses and potentially harmful cross-reactions with self-antigens, which can eventually lead to development of autoimmune diseases.

Downregulation or loss of Fas expression has also been observed in several tumors, often associated with constitutive expression of FasL. This adaptation allows the cancer cell to survive the attack by cytotoxic T lymphocytes and natural killer (NK) cells by binding Fas on their surface and inducing apoptosis, while at the same time increasing its own resistance to Fas-mediated apoptosis *(40,41)*. Consistently, the role of Fas in carcinogenesis is supported by the evidence that Fas-defective animals show increased risk of developing tumors.

On the other hand, unregulated overexpression the Fas/FasL system can be equally deleterious, causing massive tissue destruction and even leading to death of the organism. Injection of an agonistic anti-Fas antibody or recombinant FasL into mice has been shown to strongly activate the Fas system in vivo and quickly kill the animals as a consequence of acute liver failure *(42)*. The symptoms of the liver failure closely resemble those of fulminant hepatitis, which results from massive hepatocyte apoptosis caused by the reaction between abnormally activated T-cells and hepatitis B or C virus-transformed hepatocytes overexpressing Fas. Thus, overexpression of Fas may be a cause of fulminant hepatitis. The best example of this in human pathology is a disease characterized by an accumulation of copper in the liver. This transition metal promotes oxidative stress and *de novo* expression of FasL in the liver. The FasL-expressing hepatocytes induce apoptosis in Fas-expressing neighboring cells (fratricide killing) causing liver damage. Similar mechanisms may also be important in alcoholic hepatitis. Overexpression of Fas is also known to play a role in AIDS.

TUMOR NECROSIS FACTOR-RECEPTOR 1 (TNF-R1) AND TNF-α

TNF-R1

The TNF/TNF-receptor signaling system consists of two distinct receptors, TNF-R1 (also called p55 or CD120a) and TNF-R2 (also called p75 or CD120b), and three ligands, the membrane-bound TNF-α (mTNF-α), the soluble TNF-α (sTNF-α), and the soluble lymphocyte-derived cytokine (LTα, also called TNF-β) *(22)*. TNF-R1 and TNF-R2 are both type I transmembrane proteins containing an amino-terminus, disulfide-rich, extracellular domain that recognizes TNF, a transmembrane helix, and a cytoplasmic tail. However, TNF-R1 only, and not TNR-R2, possesses an intracellular DD, and, therefore, is likely to be the sole mediator of the apoptotic signal. Both receptors are expressed ubiquitously in cells; however, TNF-R1 expression seems to be constitutively low and controlled by a noninducible promoter, whereas TNF-R2 expression is regulated inducibly by a number of extracellular stimuli. TNF-R1 and TNF-R2 interact with both forms of TNF-α, as well as with LTα. Nonetheless, TNF-R1 appears to be entirely responsible for TNF signaling in most cell types, and in vivo studies in a TNF-R1-deficient mouse model showed that TNF-R1 is essential for TNF-induced apoptosis of pathogen-infected cells.

TNF-α

Tumor necrosis factor-α (TNF-α) was originally named for its ability to elicit hemorrhagic necrosis of transplanted mouse tumors and for its selective cytotoxicity for transformed cells. In the years that followed, TNF-α was found to play a key role in inflammation and immunity. Moreover, TNF-α is able to induce proliferation and differentiation of many different target cells. TNF-α is mainly produced by macrophages, monocytes, and T cells in response to infection and inflammatory conditions, but also by other cell types, such as B cells, fibroblasts, and hepatocytes. TNF-α is expressed as a 26 kDa integral type II transmembrane protein with homotrimeric structure. From this precursor, a 17 kDa soluble form is released after cleavage by the metalloprotease TNF-α-converting enzyme

(TACE) *(43)*. Both soluble and membrane-bound TNF-α are biologically active, and while the soluble form acts as an effector molecule at a distance from the producer cell, the membrane-bound form likely has a specific role in localized TNF-α responses.

TNF-R1 Signaling

Intracellular signals originating from the TNF-R1 are extremely complex and can lead to multiple, even opposite, cell responses, from cell proliferation to inflammation to cell death. Most cells treated with TNF-α, however, do not undergo apoptosis unless protein or RNA synthesis is blocked, which suggests the predominance of survival signals over death signals under normal circumstances, and the requirement of neosynthesized proteins to suppress the apoptotic stimulus. The expression of these anti-apoptotic proteins is likely to be controlled by the activity of the transcription factor NF-κB, as inhibition of NF-κB sensitizes cells to TNF-α-induced apoptosis. JNK (c-Jun NH$_2$-terminal kinase)/SAPK (stress-activated protein kinase) is also activated by NF-κB, and considerable data now suggest JNK induces the transcription of pro-apoptotic genes in many cell types *(44,45)*. Indeed, NF-κB complexes downregulate the JNK cascade through upregulation of *gadd45β/myd118*, a gene associated with cell-cycle control and DNA repair, and *xiap*, which encodes the endogenous inhibitor of apoptosis, XIAP, and promotes cell survival *(44,45)*.

Engagement of TNF-R1 by TNF-α results in conformational changes in the receptor's intracellular domain, resulting in rapid recruitment of several cytoplasmic DD-containing adapter proteins via homophilic interaction with the DD of the receptor *(22)* (Fig. 5). The unstimulated TNF-R1 has been found to be associated with the recently isolated SODD (silencer of death domains), which effectively prevents self-aggregation of the DD and spontaneous initiation of signaling *(46)*. Upon stimulation, SODD promptly dissociates from TNF-R1, allowing the adapter protein TRADD (TNFR-associated protein with death domain) to bind to the clustered-receptor DD (Fig. 6). TRADD functions as a docking protein that recruits several signaling molecules to the activated receptor, such as FADD, TRAF-2 (TNF-associated factor-2), RIP (receptor-interacting protein), and RAIDD (RIP-associated ICH-1/CED-3-homologous protein with a death domain). These proteins have no enzymatic activity, except for RIP, which possesses serine-threonine kinase activity. The role of its kinase activity in apoptosis, however, remains to be established; it may have a role, however, in mediating death receptor-induced nonapoptotic cell death *(47)*. FADD binds and activates caspase-8, promoting apoptosis through a pathway similar, though probably not identical, to that triggered by Fas. RIP binds RAIDD, which engages a death pathway by recruitment of caspase-2 through homophylic interaction between homologous sequences in its amino-terminal domain and the prodomain of caspase-2 *(48)*. RIP also associates with TRAF-2, stimulating prosurvival pathways and regulating the immune response. Two distinct pathways originate from the association of RIP and TRAF-2 to the receptor *(49)*. The first one signals through the activation of the protein kinase NIK (NF-κB-inducing kinase), which further activates the catalytic IKK complex (IκB kinase complex), comprised of the three proteins IKKα (IKK1), IKKβ (IKK2), and IKKγ (NEMO), leading to phosphorylation of the NF-κB inhibitory protein IκBα. Phospho-IκBα is then degraded via the ubiquitin-proteasome pathway, allowing NF-κB to translocate to the nucleus and initiate transcription of target genes. The second pathway involves the mitogen-activated protein (MAP) kinases, and leads to activation of JNK, via the consequential activation of MEKK-1 (mitogen-activated protein/Erk kinase kinase-1), and JNKK (JNK kinase). JNK phoshorylates and activates of a number of transcription factors, including c-Jun, ATF-2 (activating-transcription factor 2), and AP-1. Moreover, TRAF-2 has been shown to bind to the anti-apoptotic factors cIAP-1 and -2 (cellular inhibitor of apoptosis-1 and -2), and TRAF-1, to form a receptor signaling complex that inhibits the apoptotic TRADD/FADD/caspase-8 pathway, possibly by facilitating the ubiquitination and degradation of caspase-8 *(50,51)*.

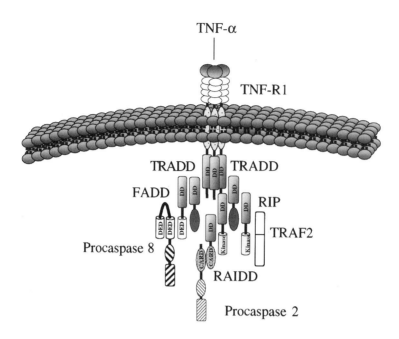

Fig. 5. The TNF-R1 DISC. Schematic representation of the death-inducing signaling complex (DISC) triggered by engagement of TNF-α to TNF-R1. See text for details.

TNF-α also activates other MAP kinases, namely p38 and ERK kinase, but much less is known about these signaling pathways in cell death. However, studies employing knockout animals suggest that MKK-3 (MAP kinase kinase-3) is required for TNF-α-induced p38 activation *(52)*.

Physiology and Pathophysiology

The TNF-α/TNF-R system is an important mediator in various physiological and pathophysiological conditions. The complicated signaling pathways originating from the receptor complex, indeed, impact different biological processes, such as cell proliferation, cell death, and inflammation.

After partial hepatectomy, biliary epithelial cells and venous endothelial cells rapidly secrete large amounts of TNF-α, thus elevating the hepatic level of the cytokine. By signaling through TNF-R1, TNF-α promotes the proliferative response of liver cells by both stimulating the transcriptional activity of AP-1 and NF-κB in the hepatocytes, and inducing secretion of another cytokine critical in hepatic regeneration, interleukin-6 (IL-6), from Kupffer cells or other nonparenchymal cells *(53)*. IL-6, in turn, binds to its receptor on the hepatocytes and induces activation of a third transcription factor, STAT-3 *(54)*. As a result of this concerted transcriptional activity, hepatocytes are forced to leave G_0 and enter the proliferative stages of the cell cycle. Thus it seems TNF-α acts as an initiator and potentiator of hepatocyte proliferation and liver regeneration. Consistently, liver regeneration is severely impaired in TNF-R1 knockout mice *(54)*.

The role of TNF-α as mediator of cell death as been described in a wide variety of liver diseases. Serum TNF-α levels are elevated in patients with alcoholic hepatitis and directly correlate with increased mortality. In these patients, TNF-α has been show to signal via the TNF-R1 pathway, inducing both apoptosis and necrosis of the hepatic parenchyma *(55)*. Serum TNF-α levels are also higher in patients with both fulminant and chronic hepatitis, and seem to correlate, at least in fulminant hepatitis, with the severity of the disease. The serum levels of soluble TNF-R1 and TNF-R2 are also significantly elevated in chronic hepatitis B. A massive production of TNF-α by blood mononuclear cells has been observed in chronic hepatitis-B patients undergoing interferon-α (IFN-α)-

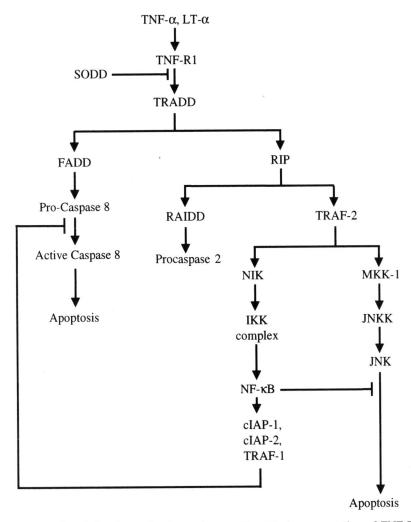

Fig. 6. TNF-R1-mediated signal transduction pathways. Simplified representation of TNF-R1-mediated apoptotic and anti-apoptotic pathways in the cell. See text for details.

treatment at the time of successful antigen seroconversion, suggesting that TNF-α may be involved in viral clearance *(56)*.

The study of genetic models of knockout animals demonstrated the crucial role of the TNF/TNF-R1 system in immune and inflammatory responses. The apparently normal phenotype of TNF-R1 and TNF knockout mice suggest that TNF is not essential for embryonic development. However, mice lacking TNF-R1 show some abnormalities in the lymphoid organs, such as lack of Peyer's patches, and impaired differentiation of follicular dentritic cells and formation of germinal centers. They also produce a reduced antibody response after immunization, and display an increased susceptibility to *Listeria monocytogenes* and *Mycobacterium tuberculosis*, infections usually controlled by the mice *(57–59)*. TNF knockout mice show essentially the same characteristics, except that they do develop Peyer's patches *(60)*. Those genetic models allowed the demonstration that TNF has a dual role during the inflammatory process: a pro-inflammatory role in the initial phase of infection and inflammation, and an anti-inflammatory/repair function after the infectious or toxic agent has been localized and controlled *(60)*.

Several human immune diseases have been found to be associated with a dysregulation of the TNF/TNF-R system. In a genetic disease called TNF receptor-associated periodic syndrome (TRAPS), heterozygous dominant alleles of TNF-R1, with amino acid changes in the extracellular domain, cause a decrease in TNF-R1 shedding, and enhancement of the pro-inflammatory effects of TNF, which results in severe localized inflammation and development of familial periodic fever *(61)*. TRAPS has been successfully treated with soluble TNFR2–Ig fusion protein constructs, which act as decoy receptors for the cytokine. Fusion constructs containing immunoglobulin or leucin zipper oligomerization domains allow to achieve a high-level activity of the soluble ligand or receptor.

Overproduction of TNF has been associated with several autoimmune conditions. Elevated levels of TNF-α have been detected in diseases such as rheumatoid arthritis (RA), inflammatory bowel disease (IBD; Crohn's disease), multiple sclerosis (MS), and systemic lupus erythematosus (SLE). Several clinical trials for the treatment of RA and Crohn's disease have utilized anti-TNF-α MAbs to neutralize the pro-inflammatory cytokine with successful results, demonstrating that this approach may be an effective therapeutic option in the treatment of chronic inflammatory diseases *(62)*.

TRAIL RECEPTORS AND TRAIL

TRAIL Receptors

Since the identification of the new member of the TNF family, TNF-related apoptosis-inducing ligand (TRAIL, also called APO-2L), five different cognate receptors have been shown to bind it (Fig. 7). Two of them, TRAIL-R1 (also called DR4) and TRAIL-R2 (also called DR5 or Killer or TRICK2) are considered actual death receptors, as engagement of TRAIL with these receptors results in apoptosis. TRAIL-R1 and TRAIL-R2 are plasma-membrane receptors consisting of an extracellular region containing two cysteine-rich domains, which enable the receptors to bind TRAIL, a transmembrane domain, and a cytoplasmic tail containing a DD essential for transducing the apoptotic signal. TRAIL-R1 is expressed in most human tissues, including spleen, thymus, liver, peripheral blood leukocytes, activated T cells, small intestine, and some tumor cell lines *(63)*. TRAIL-R2 expression has a ubiquitous distribution both in normal tissue and tumor cell lines, but is particularly high in spleen, peripheral blood leukocytes, and activated lymphocytes. Two other receptors, TRAIL-R3 (also called DcR1 or TRID or LIT) and TRAIL-R4 (DcR2/TRUNDD), are so-called decoy receptors. Although quite similar to TRAIL-R1 and TRAIL-R2 in their extracellular and transmembrane regions, TRAIL-R3 and TRAIL-R4 lack a death domain, or possess a nonfunctional death domain, respectively, therefore binding of TRAIL to these receptor fails to trigger apoptosis. TRAIL-R3 and TRAIL-R4 transcripts are almost ubiquitously expressed in healthy human tissues, but not in most cancer cell lines *(63)*. The preferential distribution of the decoy receptors in normal tissues, together with their ability to compete with the death receptors for binding to TRAIL, has been thought to account for the higher resistance of normal cells to TRAIL-induced apoptosis. However, recent studies suggest intracellular regulation of TRAIL signaling more likely accounts for the resistance of healthy cells to TRAIL-mediated apoptosis, rather than the surface density of decoy receptors. The fifth identified receptor for TRAIL is the soluble osteoprotegerin receptor (OPG), which also binds the osteoclast differentiation factor (ODF), another member of the TNF family. OPG also can act as a decoy receptor, because it efficiently binds TRAIL, but does not induce apoptosis. At 37°C, TNR-R2 has the highest affinity for TRAIL, whereas OPG has the weakest *(64)*.

TRAIL

TRAIL/APO-2 was first identified in 1995 through screening of DNA databases based on sequence homology with other members of the TNF family *(65,66)*. In particular, among the TNF family, TRAIL was found to share the highest sequence homology with FasL, but did not bind Fas or any of the other previously known receptors of the TNF-R-family. TRAIL gene is located in chromosome 3, and its mRNA is expressed constitutively in many tissues. Like the other ligands of the same family,

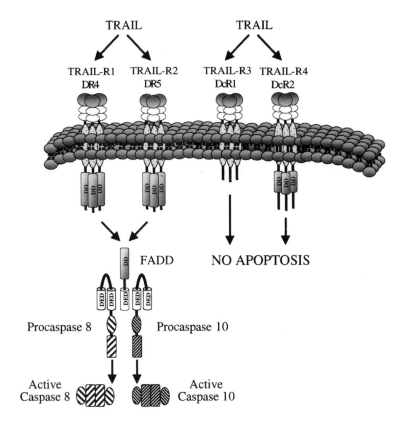

Fig. 7. TRAIL receptors and the TRAIL DISC. Schematic representation of TRAIL receptors and the signaling complex triggered by engagement of TRAIL to TRAIL-R1 or TRAIL-R2. Because TRAIL-R3 and TRAIL-R4 lack a functional death domain, they do not induce cell death and function as decoy receptors. Further details are found in the text.

TRAIL is a type II transmembrane protein (281 amino acids) that can also be cleaved by metalloproteases to generate a soluble form *(63,67)*. Its carboxy-terminal extracellular domain shows significant homology to other TNF family members, whereas the cytosolic amino-terminus is considerably shorter and not conserved among species. The biologically active form of TRAIL is a homotrimer with the cysteine residues in position 230 coordinating a zinc ion, essential for proper folding, trimer association, and activity of the cytokine itself. TRAIL seems to trigger apoptosis more specifically in tumor cell lines and tumor xenografts rather than in normal cells, although the reason for this differential sensitivity has not yet been explained.

TRAIL-R1 and TRAIL-R2 Signaling (Fig. 5)

Like Fas, activated TRAIL-R1 and TRAIL-R2 recruit FADD and caspase-8 and caspase-10 to their respective DISCs *(68)*. FADD and caspases-8/-10 not only are integral components of the TRAIL receptor DISC, but are also essential for TRAIL-induced apoptosis. Therefore, TRAIL-R1 and TRAIL-R2 seem to trigger apoptosis through a pathway similar to that activated by Fas *(69)* (Fig. 7). Like Fas, TRAIL-induced apoptosis is effectively inhibited by overexpression of c-FLIP, which appears to be the key factor in determining cell sensitivity to TRAIL-induced apoptosis *(70)*.

Several reports also show that TRAIL, in addition to inducing cell death, promotes activation of NF-κB and JNK through distinct, independent pathways. In particular, TRAIL-R1, TRAIL-R2, and

TRAIL-R4 have been shown to activate NF-κB via a TRAF-2–NIK-IKKα/β-dependent signaling cascade, whereas TRAIL-R1 induces JNK activation via a TRAF-2-MEKK1-MKK4-dependent pathway. This suggests a bifurcation in the signaling pathway at the level of TRAF-2 similar to that described for TNF-R1. Because TRAF-2 does not directly associate with any of the TRAIL receptors, it is conceivable that this pathway requires a still-unidentified adapter protein other than TRADD or FADD *(71)*. Interestingly, activation of NF-κB is not sufficient to block TRAIL-induced apoptosis, suggesting that this might be just an epiphenomenon *(70)*.

Physiology and Pathophysiology

The unavailability of genetic models of mice overexpressing or defective in the TRAIL receptors or TRAIL renders more difficult to establish the importance of the TRAIL-R/TRAIL system in the pathogenesis of human and animal disease. Unpublished data, however, suggest TRAIL knockout mice are phenotypically normal.

A recent study provided the first evidence supporting a physiological role of TRAIL as a tumor suppressor *(72)*. Indeed, TRAIL has been found to be constitutively expressed in a large number of mouse liver NK cells, and likely accounts for the anti-metastatic function of liver NK cells against TRAIL-sensitive tumor cells. As TRAIL has been shown to be regulated by endogenously produced interferon-γ (IFN-γ), it appears to be, at least in part, responsible for the IFN-γ-dependent pathway of NK-cell-mediated anti-tumor immunity.

The apparent selectivity of TRAIL to kill tumor cells renders it a promising candidate for cancer therapy. Initial studies seemed to suggest that the tumoricidal activity of TRAIL was not accompanied by any significant systemic toxicity when administered to mice and nonhuman primates *(73,74)*. However, a recent report on a study testing TRAIL cytotoxic effects in vitro on isolated hepatocytes from different species, showed significant species-related differences in TRAIL sensitivity, and warned about a possible substantial liver toxicity if TRAIL were used in humans *(75)*. These studies, however, were performed using a tagged TRAIL to induce apoptosis. Studies using nontagged, soluble zinc-replete form of TRAIL do not appear to be hepatotoxic *(76)*. Further studies are therefore required to establish whether or not TRAIL could be safely used as a cancer therapeutic.

DEATH RECEPTOR 3 (DR3/APO-3/TRAMP/WSL-1/LARD) AND APO-3L

DR3 has been identified and characterized in 1996 by its high homology with TNF-R1, particularly in its DD *(77–79)*. DR3 is a 393 amino acid protein of approx 47–54 kDa. Like the other death receptors, DR3 has an extracellular, N-terminal domain, which contains four cysteine-rich motifs, followed by a transmembrane domain, and a 193 amino acid long, C-terminal cytosolic tail. Its mRNA has been detected in spleen, thymus, and peripheral blood lymphocytes, but not in heart, brain, placenta, lung, liver, skeletal muscle, kidney, or pancreas. A natural ligand of DR3 was identified shortly after the receptor, and was named Apo3L *(80)*. Apo3L is a 249 amino acid protein with a molecular weight of approx 27 kDa, and, like the other ligands of the TNF family, is a type II transmembrane protein. Like their receptors, Apo3L and TNF also share high protein sequence identity. However, their localization is substantially different, as Apo3L is ubiquitously expressed in fetal and adult tissues, whereas TNF is expressed mainly in activated macrophages and lymphocytes. The responses and signaling, triggered either by overexpression of DR3 or binding of DR3 to Apo3L, resemble that mediated by TNF-R1. Indeed, DR3 activates NF-κB through TRADD, TRAF-2, RIP, and NIK, and also induces apoptosis, through a pathway mediated by TRADD, FADD, and caspase-8. The different expression of the ligands and receptors, however, suggests that, despite the similarities in the signaling mechanisms, DR3 and TNF-R1 likely have distinct biological roles. Indeed, negative selection and anti-CD3-induced apoptosis have been found to be significantly impaired in DR3-null mice, suggesting a unique, nonredundant role for this receptor in the removal of self-reactive T cells in the thymus *(81)*.

DEATH RECEPTOR 6 (DR6)

DR6 has been identified as a novel death receptor in 1998 based on the presence of the characteristic extracellular cysteine-rich domain and the intracellular DD *(82)*. However, unlike the other death receptors, the DD is not localized at the C-terminus, but it locates adjacent to the transmembrane domain, followed by a 150 amino acid tail. Following the DD, there is a putative leucine zipper sequence overlapping with as proline-rich region, whose functional relevance is still obscure. DR6 is highly expressed in thymus, spleen, and lymphoid cells.

The signaling pathway originating from this receptor has been poorly defined so far. No cognate ligand for DR6 has yet been identified. DR6 engages a cell-death pathway in mammalian cells different from those initiated by the other known death receptors, as it seems not to associate with DD-containing adaptor molecules such as TRADD, FADD, RAIDD, or RIP. Upon overexpression, DR6 is a potent inducer of JNK and NF-κB activation. The ability to activate the JNK pathway and the predominant expression in lymphoid organs suggest that DR6 may have a role in the immune system. In particular, studies employing DR6 knockout mice demonstrated that DR6 works through JNK to regulate the differentiation of naïve CD4$^+$ into Th1 and Th2 cells *(83)*.

CONCLUSIONS

In summary, multiple death receptors are expressed differentially in mammalian cells. Their role in maintaining tissue homeostasis through regulation of cell death and survival is crucial, as demonstrated by the apparent redundancy of their signaling pathways. The death receptors do not have catalytic activity, but they undergo oligomerization by binding to selective, noncrossreacting ligands, and recruit adapter molecules by interaction of their common cytoplasmic death domains. The adapter molecules, in turn, recruit initiator procaspases, which undergo self-processing in the receptor complex, likely, by induced proximity and weak, but intrinsic catalytic activity. The activated caspases then initiate complex cell-signaling cascades culminating in cell death by apoptosis. Several intrinsic cell proteins (i.e., cFLIPs, cIAPs) and survival signaling pathways (i.e., NF-κB) can inhibit death receptor-mediated apoptosis. The ultimate fate of the cell depends upon the cell context, simultaneous signaling events, and other stimuli. Death receptors are particularly important in immune regulation and tissue injury in disease states, and they have already been targeted for the treatment of several human inflammatory and autoimmune diseases. Purposeful induction of apoptosis by death receptors may ultimately be useful also in cancer therapy. Much remains to be learned about the complexity of death-receptor signaling in health and disease.

ACKNOWLEDGMENTS

This work was supported by grants from the National Institute of Health DK 41876 (to G.J.G.) and the Mayo Foundation, Rochester, Minnesota.

REFERENCES

1. Ashkenazi, A. and Dixit, V. M. (1998) Death receptors: signaling and modulation. *Science* **281,** 1305–1308.
2. Locksley, R. M., Killeen, N., and Lenardo, M. J. (2001) The TNF and TNF receptor superfamily: integrating mammalian biology. *Cell* **104,** 487–501.
3. Nagata, S. (1997) Apoptosis by death factor. *Cell* **88,** 355–365.
4. Cascino, I., Fiucci, G., Papoff, G., and Ruberti, G. (1995) Three functional soluble forms of the human apoptosis-inducing Fas molecule are produced by alternative splicing. *J. Immunol.* **154,** 2706–2713.
5. Bennet, M., MacDonald, K., Chan, S.-W., Luzio, J. P., Simari, R., and Weissberg, P. (1998) Cell surface trafficking of Fas: a rapid mechanism of p53-mediated apoptosis. *Science* **282,** 290–293.
6. Sodeman, T., Bronk, S. F., Roberts, P. J., Miyoshi, H., and Gores, G. J. (2000) Bile salts mediate hepatocyte apoptosis by increasing cell surface trafficking of Fas. *Am. J. Physiol. Gastrointest. Liver Physiol.* **278,** G992–G999.
7. Feng, G. and Kaplowitz, N. (2000) Colchicine protects mice from the lethal effect of an agonistic anti-Fas antibody. *J. Clin. Invest.* **105,** 329–339.
8. Peter, M. E., Hellbardt, S., Schwartz-Albiez, A., Westendorp, M. O., Walczak, H., Moldenhauer, G., et al. (1995) Cell

surface sialylation plays a role in modulating sensitivity towards APO-1–mediated apoptotic cell death. *Cell Death Diff.* **2**, 163–171.

9. Chan, H., Bartos, D. P., and Owen-Schaub, L. B. (1999) Activation-dependent transcriptional regulation of the human Fas promoter requires NF-kappaB p50–p65 recruitment. *Mol. Cell Biol.* **19**, 2098–2108.

10. Friesen, C., Herr, I., Krammer, P. H., and Debatin, K. M. (1996) Involvement of the CD95 (APO-1/Fas) receptor/ligand system in drug-induced apoptosis in leukemia cells. *Nat. Med.* **2**, 574–580.

11. Muller, M., Strand, S., Hug, H., Heinemann, E. M., Walczak, H., Hofmann, W. J., et al. (1997) Drug-induced apoptosis in hepatoma cells is mediated by the CD95 (APO-1/Fas) receptor/ligand system and involves activation of wild-type p53. *J. Clin. Invest.* **99**, 403–413.

12. Muller, M., Wilder, S., Bannasch, D., Israeli, D., Lehlbach, K., Li-Weber, M., et al. (1998) p53 activates the CD95 (APO-1/Fas) gene in response to DNA damage by anticancer drugs. *J. Exp. Med.* **188**, 2033–2045.

12a. Suda, T., Takahashi, T., Golstein, P., and Nagata, S. (1993) Molecular cloning and expression of the Fas ligand, a novel member of the tumor necrosis factor family. *Cell* **75**, 1169–1178.

13. Berke, G. (1995) The CTL's kiss of death. *Cell* **81**, 9–12.

14. Kagi, D., Vignaux, F., Ledermann, B., Burki, K., Depraetere, V., Nagata, S., et al. (1994) Fas and perforin pathways as major mechanisms of T-cell-mediated cytotoxicity. *Science* **265**, 528–530.

15. Lowin, B., Hahne, M., Mattmann, C., and Tschopp, J. (1994) Cytolytic T-cell cytotoxicity is mediated through perforin and Fas lytic pathways. *Nature* **370**, 650–652.

16. Schneider, P., Holler, N., Bodmer, J. L., Hahne, M., Frei, K., Fontana, A., and Tschopp, J. (1998) Conversion of membrane-bound Fas(CD95) ligand to its soluble form is associated with downregulation of its proapoptotic activity and loss of liver toxicity. *J. Exp. Med.* **187**, 1205–1213.

17. Tanaka, M., Suda, T., Haze, K., Nakamura, N., Sato, K., Kimura, F., et al. (1996) Fas ligand in human serum. *Nat. Med.* **2**, 317–322.

18. Tanaka, M., Suda, T., Takahashi, T., and Nagata, S. (1995) Expression of the functional soluble form of human fas ligand in activated lymphocytes. *EMBO J.* **14**, 1129–1135.

19. Shudo, K., Kinoshita, K., Imamura, R., Fan, H., Hasumoto, K., Tanaka, M., et al. (2001) The membrane-bound but not the soluble form of human Fas ligand is responsible for its inflammatory activity. *Eur. J. Immunol.* **31**, 2504–2511.

20. Suda, T., Hashimoto, H., Tanaka, M., Ochi, T., and Nagata, S. (1997) Membrane Fas ligand kills human peripheral blood T lymphocytes, and soluble Fas ligand blocks the killing. *J. Exp. Med.* **186**, 2045–2050.

21. Hohlbaum, A. M., Moe, S., and Marshak-Rothstein, A. (2000) Opposing effects of transmembrane and soluble Fas ligand expression on inflammation and tumor cell survival. *J. Exp. Med.* **191**, 1209–1220.

22. Wallach, D., Varfolomeev, E. E., Malinin, N. L., Goltsev, Y. V., Kovalenko, A. V., and Boldin, M. P. (1999) Tumor necrosis factor receptor and Fas signaling mechanisms. *Annu. Rev. Immunol.* **17**, 331–367.

23. Kischkel, F. C., Lawrence, D. A., Tinel, A., LeBlanc, H., Virmani, A., Schow, P., et al. (2001) Death receptor recruitment of endogenous caspase-10 and apoptosis initiation in the absence of caspase-8. *J. Biol. Chem.* **276**, 46639–46646.

24. Wang, J., Chun, H. J., Wong, W., Spencer, D. M., and Lenardo, M. J. (2001b) Caspase-10 is an initiator caspase in death receptor signaling. *Proc. Natl. Acad. Sci. USA* **98**, 13884–13888.

25. Scaffidi, C., Fulda, S., Srinivasan, A., Friesen, C., Li, F., Tomaselli, K. J., et al. (1998) Two CD95 (APO-1/Fas) signaling pathways. *EMBO J.* **17**, 1675–1687.

26. Susin, S. A., Zamzami, N., Castedo, M., Hirsch, T., Marchetti, P., Macho, A., et al. (1996) Bcl-2 inhibits the mitochondrial release of an apoptogenic protease. *J. Exp. Med.* **184**, 1331–1341.

27. Du, C., Fang, M., Li, Y., Li, L., and Wang, X. (2000) Smac, a mitochondrial protein that promotes cytochrome c-dependent caspase activation by eliminating IAP inhibition. *Cell* **102**, 33–42.

28. Verhagen, A. M., Ekert, P. G., Pakusch, M., Silke, J., Connolly, L. M., Reid, G. E., et al. (2000) Identification of DIABLO, a mammalian protein that promotes apoptosis by binding to and antagonizing IAP proteins. *Cell* **102**, 43–53.

29. Meinl, E., Fickenscher, H., Thome, M., and Tschopp, J. (1998) Anti-apoptotic strategies of lymphotropic viruses. *Immunol. Today* **19**, 474–479.

30. Tschopp, J., Irmler, M., and Thome, M. (1998) Inhibition of Fas death signals by FLIPs. *Curr. Opin. Immunol.* **10**, 552–558.

31. Scaffidi, C., Schmitz, I., Krammer, P. H., and Peter, M. E. (1999) The role of c-FLIP in modulation of CD95-induced apoptosis. *J. Biol. Chem.* **274**, 1541–1548.

32. Krueger, A., Baumann, S., Krammer, P. H., and Kirchhoff, S. (2001) FLICE-inhibitory proteins: regulators of death receptor-mediated apoptosis. *Mol. Cell Biol.* **21**, 8247–8254.

33. Takahashi, T., Tanaka, M., Brannan, C. I., Jenkins, N. A., Copeland, N. G., Suda, T., and Nagata, S. (1994) Generalized lymphoproliferative disease in mice, caused by a point mutation in the Fas ligand. *Cell* **76**, 969–976.

34. Watanabe-Fukunaga, R., Brannan, C. I., Copeland, N. G., Jenkins, N. A., and Nagata, S. (1992) Lymphoproliferation disorder in mice explained by defects in Fas antigen that mediates apoptosis. *Nature* **356**, 314–317.

35. Cohen, P. L. and Eisenberg, R. A. (1991) Lpr and gld: single gene models of systemic autoimmunity and lymphoproliferative disease. *Annu. Rev. Immunol.* **9**, 243–269.

36. Krammer, P. H. (2000) CD95's deadly mission in the immune system. *Nature* **407**, 789–795.

37. Nagata, S. and Golstein, P. (1995) The Fas death factor. *Science* **267**, 1449–1456.

38. Fischer, G. H., Rosenberg, F. J., Straus, S. E., Dale, J. K., Middelton, L. A., Lin, A. Y., et al. (1995) Dominant interfering Fas gene mutations impair apoptosis in a human autoimmune lymphoproliferative syndrome. *Cell* **81**, 935–946.

39. Rieux-Laucat, F., Le Deist, F., Hivroz, C., Roberts, I. A., Debatin, K. M., Fischer, A., and de Villarty, J. P. (1995) Mutations in Fas associated with human lymphoproliferative syndrome and autoimmunity. *Science* **268**, 1347–1349.

40. Hahne, M., Rimoldi, D., Schroter, M., Romero, P., Schreier, M., French, L. E., et al. (1996) Melanoma cell expression of Fas(Apo-1/CD95) ligand: implications for tumor immune escape. *Science* **274**, 1363–1366.
41. Strand, S., Hofmann, W. J., Hug, H., Muller, M., Otto, G., Strand, D., et al. (1996) Lymphocyte apoptosis induced by CD95 (APO-1/Fas) ligand-expressing tumor cells: a mechanism of immune evasion? *Nat. Med.* **2**, 1361–1366.
42. Ogasawara, J., Watanabe-Fukunaga, R., Adachi, M., Matsuzawa, A., Kasugai, T., Kitamura, Y., et al. (1993) Lethal effect of the anti-Fas antibody in mice. *Nature* **364**, 806–809.
43. Black, R. A., Rauch, C. T., Kozlosky, C. J., Peschon, J. J., Slack, J. L., Wolfson, M. F., et al. (1997) A metalloproteinase disintegrin that releases tumor necrosis factor α from cells. *Nature* **385**, 729–733.
44. De Smaele, E., Zazzeroni, F., Papa, S., Nguyen, D. U., Jin, R., Jones, J., et al. (2001) Induction of *gadd45β* by NF-κB downregulates pro-apoptotic JNK signalling. *Nature* **414**, 308–313.
45. Tang, G., Minemoto, Y., Dibling, B., Purcell, N. H., Li, Z., Karin, M., and Lin, A. (2001) Inhibition of JNK activation through NF-κB target genes. *Nature* **414**, 313–317.
46. Jiang, Y. P., Woronicz, J. D., Liu, W., and Goeddel, D. V. (1999) Prevention of constitutive TNF receptor 1 signaling by silencer of death domains. *Science* **283**, 543–546.
47. Holler, N., Zaru, R., Micheau, O., Thome, M., Attinger, A., Valitutti, S., et al. (2000) Fas triggers an alternative, caspase-8-independent cell death pathway using the kinase RIP as effector molecule. *Nat. Immunol.* **1**, 489–495.
48. Duan, H. and Dixit, V. M. (1997) RAIDD is a new "death" adaptor molecule. *Nature* **385**, 86–89.
49. Wajant, H. and Scheurich, P. (2001) Tumor necrosis factor receptor-associated factor (TRAF) 2 and its role in TNF signaling. *Int. J. Biochem. Cell Biol.* **33**, 19–32.
50. Huang, H., Joazeiro, C. A., Bonfoco, E., Kamada, S., Leverson, J. D., and Hunter, T. (2000) The inhibitor of apoptosis, cIAP2, functions as a ubiquitin-protein ligase and promotes in vitro monoubiquitination of caspase-3 and -7. *J. Biol. Chem.* **275**, 26661–26664.
51. Wang, C. Y., Mayo, M. W., Korneluk, R. G., Goeddel, D. V., and Baldwin, A. S. (1998) NF-κB antiapoptosis: induction of TRAF 1 and TRAF 2 and c-IAP1 and c-IAP2 to suppress caspase-8 activation. *Science* **281**, 1680–1683.
52. Wysk, M., Yang, D. D., Lu, H. T., Flavell, R. A., and Davis, R. J. (1999) Requirement of mitogen-activated protein kinase kinase 3 (MKK 3) for tumor necrosis factor-induced cytokine expression. *Proc. Natl. Acad. Sci. USA* **96**, 3763–3768.
53. Loffreda, S., Rai, R., Yang, S. Q., Lin, H. Z., and Diehl, A. M. (1997) Bile duct and portal and central veins are major producers of tumor necrosis factor alpha in regenerating rat liver. *Gastroenterology* **112**, 2089–2098.
54. Yamada, Y., Kirillova, I., Peschon, J. J., and Fausto, N. (1997) Initiation of liver growth by tumor necrosis factor: deficient liver regeneration in mice lacking type I tumor necrosis receptor. *Proc. Natl. Acad. Sci. USA* **94**, 1441–1446.
55. Yin, M., Wheeler, M. D., Kono, H., Bradford, B. U., Gallucci, R. M., Luster, M. I., and Thurman, R. G. (1999) Essential role of tumor necrosis factor alpha in alcohol-induced liver injury in mice. *Gastroenterology* **117**, 942–952.
56. Daniels, H., Meager, A., Goka, J., Eddlestone, A. L. W. F., Alexander, G. J. M., and Williams, R. (1990) Spontaneous production of tumor necrosis factor α and interleukin β during interferon treatment of chronic HBV infection. *Lancet* **335**, 875–877.
57. Flynn, J. L., Goldstein, M. M., Chan, J., Triebold, K. J., Pfeffer, K., Lowenstein, C. J., et al. (1995) Tumor necrosis factor-alpha is required in the protective immune response against Mycobacterium tuberculosis in mice. *Immunity* **2**, 561–572.
58. Le Hir, M., Bluethmann, H., Kosco-Vilbois, M. H., Muller, M., di Padova, F., Moore, M., et al. (1996) Differentiation of follicular dendritic cells and full antibody responses require tumor necrosis factor receptor-1 signaling. *J. Exp. Med.* **183**, 2367–2372.
59. Pfeffer, K., Matsuyama, T., Kundig, T. M., Wakeham, A., Kishihara, K., Shahinian, A., et al. (1993) Mice deficient for the 55 kd tumor necrosis factor receptor are resistant to endotoxic shock, yet succumb to L. monocytogenes infection. *Cell* **73**, 457–467.
60. Marino, M. W., Dunn, A., Grail, D., Inglese, M., Noguchi, Y., Richards, E., et al. (1997) Characterization of tumor necrosis factor-deficient mice. *Proc. Natl. Acad. Sci. USA* **94**, 8093–8098.
61. Galon, J., Aksentijevich, I., McDermott, M. F., O'Shea, J. J., and Kastner, D. L. (2000) TNFRSF1A mutations and autoinflammatory syndromes. *Curr. Opin. Immunol.* **12**, 479–486.
62. Illei, G. G. and Lipsky, P. E. (2000) Novel, non-antigen-specific therapeutic approaches to autoimmune/inflammatory diseases. *Curr. Opin. Immunol.* **12**, 712–718.
63. Golstein, P. (1997) Cell death: TRAIL and its receptors. *Curr. Biol.* **7**, R750–R753.
64. Truneh, A., Sharma, S., Silverman, C., Khandekar, S., Reddy, M. P., Deen, K. C., et al. (2000) Temperature-sensitive differential affinity of TRAIL for its receptors. DR5 is the highest affinity receptor. *J. Biol. Chem.* **275**, 23319–23325.
65. Pitti, R. M., Marsters, S. A., Ruppert, S., Donahue, C. J., Moore, A., and Ashkenazi, A. (1996) Induction of apoptosis by Apo-2 ligand, a new member of the tumor necrosis factor cytokine family. *J. Biol. Chem.* **271**, 12687–12690.
66. Wiley, S. R., Schooley, K., Smolak, P. J., Din, W. S., Huang, C. P., Nicholl, J. K., et al. (1995) Identification and characterization of a new member of the TNF family that induces apoptosis. *Immunity* **3**, 673–682.
67. Ashkenazi, A. and Dixit, V. M. (1999) Apoptosis control by death and decoy receptors. *Curr. Opin. Cell Biol.* **11**, 255–260.
68. Kuang, A. A., Diehl, G. E., Zhang, J., and Winoto, A. (2000) FADD is required for DR4- and DR5-mediated apoptosis: lack of trail-induced apoptosis in FADD-deficient mouse embryonic fibroblasts. *J. Biol. Chem.* **275**, 25065–25068.
69. Peter, M. E. (2000) The TRAIL DISCussion: it is FADD and caspase-8! *Cell Death Diff.* **7**, 759–760.
70. Walczak, H. and Krammer, P. H. (2000) The CD95 (APO-1/Fas) and the TRAIL (APO-2L) apoptosis systems. *Exp. Cell Res.* **256**, 58–66.

71. Hu, W. H., Johnson, H., and Shu, H. B. (1999) Tumor necrosis factor-related apoptosis-inducing ligand receptors signal NF-kappaB and JNK activation and apoptosis through distinct pathways. *J. Biol. Chem.* **274,** 30603–30610.
72. Takeda, K., Hayakawa, Y., Smyth, M. J., Kayagaki, N., Yamaguchi, N., Kakuta, S., et al. (2001) Involvement of tumor necrosis factor-related apoptosis-inducing ligand in surveillance of tumor metastasis by liver natural killer cells. *Nat. Med.* **7,** 94–100.
73. Ashkenazi, A., Pai, R. C., Fong, S., Leung, S., Lawrence, D. A., Marsters, S. A., et al. (1999) Safety and antitumor activity of recombinant soluble Apo2 ligand. *J. Clin. Invest.* **104,** 155–162.
74. Walczak, H., Miller, R. E., Ariail, K., Gliniak, B., Griffith, T. S., Kubin, M., et al. (1999) Tumoricidal activity of tumor necrosis factor-related apoptosis-inducing ligand in vivo. *Nat. Med.* **5,** 157–163.
75. Jo, M., Kim, T. H., Seol, D. W., Esplen, J. E., Dorko, K., Billiar, T. R., and Strom, S. C. (2000) Apoptosis induced in normal human hepatocytes by tumor necrosis factor-related apoptosis-inducing ligand. *Nat. Med.* **6,** 564–567.
76. Gores, G. J. and Kaufmann, S. H. (2001) Is TRAIL hepatotoxic? *Hepatology* **34,** 3–6.
77. Bodmer, J. L., Burns, K., Schneider, P., Hofmann, K., Steiner, V., Thome, M., et al. (1997) TRAMPS, a novel apoptosis-mediating receptor with sequence homology to Tumor Necrosis Factor Receptor 1 and Fas (Apo-1/CD95). *Immunity* **6,** 79–88.
78. Kitson, J., Raven, T., Jiang, Y. P., Goeddel, D. V., Giles, K. M., Pun, K. T., et al. (1996) A death-domain-containing receptor that mediates apoptosis. *Nature* **384,** 372–375.
79. Marsters, S. A., Sheridan, J. P., Donahue, C. J., Pitti, R. M., Gray, C. L., Goddard, A. D., et al. (1996) Apo-3, a new member of the tumor necrosis factor receptor family, contains a death domain and activates apoptosis and NF-κB. *Curr. Biol.* **6,** 1669–1676.
80. Marsters, S. A., Sheridan, J. P., Pitti, R. M., Brush, J., Goddard, A. D., and Ashkenazi, A. (1998) Identification of a ligand for the death-domain-containing receptor Apo3. *Curr. Biol.* **8,** 525–528.
81. Wang, E. C. Y., Thern, A., Denzel, A., Kitson, J., Farrow, S. N., and Owen, M. J. (2001a) DR3 regulates negative selection during thymocyte development. *Mol. Cell Biol.* **21,** 3451–3461.
82. Pan, G., Bauer, J. H., Haridas, V., Wang, S., Liu, D., Yu, G., et al. (1998) Identification and functional characterization of DR6, a novel death domain-containing TNF receptor. *FEBS Lett.* **431,** 351–356.
83. Zhao, H., Yan, M., Wang, H., Erickson, S., Grewal, I. S., and Dixit, V. M. (2001) Impaired c-Jun amino terminal kinase activity and T cell differentiation in death receptor 6-deficient mice. *J. Exp. Med.* **194,** 1441–1448.

6

The Mitochondrial Apoptosis Pathway

Bruno Antonsson

INTRODUCTION

Multicellular organisms are dependent on the removal of damaged or unwanted cells both during their development and in the adult life. This is ensured by apoptosis or programmed cell death. Apoptosis is an active energy (ATP) requiring process where cells are eliminated in an ordered manner. A large variety of different stimuli, both extracellular signals and signals generated from inside the cell itself can initiate the apoptosis signaling cascades. Two intracellular apoptosis signaling pathways have been identified, the death-receptor pathway and the mitochondrial pathway. Apoptosis is accompanied by characteristic morphological and biochemical changes, such as chromatin condensation, DNA degradation, nuclear fragmentation, exposure of phosphatidylserine on the outside of the plasma membrane, and finally cellular fragmentation into apoptotic bodies. Although, apoptosis is an essential process, it is also involved in a wide range of pathologies. Increased apoptosis has been associated with stroke; myocardial infarct; reperfusion injury; arteriosclerosis; heart failure; infertility; diabetes; AIDS; hepatitis; renal failure; and neurodegenerative diseases like multiple sclerosis (MS); amyotropic lateral sclerosis (ALS); and Alzheimer's, Huntington's, and Parkinson's diseases (1–5). On the other hand, impaired apoptosis is associated with various forms of cancer and autoimmune diseases (6).

THE MITOCHONDRIA

Mitochondria are intracellular organelles whose primary function is to provide energy for the cellular functions. Without the energy, in the form of ATP, produced by the mitochondrial respiratory chain, cells are not able to survive under normal conditions. The mitochondria has a complex structure, they are composed of an outer membrane, an inner membrane, an intermembrane space, cristae structures formed by the folding of the inner membrane, and the mitochondrial matrix inside the inner membrane (Fig. 1). The inner and outer membranes are different in terms of lipid and protein composition. The outer membrane is a fairly permeable membrane allowing free passage of molecules of a molecular weight of less than 1.5 kDa. The inner membrane, on the contrary, is a tightly sealed barrier. However, both membranes contain large channels or pores, which allow for a selective passage of large molecules, including nucleotides and proteins, into the intermembrane space and the matrix. The two main channels in the outer membrane are the voltage-dependent anion channel (VDAC) and the protein import channel, the translocase of the outer membrane (TOM). The

From: *Essentials of Apoptosis: A Guide for Basic and Clinical Research*
Edited by: X-M. Yin and Z. Dong © Humana Press Inc., Totowa, NJ

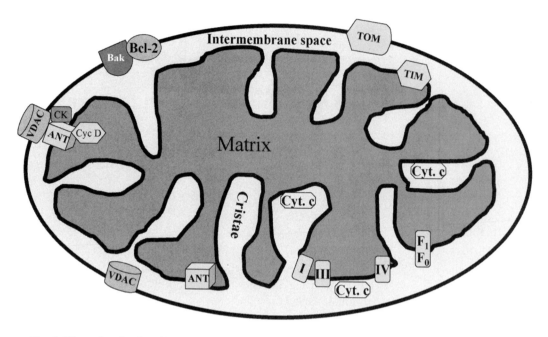

Fig. 1. The main mitochondrial structures. The outer membrane containing channel-forming proteins, such as VDAC, TOM, Bak, and Bcl-2. The inner membrane containing ANT; TIM; the complexes I, III, and IV of the respiratory electron-transfer chain; and the F_0F_1-ATPase. The permeability transition pore complex formed by VDAC, creatine kinase, ANT, and cyclophilin D is formed at the contact sites between the outer and the inner membranes. The intermembrane space with cytochrome c attached to the outside of the inner membrane. The cristae structures formed by the folding of the inner membrane and the matrix inside the inner membrane.

inner membrane contains the adenine nucleotide translocator (ANT) and the protein translocase of the inner membrane (TIM) *(7)*.

Normal functioning mitochondria have an electrochemical gradient ($\Delta\Psi$m) across their inner membrane. The gradient is created through the efflux of H^+ ions from the matrix to the intermembrane space driven by the complexes of the respiratory electron transfer chain in the inner membrane. The disequilibration in H^+ ions concentration results in a pH and voltage gradient across the membrane. The F_0F_1-ATPase/H^+ pump is also contributing to the H^+ ions balance across the membrane. Under normal conditions, the ATPase converts ADP into ATP, whereby the driving force is the flux of H^+ ions from the intermembrane space into the matrix. However, in the absence of a H^+ ions gradient, the enzyme works in the reverse direction, hydrolyzing ATP into ADP, thereby pumping H^+ ions from the matrix to the intermembrane space. In addition, several other ion channels have been identified in the inner membrane, including the H^+/K^+ antiporter, Cl^-/HCO_3^- antiporter, Na^+/H^+ exchanger and uncoupling proteins, which also contribute to maintaining the ion-flux balance *(8,9)*.

Mitochondrial dysfunction is associated with both necrotic and apoptotic cell death. During necrosis, mitochondrial function is compromised through the loss of the mitochondrial membrane potential, leading to an incapacity of the organelle to synthesis ATP and provide the cell with energy. The loss of the membrane potential is a result of opening of large pores across the inner membrane, resulting in mitochondrial permeability transition. The mitochondrial permeability transition pore (PTP) is a multiprotein complex present at the contact sites between the outer and the inner membranes. The complex is spanning from the matrix to the cytosol and it is thought to contain VDAC in the outer membrane, ANT in the inner membrane, the matrix protein cyclophilin D, creatine kinase from the intermembrane space, and cytosolic hexokinase *(10,11)*. However, VDAC, ANT, and cyclophilin D are often considered the core proteins of the pore. In apoptosis, mitochondrial dysfunction appears to be mainly the result of a specific permeabilization of the outer mitochondrial membrane to large molecules, including cytochrome *c*. The loss of cytochrome *c* from the outer side of the inner membrane results in impairment of mitochondrial respiration, effectively blocking the electron transport between complex III (cytochrome *c* reductase) and complex IV (cytochrome *c* oxidase). This not only results in a disturbance of the membrane potential across the inner membrane, but it also increases the production of reactive oxygen species (ROS), which results in increased lipid peroxidation.

APOPTOSIS SIGNALING PATHWAYS

Two intracellular apoptosis-signaling pathways have been identified, the death-receptor pathway and the mitochondrial pathway. The receptor pathway is activated through ligand binding to receptors in the plasma membrane, such as the Fas/CD95, tumor necrosis factor-α (TNF-α) and TNF-related apoptosis-inducing ligand (TRAIL) receptors *(12)* (*see* Chapter 5). The mitochondrial pathway is controlled by the Bcl-2 proteins. When activated, the multidomain pro-apoptotic proteins permeabilize the outer mitochondrial membrane and trigger the release of pro-apoptotic proteins, including cytochrome *c*, from the intermembrane space. Several stimuli, including hypoxia, ROS, ultraviolet (UV) or gamma irradiation, growth-factor deprivation, kinase inhibitors like staurosporine, and several cytotoxic compounds have all been shown to activate this pathway. However, how the individual stimuli, which are very different in nature, at the molecular level initiate and activate the signaling cascade upstream of the Bcl-2 proteins remains unknown.

The receptor and the mitochondrial pathways initially appeared to be two independent pathways. However, it is now clear that a crosstalk between the two pathways exist. This is mediated by a member of the Bcl-2 protein family, the "BH3 domain only" protein Bid *(13)*.

STRUCTURE AND CHANNEL FORMING ACTIVITY OF THE BCL-2 PROTEINS

The Bcl-2 protein family contains over 20 members (*see* Chapter 2). The proteins have either pro- or anti-apoptotic activity and they control the mitochondrial apoptosis pathway. The three-dimensional structures are available for four Bcl-2 family proteins. Two anti-apoptotic members (Bcl-X_L and Bcl-2) and two pro-apoptotic proteins, one from the multidomain subgroup (Bax) and one of the "BH3 domain only" group (Bid) *(14–18)*. The overall structures of all the proteins are very similar, despite their totally opposite activity in regulation of apoptosis. The proteins show a fairly compact globular structure with two hydrophobic central helixes (α5 and α6) surrounded by amphipathic helixes. The helixes are connected by flexible loop structures, which show a larger variation among the proteins.

The structural similarity of the Bcl-2 proteins to diphtheria toxin and colicins suggested that the proteins could possess channel-forming activity. Surprisingly, both the pro- and the anti-apoptotic proteins have channel-forming activity in artificial lipid membranes *(19)*. The anti-apoptotic proteins have channel-forming activity only at low pH, below pH 5.5 *(20,21)*. The pro-apoptotic proteins have channel-forming activity at neutral pH, although the activity is enhanced at lower pH.

Channels formed by oligomeric Bax have multiconductance levels, ranging from a few pS up to 4–5 nS, are pH-sensitive, slightly cation-selective, and Ca^{2+}-insensitive *(19,20)*. Monomeric Bax does not possess channel-forming activity. Oligomerization of Bax can be induced artificially by several detergents *(22)*. Recombinant Bax or cytosolic Bax exposed to Triton X-100 or octyl glucoside form oligomers with a molecular weight of 80 kDa *(23)*. These oligomers are able to form at least dimers resulting in oligomers of 160 kDa. This would correspond to Bax tetramers and octamers. Oligomerization only appears to take place when the detergent is present over the critical micelle concentration. This is supported by the nuclear magnetic resonance (NMR) structure study where a low octyl glucoside concentration did not induce any significant changes in the structure; however, at a concentration of 0.6%, Bax structure changed dramatically, indicating aggregation or oligomer formation *(17)*. In a study by Saito et al., it was shown that Bax oligomers were able to release cytochrome *c* from liposomes *(24)*. The structure was estimated to be Bax tetramers. In a recent study monomeric Bax incubated with caspase-8 cleaved Bid in the presence of artificial lipid membranes formed low-conductance, highly cation-selective channels, which were permeable to small molecules but not to cytochrome *c (25)*. One study indicated that Bax could destabilize lipid membranes through reducing the linear tension without forming ion channels *(26)*. These might indicate that Bax, and maybe other multidomain proteins, can form different types of channels and permeabilize the outer mitochondrial membrane by various mechanisms.

ACTIVATION OF THE PRO-APOPTOTIC MULTIDOMAIN BCL-2 PROTEINS

Regulation of the pro-apoptotic multidomain protein activities appears to be mainly on the post-translational level, although Bax levels have been reported to change during apoptosis *(27,28)*. However, in most, if not all cells, the multidomain proteins are present as inactive forms.

In normal cells or tissues, Bax is predominantly localized in the cytosol as a monomer *(22,23)*. After apoptotic stimulation, Bax translocates from the cytosol specifically to the mitochondria *(29)*. Activation of Bax has been shown to be accompanied by conformational changes in both the C- and the N-terminal domains of the protein resulting in changes in its quaternary structure *(17,30,31)*. After activation Bax is found inserted into the outer mitochondrial membrane as large oligomers *(32,33)*. Bcl-2 inhibits Bax activation and oligomerization *(32–34)*.

Bak is found inserted in the outer mitochondrial membrane of normal cells *(35)*. Similar to Bax, activation of Bak has been shown to be associated with changes in its tertiary and quaternary structure *(35,36)*. Combined, these results show that activation of the pro-apoptotic multidomain proteins is associated with conformational changes in the proteins.

The "BH3 domain only" proteins appear to function mainly through activation of the multidomain proteins *(see* Chapter 2). A few years ago, Bid was shown to provide a shunt between the death receptor and the mitochondrial pathway *(37)*. Caspase-8 activated through the receptor pathway cleaves Bid generating a C-terminal fragment (t^{c}Bid), which interacts with Bax and Bak triggering activation of these proteins and the mitochondrial pathway. This mechanism functions as an amplification loop of the receptor pathway. Fibroblasts from Bax and Bak double-deficient mice were completely resistant to t^{c}Bid-induced cell death *(38)*. Thus, at least in this system Bid acts exclusively through activation of the multidomain proteins. However, in artificial membrane systems t^{c}Bid has a membrane-destabilizing effect *(39)*. Whether such an activity is also present under physiological conditions remains to be elucidated. It is conceivable that t^{c}Bid, through destabilization of the mitochondrial membrane, could facilitate insertion of the multidomain proteins and so promote their activity.

Other "BH3 domain only" proteins have been shown to activate the multidomain proteins in an indirect way. For example, unphosphorylated Bad binds to Bcl-2 and Bcl-X$_L$ neutralizing their anti-apoptotic activity *(40)*. Survival factors like interleukin-3 or NGF have been shown to induce phosphorylation of Bad, which then dissociates from Bcl-2 and Bcl-X$_L$ allowing the proteins to interact with the multidomian proteins, thereby preventing their pro-apoptotic activity. Another "BH3 domain only" protein Bik has also been found to be phosphorylated and in this case phosphorylation reduced its pro-apoptotic activity *(41)*. The "BH3 domain only" protein Bim is bound to the microtubulin-associated dynein complex through interactions with the LC8 dynein light chain in normal cells. After induction of apoptosis by various stimuli, including UV irradiation, staurosporine, γ-irradiation, or growth-factor deprivation, Bim was released from the microtubuli complex *(42)*. In the cytosol Bim is thought to bind to anti-apoptotic proteins inhibiting their activities.

HOW ARE PROTEINS RELEASED FROM THE MITOCHONDRIAL INTERMEMBRANE SPACE?

The first report indicating that mitochondria are essential for apoptosis was by Newmeyer et al. using a cell-free model system *(43)*. Subsequently, using cell extracts, Liu et al. demonstrated that cytochrome *c* could activate the apoptosis-signaling cascade in vitro. They also showed an increased cytochrome *c* concentration in the cytosol from apoptotic cells *(44)*. It is now clear that the pro-apoptotic Bcl-2 proteins execute their function at the mitochondria, although controversy persists over the molecular mechanisms. At the mitochondria, the multidomain proteins trigger the release of several proteins from the intermembrane space, including cytochrome *c*, apoptosis-inducing factor (AIF), adenylate kinase, endonuclease G, Smac/Diablo, and HtrA2/Omi *(45–51)*.

The controversial question is how Bax and other multidomain proteins induce the release of cytochrome *c* and the other proteins. Two mechanisms have been suggested that would result in permeabilization of the outer mitochondrial membrane: 1) formation of specific channels or pores in the outer membrane; or 2) opening of the permeability transition pore (PTP), resulting in mitochondria matrix swelling and rupture of the outer membrane. Several models have been proposed (Fig. 2): 1) Bax and Bak form channels themselves; 2) Bax forms chimeric channels with VDAC; 3) Bax destabilizes the mitochondrial membrane inducing "lipid holes" *(26)*; and 4) Bax triggers opening of the mitochondrial PTP.

Bax Channels

As described earlier, recombinant Bax is able to form channels in artificial membranes without any additional proteins *(19)*. These "Bax alone" channels are able to permeabilize lipid membranes to cytochrome *c (24)*. During apoptosis, Bax specifically translocates to the outer mitochondrial membrane. In the mitochondrial membrane, both Bax and Bak have been shown to form large oligomers *(36)*. Bax oligomers extracted from mitochondria of apoptotic HeLa or HEK cells have molecular masses of 96k and 250 kDa *(32)*. The Bax oligomers did not comigrate with either VDAC or ANT, showing that these proteins are not part of the stable Bax oligomers. In a study by Mikhailov et al., Bax oligomers, up to hexamers, were identified in rat kidney cells deprived of ATP to induce apoptosis *(34)*. These results show that Bax oligomers formed in the outer mitochondrial membrane in cells undergoing apoptosis are larger than those shown to be required to permeabilize liposomes to cytochrome *c*, and could be the cytochrome *c*-conducting channel in the outer mitochondrial membrane.

In a recent study, a new high-conductance channel was identified in mitochondria from apoptotic cells *(52)*. The channel was localized in the outer mitochondrial membrane and its activity correlated with the onset of apoptosis. The mitochondrial apoptosis-induced channel (MAC) has a conductance of 2.5 nS, shows cation selectivity and the pore size was estimated to 4 nm. This size would allow the passage of cytochrome *c* and even larger proteins. The electrophysiological properties of the channel resemble those of the high-conductance Bax channel, but are clearly distinct from the main outer membrane channels VDAC and TOM. These results provide the first proof for a specific apoptosis related channel in the outer mitochondrial membrane.

Several studies have indicated that the mitochondrial structures are not damaged during the release of proteins from the intermembrane space during apoptosis. When Bax is activated through NGF deprivation in SCG neurons, cytochrome *c* release is induced without mitochondria swelling; on the contrary, the mitochondria are smaller compared to untreated cells *(53)*. Furthermore, in the presence of caspase inhibitors, the cells are rescued and, after re-addition of growth factor, the mitochondria regain their normal cytochrome *c* content and size, indicating that no irreversible damages, such as rupture of the outer membrane, have been inflicted. Finucane et al. showed that Bax induced cytochrome *c* release when overexpressed in cells or added to isolated mitochondria. However, no mitochondrial swelling was detected, demonstrating that opening of PTP was not involved *(54)*.

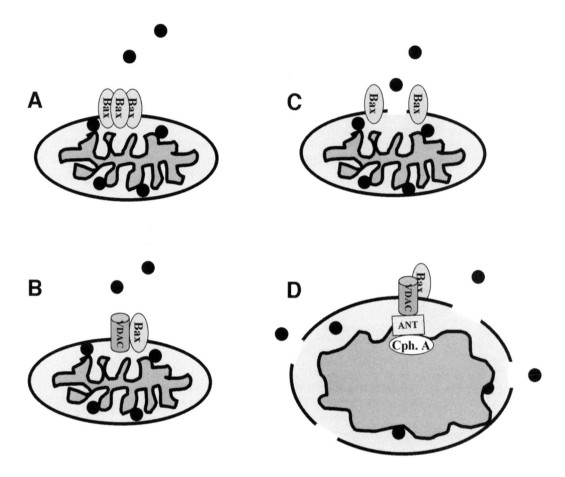

Fig. 2. Models for permeabilization of the outer mitochondrial membrane during apoptosis. Several models have been proposed for how the pro-apoptotic multidomain proteins, such as Bax and Bak, induce the release of proteins, including cytochrome *c*, from the intermembrane space. **(A)** A channel is formed by Bax oligomers. Whether these oligomers in the mitochondria are composed of Bax only or contain other not-yet-identified proteins is unclear. However, the Bax oligomers do not contain VDAC or ANT. **(B)** Bax forms a chimeric channel with VDAC. **(C)** Insertion of Bax destabilizes the membrane, inducing "lipid holes" in the membrane. **(D)** The multidomain proteins trigger opening of the permeability transition pore, which results in matrix swelling and rupture of the outer membrane. The models (A) and (B) propose formation of specific channels, which could induce release of specific proteins, whereas models (C) and (D) would permeabilize the membrane in an unspecific way, releasing proteins in an uncontrolled manner.

In a study by von Ahsen et al., isolated mitochondria treated with Bax or Bid and cells treated with UV-irradiation or staurosporine were depleted of cytochrome *c*; however, the mitochondria retained a fully intact protein import machinery *(55)*. Mitochondrial protein import is dependent on the $\Delta\psi m$ and ATP. Thus, these results show that although the outer membrane has been permeabilized to allow the passage of cytochrome *c*, the inner membrane remains intact. By single-cell analysis it has been shown that in the presence of caspase inhibitors, the release of cytochrome *c* results in a drop in $\Delta\psi m$. However, over the following 30–60 min the potential recovered to its original level *(56)*. When the downstream caspase cascade is inhibited, the mitochondria remain functional long after cytochrome *c* has been released, indicating intact mitochondrial structures. Furthermore, the membrane-permeabilizing activity of Bax is not inhibited by the PTP inhibitors cyclosporin A, EDTA, or Mg^{2+}. In fact, Mg^{2+} enhances the cytochrome *c*-releasing activity of Bax *(57)*.

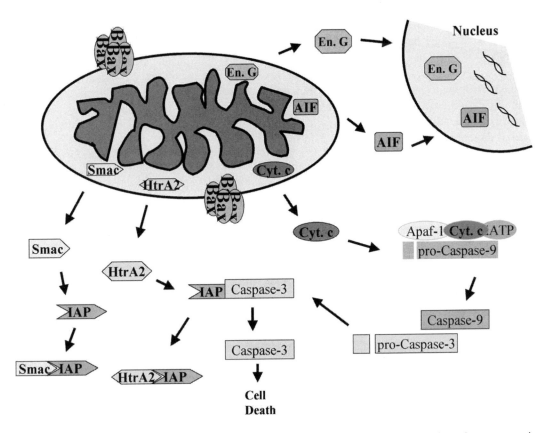

Fig. 3. Model for the release of mitochondrial proteins during apoptosis. The activated pro-apoptotic multidomain Bcl-2 proteins (Bax) form high-conductance channels in the outer mitochondrial membrane. This results in permeabilization of the outer membrane and release of proteins normally residing in the intermembrane space into the cytosol. In the cytosol, these proteins induce apoptosis-signaling cascades. Several proteins have been shown to be released. In the cytosol, cytochrome *c* forms a complex with Apaf-1, dATP and pro-caspase-9. Upon complex formation, caspase-9 is activated. Caspase-9 subsequently activates executioner caspases, such as caspase-3, ultimately leading to apoptotic cell death. In the cytosol, a group of proteins known as IAPs binds to activated caspases and inhibit their activity. Two proteins released from the mitochondrial intermembrane space, Smac/DIABLO and HtrA2, bind to IAPs, preventing them from binding to the active caspases, thus promoting caspase activity and apoptosis. AIF and endonuclease G are also released from the mitochondria. These two proteins translocate to the nucleus after their release, where they induce DNA degradation in a caspase-independent manner.

Combined, these results strongly suggest that Bax and Bak can form ion channels in the outer mitochondrial membrane and trigger cytochrome *c* release independent of PTP or its components (Fig. 2).

Bax-VDAC Channels

The involvement of VDAC in Bax-induced apoptosis has been suggested by some studies. Shimizu et al. showed that Bax and Bak could induce cytochrome *c* release from liposomes in which VDAC had been incorporated, whereas neither of the proteins was active alone *(58)*. Conversely, Sato et al. have shown that Bax oligomers alone are able to form cytochrome *c*-conducting channels in liposomes *(24)*. The difference might be due to the quaternary structure of the Bax protein, because Bax monomers do not have channel-forming activity. Priault et al. showed that Bax induced cytochrome *c* release in yeast deficient of VDAC as efficient as in wild-type yeast. The release was prevented by Bcl-X$_L$ *(59)*. A study by Gross et al. also concluded that VDAC was not required for Bax killing activity in yeast *(60)*. In a recent study, antibodies against VDAC were shown to prevent Bax-induced apoptosis. It was further shown that binding of Bax and Bak to red blood cells was dependent on a plasma membrane VDAC protein *(61)*. Although these results suggest the involvement of VDAC in multidomain protein activity, they do not show that VDAC is part of the channel-forming structure. VDAC might function as a receptor protein in the mitochondria membrane. This could explain the specific targeting of Bax to mitochondria during apoptosis.

The Permeability Transition Pore

The mitochondrial permeability transition pore (PTP) is a multiprotein complex spanning over the two mitochondrial membranes. The core proteins of the pore are thought to be VDAC in the outer membrane, ANT in the inner membrane, and the matrix protein cyclophilin D *(10)*. Ca^{2+} ions trigger opening of the pore, whereas the ANT ligand bongkrekate, cyclosporin A, which is interacting with cyclophilin D, Mg^{2+} ions, ADP, and ATP, block the pore *(11)*. PTP opening permeabilize the inner mitochondrial membrane to molecules up to 1500 Da. This results in matrix swelling, leading to rupture of the outer mitochondrial membrane and an unspecific release of proteins, including cytochrome *c*, from the intermembrane space *(62)*.

The main arguments for the involvement of PTP in apoptosis are early observations that apoptosis is accompanied by a loss of the transmembrane potential across the inner mitochondrial membrane, and that cytochrome *c* release could be inhibited by the PTP inhibitor cyclosporin A *(63,64)*. Furthermore, Bax was found to co-immunoprecipitate with ANT and interactions between the two proteins were detected in yeast two-hybrid experiments *(65)*. However, the decrease in membrane potential detected during apoptosis is merely a result of loss of cytochrome *c* and it appears to be transient *(56,66)*. Addition of exogenous cytochrome *c* can re-establish the potential, indicating that the inner membrane remains intact *(67)*. The PTP inhibitors cyclosporin A and Mg^{2+} have now been shown not to inhibit cytochrome *c* release *(57)*. Furthermore, it now appears clear that, at least in most forms of apoptosis, the mitochondrial structures remain intact. These would suggest that activation of the PTP is not involved. However, the intracellular ATP concentration has been shown to influence whether a cell is dying by apoptosis or necrosis *(68)*. It is therefore possible that cells starting to die by the apoptosis pathway, if depleted of ATP, could switch to necrotic death, thus activating the PTP.

ACTIVITIES OF MITOCHONDRIAL PROTEINS RELEASED

Several proteins normally residing in the mitochondrial intermembrane space are released during apoptosis (Fig. 3). Although the mitochondrial activity of some of these proteins remains unclear, others such as cytochrome *c*, have well-defined essential functions. However, once in the cytosol, all these proteins initiate or promote apoptosis-signaling cascades. The proteins released can be divided into two groups. One group, cytochrome *c*, Smac/DIABLO, and HtrA2/Omi, activate or promote the activity of caspases. The second group, endonuclease G and AIF, translocate to the nucleus and induce DNA degradation in a caspase-independent manner. Thus, it appears that the mitochondrial proteins released can trigger at least two different death-signaling pathways.

Cytochrome c

Cytochrome *c* is in the cells found attached to the outer surface of the inner mitochondrial membrane where the protein functions in the respiratory electron transfer chain. Several pools of cytochrome *c* with different conformations have been described *(69)*. One conformation of cytochrome *c* is loosely attached to the lipids, primary cardiolipin, through electrostatic interactions, whereas other pools are more tightly bound involving hydrophobic interactions or they are partly inserted into the membrane. Cytochrome *c* is, to a great extent (80–90%), localized in the cristae structures, which are connected with the narrow intermembrane space directly facing the outer membrane through narrow junctions *(70,71)*. During apoptosis, cytochrome *c* is released into the cytosol. Using green fluorescent protein (GFP)-tagged cytochrome *c*, Goldstein et al. showed that cytochrome *c* release is a fast process. Once the release has been initiated, all mitochondria in the cell lose their cytochrome *c* within approx 5 min *(72)*. Recent studies have suggested that cytochrome *c* release is a two-step process. First, cytochrome *c* is detached from the inner membrane and then the solubilized protein is subsequently released after permeabilization of the outer membrane *(73)*. An intriguing question is how can cytochrome *c* residing in the cristae structures be released from the mitochondria within a few minutes. A recent paper by Scorrano et al. shows that mitochondria exposed to t^cBid undergoes fast morphological rearrangements of the cristae membrane, allowing the diffusion and release of cristael cytochrome *c (74)*. The t^cBid-induced cristae remodeling was not BH3 domain-dependent, and Bak was not required.

The central role of cytochrome *c* in the mitochondrial apoptosis-signaling pathway has been shown in cell-free systems in vitro, through injection of cytochrome *c* into cells, and in apoptotic cells *(44,75,76)*. Murine embryos deprived of cytochrome *c* are, as expected, not viable and die by midgestation. However, cell lines derived from the cytochrome *c* knockout embryos can be maintained in culture. These cells show decreased activation of caspase-3 compared to wild-type controls. They are resistant to stress-induced apoptosis, such as UV irradiation, serum withdrawal, or staurosporine *(77)*. On the contrary, they are more sensitive to apoptosis induced through the death-receptor pathway.

In the cytosol cytochrome *c* binds to Apaf-1, a homolog to the *C. elegans* Ced-4 protein *(78)*. Apaf-1 is a cytosolic protein containing a caspase-recruitment domain (CARD) and a nucleotide binding site. Complex formation with cytochrome *c* increases the affinity for dATP, which now binds to Apaf-1, and induces conformational changes in the protein leading to oligomerization and exposure of the caspase recruitment domain (CARD). This result in the recruitment of procaspase-9 to the complex, which is referred to as the apoptosome *(79)*. Once bound to the complex procaspase-9 is proteolyticly autoactivated *(80)*. The active caspase-9 then activates downstream executioner caspases, such as caspases-3 and -7. These caspases subsequently cleave various substrates, leading to the characteristic morphological changes associated with apoptosis, such as DNA fragmentation, chromatin condensation, externalization of phosphatidylserine, and ultimately formation of apoptotic bodies and cell death. In contrast to cytochrome *c* knockouts, mice deficient in Apaf-1 developed normally to adult life, apart from an increased number of neurons and male sterility *(81,82)*. Cells isolated from these animals died when exposed to stress-apoptotic signals *(81)*, although no activation of caspase-3 was detected.

Smac/DIABLO

Smac/DIABLO is a 25 kDa protein, which, when released into the cytosol, binds to inhibitor of apoptosis proteins (IAPs) *(48,49)*. IAPs are a group of proteins that binds to both procaspases and activated caspases to inhibit their activation and activity, respectively *(83)* (*see* Chapter 3). The IAP proteins contain at least one baculoviral IAP repeat (BIR). These are 70–80 amino acid stretches containing a zinc-finger motive that coordinates a zinc atom *(84)*. The BIR domains are essential for interactions with active caspases and with Smac/DIABLO *(85)*. Smac/DIABLO is nuclear-encoded, synthesized in the cytosol, and imported into the mitochondria. The protein contains a 55 amino acid mitochondrial-signal sequence that is removed during the mitochondrial import process, yielding the mature protein *(49)*. The mature protein migrates as a 100 kDa protein on gel filtration, indicating that the protein forms multimers. Only the mature protein binds to active caspases. The first four N-terminal amino acids (Ala-Val-Pro-Ile) of mature Smac/DIABLO have been shown to interact with the BIR domains of the IAP proteins *(86–87)*. Modifications of the N-terminal amino acid sequence abolished binding to XIAP. On the contrary, deletions at the C-terminal did not affect the activity *(86)*.

XIAP contains 3 BIR domains with a high sequence homology. However, Smac/DIABLO was shown to interact with BIR 2 and 3 only *(86)*. Furthermore, it has been shown that XIAP binds to the N-terminal tetrapeptide of the caspase-9 linker peptide after activation of the enzyme *(85)*. The four amino acids of the N-terminus of the linker peptide is highly homology to the N-terminal sequence of Smac/DIABLO. Binding of Smac/DIABLO to the BIR3 domain of XIAP prevents the latter from binding to the linker peptide. This would suggest that Smac/DIABLO enhance caspase-9 activity through sequestering XIAP, preventing its binding to the activated caspase-9.

The presence of IAPs in the cytosol ensures that the apoptosis-signaling cascade is not accidentally induced, for example, by damaged mitochondria in a cell. However, when a massive release of cytochrome *c* and Smac/DIABLO from the mitochondria occurs during apoptosis, inhibition of IAPs by Smac/DIABLO ensures a fast propagation of the caspase cascade and execution of the apoptosis death program.

HtrA2/Omi

HtrA2/Omi is a highly conserved protein with serine protease activity present from bacteria to mammals *(88)*. The protein has been found in the endoplasmic reticulum, nucleus, and most recently in the mitochondria. The protein is synthesized as a 50 kD precursor, which is processed to the 36 kD mature protein in the mitochondria through removal of the 133 N-terminal amino acid peptide *(45)*. The mature protein was shown to be released from mitochondria in cell undergoing apoptosis induced by anti-Fas antibodies *(89)*. The release was completely blocked in cells overexpressing Bcl-2. Neither was any release detected in cells dying by necrosis after TNF stimulation. Furthermore, when isolated mitochondria were treated with t^{c}Bid, the protein was released. The N-terminal amino acid sequence of the mature HtrA2/Omi protein is AVPS, which is almost identical to the N-terminal sequence of Smac/DIABLO *(45,90)*. The protein was shown to bind to and inactivate IAPs in a way similar to Smac/DIABLO *(45,89)*. However, binding to IAPs did not interfere with the serine protease activity of the protein. Both binding to IAPs and serine protease activity appears to be associated with the proapoptotic activity of HtrA2/Omi *(91)*. N-terminal deletional mutation, which abolished IAP binding, and mutation of the catalytic serine residue both reduced the activity. However, the double mutant was completely inactive. When the mature protein was overexpressed extramitochondrially in cells, no increase in caspase activity was detected and the cells were dying by a caspase-independent pathway *(45)*. A mutant lacking serine protease activity was inactive, indicating that the protease activity was required for induction of this atypical cell death.

Endonuclease G

Endonuclease G is a nuclear encoded protein with a molecular mass of 30 kDa *(50)*. The protein is synthesized in the cytosol with a 48 amino-acid mitochondrial signal sequence and subsequently imported to the mitochondria, where the signal sequence is removed, generating the mature protein. It has been suggested that the protein is normally involved in mitochondrial DNA processing by generation of RNA primers required for DNA synthesis *(92)*. During apoptosis, endonuclease G is released from the mitochondria together with other apoptogenic factors. This was demonstrated through UV irradiation of isolated cells and treatment of isolated mitochondria with t^{c}Bid *(50)*. Furthermore, injection of anti-Fas antibodies into mice triggered the release of endonuclease G in liver cells *(93)*. The release was prevented in mice overexpressing Bcl-2. t^{c}Bid treatment induced the release of endonuclease G without release of Hsp70, a matrix protein. This indicates that a substantial amount of the protein is present in the intermembrane space and not in the matrix where the DNA processing occurs. After the protein has been released into the cytosol, it translocates to the nucleus. In the nucleus, the endonuclease cleaves chromatin DNA into fragments *(50,93)*. In classical apoptosis, DNA is fragmented by a nuclease called CAD (caspase-activated DNAse) *(94)*. In normal cells, CAD is bound to an inhibitor ICAD. During apoptosis activated caspase-3 cleaves ICAD, resulting in the release of active CAD *(95)*. CAD is sufficient to induce DNA fragmentation and nuclear chromatin condensation in isolated nucleus. In contrast to CAD, endonuclease G does not require caspase activation. It induces DNA fragmentation in a caspase-independent manner *(93)*. This suggests that the mitochondrial apoptosis pathway can activate a parallel caspase-independent pathway.

Recent studies have identified a gene in *Caenorhabditis elegans*, cps-6, with strong similarities to the mammalian endonuclease G *(96,97)*. Mutations in the protein affected DNA degradation and the development of the worm. Cps-6 is the first mitochondrial protein shown to be involved in apoptosis in *C. elegans*, reenforcing the importance of mitochondria in apoptosis signaling.

AIF

The apoptosis-inducing factor (AIF) is a 57 kDa flavoprotein normally residing in the mitochondrial intermembrane space, although its natural function remains unknown *(98)*. During apoptosis, AIF is released into the cytosol. Once in the cytosol, the protein translocates to the nucleus, where it has been shown to induce large-scale DNA fragmentation and chromatin condensation. The protein possesses NADH oxidase activity. However, this activity is not required for its apoptotic function *(97)*. Neither is AIF induced DNA fragmentation caspase-dependent.

Disruption of the *aif* gene in mice resulted in a disturbed early embryonic development *(99)*. Furthermore, embryonic stem cells lacking AIF were resistant to cell death induced by growth-factor deprivation. However, whether these effects are attributed to its apoptogenic activity only is unclear, because the protein possesses both oxoreductase and apoptogenic activities. AIF does not exhibit any detectable DNAse activity. It thus remains unclear how the protein induce DNA fragmentation and chromatin condensation. Whether the protein interacts with or activates other proteins, such as nucleases, is unknown.

EFFECTS ON MITOCHONDRIAL RESPIRATION DURING APOPTOSIS

Besides inducing permeabilization of the outer mitochondrial membrane, Bax has been suggested to have effects on components of the respiratory chain in the inner mitochondrial membrane. Activation of the mitochondrial apoptosis pathway induces matrix alkalinization and cytosolic acidification. This could be a result of a disturbance of the mitochondrial respiration, or the ATPase activity. Oligomycin, an inhibitor of F_0F_1-ATPase, has been shown to inhibit apoptosis involving the mitochondrial pathway, suggesting that the ATPase could be involved *(100,100)*. Furthermore, when Bax was expressed in wild-type yeast, cytosolic acidification, matrix alkalinization, and cell death was seen, but no effects were detected in F_0F_1-ATPase/H^+ pump-deficient yeast *(101)*.

Other components of the respiratory electron transfer chain also might be affected in Bax triggered apoptosis. In a study of Bax toxicity on yeast, it was shown that mutations of mitochondrial proteins required for oxidative phosphorylation, which make the cells respiratory incompetent, decreased Bax toxicity *(102)*. On the contrary, mutations in mitochondrial proteins unrelated to the oxidative phosphorylation machinery did not affect Bax toxicity. A study by Mootha et al. showed that in t^{c}Bid-induced apoptosis, the initial respiratory dysfunction could be restored by addition of exogenous cytochrome *c* *(103)*. However, over time an irreversible respiratory impairment developed. Taken together, these results suggest that Bax, and perhaps other multidomain proteins, in addition to their permeabilizing effect on the outer mitochondrial membrane, also have effects on components of the inner membrane, interfering with the respiration and/or ATPase activity. Whether these effects are reversible remains unclear. In a recent study, neurons and HeLa cells were exposed to apoptotic stimuli in the presence of BAF, a broad-spectrum caspase inhibitor *(104)*. Although, the caspase inhibitor assured short-term survival, mitochondria were selectively eliminated from the cells without any apparent effects on other intracellular structures. Cells deprived of mitochondria were irreversibly committed to death. In neurons, elimination of mitochondria was completely prevented by expression of Bcl-2, which works upstream of the mitochondria.

CONCLUSION

The role of mitochondria in apoptosis is now well-established. That the Bcl-2 proteins control the mitochondrial pathway and that permeabilization of the outer mitochondrial membrane is triggered by the multidomain proteins is also undisputed. However, at the molecular level, how the mitochondrial proteins are getting out from the intermembrane space remains controversial. The events downstream of the mitochondria are well-characterized. On the contrary, the upstream pathways between the initiating stimuli and activation of the multidomain proteins remain largely unknown. The mitochondrial pathway can be activated by a wide range of stimuli. A number of key questions remain to be answered. How are these signals detected by cells and how are they propagated to activate the multidomain proteins? Are the various multidomain proteins involved in specific apoptosis-signaling pathways, activated by specific stimuli, or are they redundant proteins with identical function? It would be surprising to find that these proteins are actually only redundant.

REFERENCES

1. Yuan, J. and Yankner, B. A. (2000) Apoptosis in the nervous system. *Nature* **407,** 802–809.
2. Kuhlmann, T., Lucchinetti, C., Zettl, U. K., Bitsch, A., Lassmann, H., and Bruck,W. (1999) Bcl-2-expressing oligodendrocytes in multiple sclerosis lesions. *GLIA* **28,** 34–39.
3. Kockx, M. M. and Herman, A. G. (2000) Apoptosis in atherosclerosis: beneficial or detrimental? *Cardiovasc. Res.* **45,** 736–746.
4. Yaoita, H., Ogawa, K., Maehara, K., and Maruyama, Y. (2000) Apoptosis in relevant clinical situations: contribution of apoptosis in myocardial infarction. *Cardiovasc. Res.* **45,** 630–641.
5. Chandra, J., Zhivotovsky, B., Zaitsev, S., Juntti-Berggren, L., Berggren, P. O., and Orrenius, S. (2001) Role of apoptosis in pancreatic beta-cell death in diabetes. *Diabetes* **50 (Suppl 1),** S44–S47.
6. Krammer, P. H. (2000) CD95's deadly mission in the immune system. *Nature* **407,** 789–795.
7. Sherratt, H. S. (1991) Mitochondria: structure and function. *Rev. Neurol. (Paris)* **147,** 417–430.
8. Brierley, G. P., Baysal, K., and Jung, D. W. (1994) Cation transport systems in mitochondria: Na+ and K+ uniports and exchangers. *J. Bioenerg. Biomembr.* **26,** 519–526.
9. Garlid, K. D., Jaburek, M., and Jezek, P. (1998) The mechanism of proton transport mediated by mitochondrial uncoupling proteins. *FEBS Lett.* **438,** 10–14.
10. Beutner, G., Ruck, A., Riede, B., and Brdiczka, D. (1998) Complexes between porin, hexokinase, mitochondrial creatine kinase and adenylate translocator display properties of the permeability transition pore. Implication for regulation of permeability transition by the kinases. *Biochim. Biophys. Acta* **1368,** 7–18.
11. Crompton, M., Virji, S., Doyle, V., Johnson, N., and Ward, J. M. (1999) The mitochondrial permeability transition pore. *Biochem. Soc. Symp.* **66,** 167–179.
12. Schmitz, I., Kirchhoff, S., and Krammer, P. H. (2000) Regulation of death receptor-mediated apoptosis pathways. *Int. J. Biochem. Cell Biol.* **32,** 1123–1136.
13. Li, H., Zhu, H., Xu, C. J., and Yuan, J. (1998) Cleavage of BID by caspase-8 mediates the mitochondrial damage in the Fas pathway of apoptosis. *Cell* **94,** 491–501.
14. Muchmore, S. W., Sattler, M., Liang, H., Meadows, R. P., Harlan, J. E., Yoon, H. S., et al. (1996) X-ray and NMR structure of human Bcl-xL, an inhibitor of programmed cell death. *Nature* **381,** 335–341.
15. McDonnell, J. M., Fushman, D., Milliman, C. L., Korsmeyer, S. J., and Cowburn, D. (1999) Solution structure of the proapoptotic molecule BID: a structural basis for apoptotic agonists and antagonists. *Cell* **96,** 625–634.
16. Chou, J. J., Li, H., Salvesen, G. S., Yuan, J., and Wagner, G. (1999) Solution structure of BID, an intracellular amplifier of apoptotic signaling. *Cell* **96,** 615–624.
17. Suzuki, M., Youle,R. J., and Tjandra, N. (2000) Structure of Bax: coregulation of dimer formation and intracellular localization. *Cell* **103,** 645–654.
18. Petros, A. M., Medek, A., Nettesheim, D. G., Kim, D. H., Yoon, H. S., Swift, K., et al. (2001) Solution structure of the antiapoptotic protein bcl-2. *Proc. Natl. Acad. Sci. USA* **98,** 3012–3017.
19. Antonsson, B., Conti, F., Ciavatta, A., Montessuit, S., Lewis, S., Martinou, I., et al. (1997) Inhibition of Bax channel-forming activity by Bcl-2. *Science* **277,** 370–372.
20. Schlesinger, P. H., Gross, A., Yin, X. M., Yamamoto, K., Saito, M., Waksman, G., and Korsmeyer, S. J. (1997) Comparison of the ion channel characteristics of proapoptotic BAX and antiapoptotic BCL-2. *Proc. Natl. Acad. Sci. USA* **94,** 11357–11362.
21. Minn, A. J., Velez, P., Schendel, S. L., Liang, H., Muchmore, S. W., Fesik, S. W., Fill, M., and Thompson, C. B. (1997) Bcl-x(L) forms an ion channel in synthetic lipid membranes. *Nature* **385,** 353–357.
22. Hsu, Y. T. and Youle, R. J. (1998) Bax in murine thymus is a soluble monomeric protein that displays differential detergent-induced conformations. *J. Biol. Chem.* **273,** 10777–10783.
23. Antonsson, B., Montessuit, S., Lauper, S., Eskes, R., and Martinou, J. C. (2000) Bax oligomerization is required for channel-forming activity in liposomes and to trigger cytochrome *c* release from mitochondria. *Biochem. J.* **345 (Pt 2),** 271–278.

24. Saito, M., Korsmeyer, S. J., and Schlesinger, P. H. (2000) BAX-dependent transport of cytochrome *c* reconstituted in pure liposomes. *Nat. Cell Biol.* **2,** 553–555.
25. Roucou, X., Rostovtseva, T., Montessuit, S., Martinou, J.-C., and Antonsson, B. (2002) Bid induces cytochrome c impermeable Bax channels in liposomes. *Biochem. J.* **363,** 547–552.
26. Basanez, G., Nechushtan, A., Drozhinin, O., Chanturiya, A., Choe, E., Tutt, S., et al. (1999) Bax, but not Bcl-xL, decreases the lifetime of planar phospholipid bilayer membranes at subnanomolar concentrations. *Proc. Natl. Acad. Sci. USA* **96,** 5492–5497.
27. Krajewski, S., Mai, J. K., Krajewska, M., Sikorska, M., Mossakowski, M. J., and Reed, J. C. (1995) Upregulation of bax protein levels in neurons following cerebral ischemia. *J. Neurosci.* **15,** 6364–6376.
28. Ekegren, T., Grundstrom, E., Lindholm, D., and Aquilonius, S. M. (1999) Upregulation of Bax protein and increased DNA degradation in ALS spinal cord motor neurons. *Acta Neurol. Scand.* **100,** 317–321.
29. Gross, A., Jockel, J., Wei, M. C., and Korsmeyer, S. J. (1998) Enforced dimerization of BAX results in its translocation, mitochondrial dysfunction and apoptosis. *EMBO J.* **17,** 3878–3885.
30. Desagher, S., Osen-Sand, A., Nichols, A., Eskes, R., Montessuit, S., Lauper, S., et al. (1999) Bid-induced conformational change of Bax is responsible for mitochondrial cytochrome *c* release during apoptosis. *J. Cell Biol.* **144,** 891–901.
31. Nechushtan, A., Smith, C. L., Hsu, Y. T., and Youle, R. J. (1999) Conformation of the Bax C-terminus regulates subcellular location and cell death. *EMBO J.* **18,** 2330–2341.
32. Antonsson, B., Montessuit, S., Sanchez, B., and Martinou, J. C. (2001) Bax is present as a high molecular weight oligomer/complex in the mitochondrial membrane of apoptotic cells. *J. Biol. Chem.* **276,** 11615–11623.
33. Sundararajan, R. and White, E. (2001) E1b 19k blocks bax oligomerization and tumor necrosis factor alpha-mediated apoptosis. *J. Virol.* **75,** 7506–7516.
34. Mikhailov, V., Mikhailova, M., Pulkrabek, D. J., Dong, Z., Venkatachalam, M. A., and Saikumar, P. (2001) Bcl-2 prevents bax oligomerization in the mitochondrial outer membrane. *J. Biol. Chem.* **276,** 18361–18374.
35. Griffiths, G. J., Dubrez, L., Morgan, C. P., Jones, N. A., Whitehouse, J., Corfe, B. M., et al. (1999) Cell damage-induced conformational changes of the pro-apoptotic protein Bak in vivo precede the onset of apoptosis. *J. Cell Biol.* **144,** 903–914.
36. Wei, M. C., Lindsten, T., Mootha, V. K., Weiler, S., Gross, A., Ashiya, M., et al. (2000) tBID, a membrane-targeted death ligand, oligomerizes BAK to release cytochrome c. *Genes Dev.* **14,** 2060–2071.
37. Schmitz, I., Walczak, H., Krammer, P. H., and Peter, M. E. (1999) Differences between CD95 type I and II cells detected with the CD95 ligand. *Cell Death Diff.* **6,** 821–822.
38. Wei, M. C., Zong, W. X., Cheng, E. H., Lindsten, T., Panoutsakopoulou, V., Ross, A. J., et al. (2001) Proapoptotic BAX and BAK: a requisite gateway to mitochondrial dysfunction and death. *Science* **292,** 727–730.
39. Kudla, G., Montessuit, S., Eskes, R., Berrier, C., Martinou, J. C., Ghazi, A., and Antonsson, B. (2000) The destabilization of lipid membranes induced by the C-terminal fragment of caspase-8-cleaved bid is inhibited by the N-terminal fragment. *J. Biol. Chem.* **275,** 22713–22718.
40. Zha, J., Harada, H., Osipov, K., Jockel, J., Waksman, G., and Korsmeyer, S. J. (1997) BH3 domain of BAD is required for heterodimerization with BCL-XL and pro-apoptotic activity. *J. Biol. Chem.* **272,** 24101–24104.
41. Verma, S., Zhao, L. J., and Chinnadurai, G. (2001) Phosphorylation of the pro-apoptotic protein BIK: mapping of phosphorylation sites and effect on apoptosis. *J. Biol. Chem.* **276,** 4671–4676.
42. Puthalakath, H., Huang, D. C., O'Reilly, L. A., King, S. M., and Strasser, A. (1999) The proapoptotic activity of the Bcl-2 family member Bim is regulated by interaction with the dynein motor complex. *Mol. Cell* **3,** 287–296.
43. Newmeyer, D. D., Farschon, D. M., and Reed, J. C. (1994) Cell-free apoptosis in Xenopus egg extracts: inhibition by Bcl-2 and requirement for an organelle fraction enriched in mitochondria. *Cell* **79,** 353–364.
44. Liu, X., Kim, C. N., Yang, J., Jemmerson, R., and Wang, X. (1996) Induction of apoptotic program in cell-free extracts: requirement for dATP and cytochrome c. *Cell* **86,** 147–157.
45. Suzuki, Y., Imai, Y., Nakayama, H., Takahashi, K., Takio, K., and Takahashi, R. (2001) A serine protease, HtrA2, is released from the mitochondria and interacts with XIAP, inducing cell death. *Mol. Cell* **8,** 613–621.
46. Kluck, R. M., Bossy-Wetzel, E., Green, D. R., and Newmeyer, D. D. (1997) The release of cytochrome *c* from mitochondria: a primary site for Bcl-2 regulation of apoptosis. *Science* **275,** 1132–1136.
47. Daugas, E., Susin, S. A., Zamzami, N., Ferri, K. F., Irinopoulou, T., Larochette, N., et al. (2000) Mitochondrio-nuclear translocation of AIF in apoptosis and necrosis. *FASEB J.* **14,** 729–739.
48. Verhagen, A. M., Ekert, P. G., Pakusch, M., Silke, J., Connolly, L. M., Reid, G. E., et al. (2000) Identification of DIABLO, a mammalian protein that promotes apoptosis by binding to and antagonizing IAP proteins. *Cell* **102,** 43–53.
49. Du, C., Fang, M., Li, Y., Li, L., and Wang, X. (2000) Smac, a mitochondrial protein that promotes cytochrome c-dependent caspase activation by eliminating IAP inhibition. *Cell* **102,** 33–42.
50. Li, L. Y., Luo, X., and Wang, X. (2001) Endonuclease G is an apoptotic DNase when released from mitochondria. *Nature* **412,** 95–99.
51. Kohler, C., Gahm, A., Noma, T., Nakazawa, A., Orrenius, S., and Zhivotovsky, B. (1999) Release of adenylate kinase 2 from the mitochondrial intermembrane space during apoptosis. *FEBS Lett.* **447,** 10–12.
52. Pavlov, E. V., Priault, M., Pietkiewicz, D., Cheng, E. H., Antonsson, B., Manon, S., et al. (2001) A novel, high conductance channel of mitochondria linked to apoptosis in mammalian cells and Bax expression in yeast. *J. Cell Biol.* **155,** 725–731.
53. Martinou, I., Desagher, S., Eskes, R., Antonsson, B., Andre, E., Fakan, S., and Martinou, J. C. (1999) The release of

cytochrome *c* from mitochondria during apoptosis of NGF-deprived sympathetic neurons is a reversible event. *J. Cell Biol.* **144**, 883–889.

54. Finucane, D. M., Bossy-Wetzel, E., Waterhouse, N. J., Cotter, T. G., and Green, D. R. (1999) Bax-induced caspase activation and apoptosis via cytochrome *c* release from mitochondria is inhibitable by Bcl-xL. *J. Biol. Chem.* **274**, 2225–2233.

55. von Ahsen, O., Renken, C., Perkins, G., Kluck, R. M., Bossy-Wetzel, E., and Newmeyer, D. D. (2000) Preservation of mitochondrial structure and function after Bid- or Bax-mediated cytochrome *c* release. *J. Cell Biol.* **150**, 1027–1036.

56. Waterhouse, N. J., Goldstein, J. C., von Ahsen, O., Schuler, M., Newmeyer, D. D., and Green, D. R. (2001) Cytochrome *c* maintains mitochondrial transmembrane potential and ATP generation after outer mitochondrial membrane permeabilization during the apoptotic process. *J. Cell Biol.* **153**, 319–328.

57. Eskes, R., Antonsson, B., Osen-Sand, A., Montessuit, S., Richter, C., Sadoul, R.,et al. (1998) Bax-induced cytochrome *c* release from mitochondria is independent of the permeability transition pore but highly dependent on Mg2+ ions. *J. Cell Biol.* **143**, 217–224.

58. Shimizu, S., Narita, M., and Tsujimoto, Y. (1999) Bcl-2 family proteins regulate the release of apoptogenic cytochrome *c* by the mitochondrial channel VDAC. *Nature* **399**, 483–487.

59. Priault, M., Chaudhuri, B., Clow, A., Camougrand, N., and Manon, S. (1999) Investigation of bax-induced release of cytochrome *c* from yeast mitochondria permeability of mitochondrial membranes, role of VDAC and ATP requirement. *Eur. J. Biochem.* **260**, 684–691.

60. Gross, A., Pilcher, K., Blachly-Dyson, E., Basso, E., Jockel, J., Bassik, M. C., et al. (2000) Biochemical and genetic analysis of the mitochondrial response of yeast to BAX and BCL-X(L). *Mol. Cell Biol.* **20**, 3125–3136.

61. Shimizu, S., Matsuoka, Y., Shinohara, Y., Yoneda, Y., and Tsujimoto, Y. (2001) Essential role of voltage-dependent anion channel in various forms of apoptosis in mammalian cells. *J. Cell Biol.* **152**, 237–250.

62. Marzo, I., Brenner, C., Zamzami, N., Susin, S. A., Beutner, G., Brdiczka, D., Remy, R., et al. (1998) The permeability transition pore complex: a target for apoptosis regulation by caspases and bcl-2-related proteins. *J. Exp. Med.* **187**, 1261–1271.

63. Halestrap, A. P., Connern, C. P., Griffiths, E. J., and Kerr, P. M. (1997) Cyclosporin A binding to mitochondrial cyclophilin inhibits the permeability transition pore and protects hearts from ischaemia/reperfusion injury. *Mol. Cell Biochem.* **174**, 167–172.

64. Narita, M., Shimizu, S., Ito, T., Chittenden, T., Lutz, R. J., Matsuda, H., and Tsujimoto, Y. (1998) Bax interacts with the permeability transition pore to induce permeability transition and cytochrome *c* release in isolated mitochondria. *Proc. Natl. Acad. Sci. USA* **95**, 14681–14686.

65. Marzo, I., Brenner, C., Zamzami, N., Jurgensmeier, J. M., Susin, S. A., Vieira, H. L., et al. (1998) Bax and adenine nucleotide translocator cooperate in the mitochondrial control of apoptosis. *Science* **281**, 2027–2031.

66. Madesh, M. A. (2002) Rapid kinetics of tBid-induced cytochrome *c* and Smac/DIABLO release and mitochondrial depolarization. *J. Biol. Chem.* **277**, 5651–5659.

67. Mootha, V. K., Wei, M. C., Buttle, K. F., Scorrano, L., Panoutsakopoulou, V., Mannella, C. A., and Korsmeyer, S. J. (2001) A reversible component of mitochondrial respiratory dysfunction in apoptosis can be rescued by exogenous cytochrome *c*. *EMBO J.* **20**, 661–671.

68. Nicotera, P., Leist, M., and Ferrando-May, E. (1998) Intracellular ATP, a switch in the decision between apoptosis and necrosis. *Toxicol. Lett.* **102–103**, 139–142.

69. Cortese, J. D., Voglino, A. L., and Hackenbrock, C. R. (1998) Multiple conformations of physiological membrane-bound cytochrome c. *Biochemistry* **37**, 6402–6409.

70. Bernardi, P. and Azzone, G. F. (1981) Cytochrome *c* as an electron shuttle between the outer and inner mitochondrial membranes. *J. Biol. Chem.* **256**, 7187–7192.

71. Frey, T. G. and Mannella, C. A. (2000) The internal structure of mitochondria. *Trends Biochem. Sci.* **25**, 319–324.

72. Goldstein, J. C., Waterhouse, N. J., Juin, P., Evan, G. I., and Green, D. R. (2000) The coordinate release of cytochrome *c* during apoptosis is rapid, complete and kinetically invariant. *Nat. Cell Biol.* **2**, 156–162.

73. Ott, M., Robertson, J. D., Gogvadze, V., Zhivotovsky, B., and Orrenius, S. (2002) Cytochrome *c* release from mitochondria proceeds by a two-step process. *Proc. Natl. Acad. Sci. USA* **99**, 1259–1263.

74. Scorrano, L., Ashiya, M., Buttle, K., Weiler, S., Oakes, S. A., Mannella, C. A., and Korsmeyer, S. J. (2002) A distinct pathway remodels mitochondrial cristae and mobilizes cytochrome *c* during apoptosis. *Dev. Cell* **2**, 55–67.

75. Zhivotovsky, B., Orrenius, S., Brustugun, O. T., and Doskeland, S. O. (1998) Injected cytochrome *c* induces apoptosis. *Nature* **391**, 449–450.

76. Bossy-Wetzel, E., Newmeyer, D. D., and Green, D. R. (1998) Mitochondrial cytochrome *c* release in apoptosis occurs upstream of DEVD-specific caspase activation and independently of mitochondrial transmembrane depolarization. *EMBO J.* **17**, 37–49.

77. Li, K., Li, Y., Shelton, J. M., Richardson, J. A., Spencer, E., Chen, Z. J., Wang, X., and Williams, R. S. (2000) Cytochrome *c* deficiency causes embryonic lethality and attenuates stress-induced apoptosis. *Cell* **101**, 389–399.

78. Zou, H., Henzel, W. J., Liu, X., Lutschg, A., and Wang, X. (1997) Apaf-1, a human protein homologous to C. elegans CED-4, participates in cytochrome c-dependent activation of caspase-3. *Cell* **90**, 405–413.

79. Li, P., Nijhawan, D., Budihardjo, I., Srinivasula, S. M., Ahmad, M., Alnemri, E. S., and Wang, X. (1997) Cytochrome *c* and dATP-dependent formation of Apaf-1/caspase-9 complex initiates an apoptotic protease cascade. *Cell* **91**, 479–489.

80. Zou, H., Li, Y., Liu, X., and Wang, X. (1999) An APAF-1.cytochrome *c* multimeric complex is a functional apoptosome that activates procaspase-9. *J. Biol. Chem.* **274**, 11549–11556.

81. Yoshida, H., Kong, Y. Y., Yoshida, R., Elia, A. J., Hakem, A., Hakem, R., Penninger, J. M., and Mak, T. W. (1998) Apaf1 is required for mitochondrial pathways of apoptosis and brain development. *Cell* **94**, 739–750.

82. Honarpour, N., Du, C., Richardson, J. A., Hammer, R. E., Wang, X., and Herz, J. (2000) Adult Apaf-1-deficient mice exhibit male infertility. *Dev. Biol.* **218**, 248–258.

83. Deveraux, Q. L., Roy, N., Stennicke, H. R., Van Arsdale, T., Zhou, Q., Srinivasula, S. M., et al. (1998) IAPs block apoptotic events induced by caspase-8 and cytochrome *c* by direct inhibition of distinct caspases. *EMBO J.* **17**, 2215–2223.

84. Crook, N. E., Clem, R. J., and Miller, L. K. (1993) An apoptosis-inhibiting baculovirus gene with a zinc finger-like motif. *J. Virol.* **67**, 2168–2174.

85. Srinivasula, S. M., Hegde, R., Saleh, A., Datta, P., Shiozaki, E., Chai, J., et al. (2001) A conserved XIAP-interaction motif in caspase-9 and Smac/DIABLO regulates caspase activity and apoptosis. *Nature* **410**, 112–116.

86. Chai, J., Du, C., Wu, J. W., Kyin, S., Wang, X., and Shi, Y. (2000) Structural and biochemical basis of apoptotic activation by Smac/DIABLO. *Nature* **406**, 855–862.

87. Wu, G., Chai, J., Suber, T. L., Wu, J. W., Du, C., Wang, X., and Shi, Y. (2000) Structural basis of IAP recognition by Smac/DIABLO. *Nature* **408**, 1008–1012.

88. Hu, S. I., Carozza, M., Klein, M., Nantermet, P., Luk, D., and Crowl, R. M. (1998) Human HtrA, an evolutionarily conserved serine protease identified as a differentially expressed gene product in osteoarthritic cartilage. *J. Biol. Chem.* **273**, 34406–34412.

89. van Loo, G., van Gurp, M., Depuydt, B., Srinivasula, S. M., Rodriguez, I., Alnemri, E. S., et al. (2002) The serine protease Omi/HtrA2 is released from mitochondria during apoptosis. Omi interacts with caspase-inhibitor XIAP and induces enhanced caspase activity. *Cell Death Diff.* **9**, 20–26.

90. Hegde, R., Srinivasula, S. M., Zhang, Z., Wassell, R., Mukattash, R., Cilenti, L., et al. (2002) Identification of Omi/HtrA2 as a mitochondrial apoptotic serine protease that disrupts inhibitor of apoptosis protein-caspase interaction. *J. Biol. Chem.* **277**, 432–438.

91. Verhagen, A. M., Silke, J., Ekert, P. G., Pakusch, M., Kaufmann, H., Connolly, L. M., et al. (2002) HtrA2 promotes cell death through its serine protease activity and its ability to antagonize inhibitor of apoptosis proteins. *J. Biol. Chem.* **277**, 445–454.

92. Cote, J. and Ruiz-Carrillo, A. (1993) Primers for mitochondrial DNA replication generated by endonuclease G. *Science* **261**, 765–769.

93. van Loo, G., Schotte, P., van Gurp, M., Demol, H., Hoorelbeke, B., Gevaert, K., et al. (2001) Endonuclease G: a mitochondrial protein released in apoptosis and involved in caspase-independent DNA degradation. *Cell Death Diff.* **8**, 1136–1142.

94. Enari, M., Sakahira, H., Yokoyama, H., Okawa, K., Iwamatsu, A., and Nagata, S. (1998) A caspase-activated DNase that degrades DNA during apoptosis, and its inhibitor ICAD. *Nature* **391**, 43–50.

95. Liu, X., Zou, H., Slaughter, C., and Wang, X. (1997) DFF, a heterodimeric protein that functions downstream of caspase-3 to trigger DNA fragmentation during apoptosis. *Cell* **89**, 175–184.

96. Parrish, J., Li, L., Klotz, K., Ledwich, D., Wang, X., and Xue, D. (2001) Mitochondrial endonuclease G is important for apoptosis in C. elegans. *Nature* **412**, 90–94.

97. Miramar, M. D., Costantini, P., Ravagnan, L., Saraiva, L. M., Haouzi, D., Brothers, G., et al. (2001) NADH oxidase activity of mitochondrial apoptosis-inducing factor. *J. Biol. Chem.* **276**, 16391–16398.

98. Susin, S. A., Lorenzo, H. K., Zamzami, N., Marzo, I., Snow, B. E., et al. (1999) Molecular characterization of mitochondrial apoptosis-inducing factor. *Nature* **397**, 441–446.

99. Joza, N., Susin, S. A., Daugas, E., Stanford, W. L., Cho, S. K., Li, C. Y., et al. (2001) Essential role of the mitochondrial apoptosis-inducing factor in programmed cell death. *Nature* **410**, 549–554.

100. Leist, M., Single, B., Castoldi, A. F., Kuhnle, S., and Nicotera, P. (1997) Intracellular adenosine triphosphate (ATP) concentration: a switch in the decision between apoptosis and necrosis. *J. Exp. Med.* **185**, 1481–1486.

101. Matsuyama, S., Llopis, J., Deveraux, Q. L., Tsien, R. Y., and Reed, J. C. (2000) Changes in intramitochondrial and cytosolic pH: early events that modulate caspase activation during apoptosis. *Nat. Cell Biol.* **2**, 318–325.

102. Harris, M. H., Vander Heiden, M. G., Kron, S. J., and Thompson, C. B. (2000) Role of oxidative phosphorylation in Bax toxicity. *Mol. Cell Biol.* **20**, 3590–3596.

103. Mootha, V. K., Wei, M. C., Buttle, K. F., Scorrano, L., Panoutsakopoulou, V., Mannella, C. A., and Korsmeyer, S. J. (2001) A reversible component of mitochondrial respiratory dysfunction in apoptosis can be rescued by exogenous cytochrome c. *EMBO J.* **20**, 661–671.

104. Xue, L., Fletcher, G. C., and Tolkovsky, A. M. (2001) Mitochondria are selectively eliminated from eukaryotic cells after blockade of caspases during apoptosis. *Curr. Biol.* **11**, 361–365.

7

From Caspases to Alternative Cell-Death Mechanisms

Marja Jäättelä and Marcel Leist

INTRODUCTION

Programmed cell death (PCD) is essential for the development and maintenance of multicellular organisms. Many eukaryotic cells that die and are removed in a programmed way undergo an astonishingly stereotypical series of biochemical and morphological changes, the most defining features of which are the activation of caspases, chromatin condensation, and the display of phagocytosis markers on the cell surface (1–4). The underlying death process has been called apoptosis to delineate it clearly from other death programs. A single family of proteases, the caspases, has been considered the pivotal executioner of all programmed cell death. However, recent findings of evolutionary-conserved, caspase-independent, controlled-death mechanisms have opened new perspectives on the biology of cell demise, with implications in particular for neurobiology, cancer research, and immunological processes (4–7).

DEFINITIONS OF PROGRAMMED CELL DEATH

The unclear definitions of the alternative death pathways have been the major obstacles to their better understanding. When PCD is used as a synonym of apoptosis and defined by caspase activation (8), then alternative caspase-independent PCD pathways are evidently not possible. In contrast, the below classification sorts the described modes of death into three major categories of PCD, all clearly distinct from necrosis or accidental lysis. The intention is to facilitate the cell-death discussion while taking into account the implications of the death mode for the surrounding tissue, and leaving space for different mechanistic observations and alternative interpretations. Remarkably, modifications in the mode of death do not necessarily affect the efficient removal of dying cells (9,10). Thus, some of the alternative caspase-independent death pathways may have evolved to serve the same purpose as proposed for the classical apoptosis; that is, to guarantee a safe and noninflammatory removal of corpses. In this vein, we present here a differentiated view on PCD based not on the activation of caspases, but rather the morphology and fate of dying cells.

Apoptosis

Apoptosis is defined by stereotypic morphological changes, especially evident in the nucleus, where the chromatin condenses to very compact and apparently simple geometric (globular, crescent-shaped) figures (1,4). Other typical features include phosphatidylserine exposure, cytoplasmic

From: *Essentials of Apoptosis: A Guide for Basic and Clinical Research*
Edited by: X-M. Yin and Z. Dong © Humana Press Inc., Totowa, NJ

shrinkage, zeiosis, and formation of apoptotic bodies (with nuclear fragments). In its most classical form, apoptosis is observed almost exclusively when caspases, in particular caspase-3, are activated. When death can be blocked by inhibition of any signal or activity (e.g., caspases) within the target cell, then the simplest (but most essential) condition for PCD is met. Apoptotic morphology results from one of the most elaborate forms of PCD, and it may be viewed as a far end of a continuum of death modes, with varying contributions of the cellular machinery.

Apoptosis-Like PCD

Apoptosis-like PCD is used here to describe forms of PCD with chromatin condensation that is less compact/complete than in apoptosis (geometrically more complex and lumpier shapes), and with display of phagocytosis-recognition molecules before lysis of the plasma membrane. Any degree and combination of other apoptotic features can be found. Most published forms of caspase-independent apoptosis fall into this class. Notably, also some of the classic caspase-dependent apoptosis models, such as tumor necrosis factor (TNF)-induced death of MCF-7 cells, have this morphology. For comparative examples, *see* ref. *(4)*.

Necrosis-Like PCD

Necrosis-like PCD is used here to define PCD in the absence of chromatin condensation, or at best with chromatin clustering to lose speckles *(11–16)*. Varying degrees of apoptosis-like cytosolic features, including externalization of phosphatidylserine, may occur before the lysis *(14,15)*. Necrotic PCD usually involves specialized caspase-independent signaling pathways. However, caspase-8 may be activated *(17)*, and caspase-1-driven necrosis has also been observed *(18)*. A subgroup of necrotic PCD models are often classified as aborted apoptosis; that is, a standard apoptosis program is initiated, then blocked at the level of caspase activation, and finally terminated by alternative, caspase-independent routes *(19)*.

Accidental Necrosis/Cell Lysis

Accidental necrosis/cell lysis is the conceptual counterpart to PCD, because it is prevented only by removal of the stimulus. It occurs after exposure to high concentrations of detergents, oxidants, ionophores, or high intensities of pathologic insult *(19)*. Necrosis is often associated with cellular "edema" (organelle swelling). The necrotic tissue morphology is in large part due to post-mortem events (occurring after the lysis of the plasma membrane) *(19)*.

THE GRAIN OF DOUBT: IS CASPASE ACTIVATION REALLY IDENTICAL WITH APOPTOSIS?

The unexpected ability of certain cells to survive the activation of pro-apoptotic caspases *(20–28)* demonstrates a remarkable plasticity of the cellular death program, and argues against the idea that caspases alone are sufficient for the induction of mammalian PCD (Fig. 1). Furthermore, recent evidence indicates a diversification of the apoptosis program in higher eukaryotes with respect to the necessity and role of caspases. Apoptosis-like cell death can occur in the absence of effector caspase activation *(29–36)*, and signals emanating from the established key players of apoptosis, including death receptors and caspases themselves, may result in a nonapoptotic death *(11–13,17)* (Fig. 1).

EVOLUTION OF CELL DEATH PRINCIPLES

The driving evolutionary pressures for the development of multiple cell-death programs have been increasing in parallel with the increased complexity and lifespan of the organisms *(37)*. But when in the evolution did the caspase-independent death mechanisms arise? Caspase-coding sequences are absent from the known genomes of many nonanimal species *(37)*. Nevertheless, such organisms— including plants and a number of single-celled eukaryotes—undergo PCD under conditions of stress *(38,39)* (Table 1) *(see* Chapter 8).

Cells survive caspase activation:

Ap24 inhibition
HSP70 overexpression
Cathepsin B inhibition
CDC2 inhibition
ALG-2 depletion
Erythropoietin exposure
TNF + E1B19K (caspase-8 on)
TNF + permeability transition block (caspase-8 on)
T cell activation
C. elegans (phagocytosis-defect)

Caspases/Death receptors trigger non-apoptotic death:

CD95/fas (ATP-depletion)
TNF (L9292 cells)
CD95 (caspase inhibition)

Other proteases/factors with caspase-independent essential role:

Apoptosis-inducing factor (AIF)	> e.g. Blastulation (Knock-out, Microinjection)
Endonuclease G	> e.g. *C. elegans* (RNAi; overexpression)
Serine proteases (Omi, Ap24)	> e.g. Overexpression/inhibition (TNF)
Calpains	> e.g. Neurons; Cancer cells with vitamin D
Cysteine cathepsins	> e.g. Hepatocytes (TNF); cancer cells (TNF)
Cathepsin D	> e.g. Antisense, Inhibition, Microinjection
Proteasome	> e.g. Lymphocytes, Neurons

Apoptosis-like/programmed cell death without caspases:

Adenoviral E4orf expression (apoptotic morphology)
Oncogenic ras (necrotic morphology)
CD4/CXC4 or CD99 stimulation (PS exposure)
Vitamin D (apoptotic morphology)
HSP70 depletion (apoptotic morphology)
Bin1 (DNA-fragmentation)
Bcl-2 depletion (autophagy)
Cep-1 in *C. elegans* (Ced-3 independent)

Programmed cell death triggers acting in the presence of pan-caspase inhibition:

Bax
Bak
TNF (L929, WEHI164, U937, ME-180)
CD95L (Jurkat)
GD3 (oligodendrocytes)
Staurosporine (Jurkat)
Anti-CD2 (lymphocytes)

HIV-infection of CD4-positive cells
spontaneous neutrophil apoptosis
B cell receptor activation
....

Problem with inhibitor specificity:

Inhibition [at 50 µM] protease:	caspase-3	caspase-8	cathepsin B	cathepsin L	calpain
zVAD-fmk	YES	YES	YES	NO	Weak
z-D-fmk/BAF	YES	YES	YES	NO	NO
DEVD-fmk	YES	YES	YES	?	YES
DEVD-CHO	YES	YES	Weak	?	NO

Fig. 1. Are caspases always the mediation of PCD?

For instance, in yeast this apoptosis-like death is associated with DNA-fragmentation, zeiosis, phosphatidylserine exposure, and chromatin condensation *(38)*, and can be selectively triggered or blocked by Bax-like or *ced-9* related genes, respectively. Furthermore, programmed paraptosis-like death is well-characterized in caspase-deficient slime molds *(45)*.

The introduction of the caspases, and especially of the mitochondrial CED-9/Bcl-2-related death switches *(37,39)*, may represent a decisive refinement of the old caspase-independent death programs. The relative importance of different death mechanisms seems to have been optimized subsequently in various ways. One form of extreme specialization is exemplified by the somatic cell death in the nematode *Caenorhabditis elegans*. The requirements for PCD in *C. elegans* are adapted to its

Table 1
Programmed Cell Death in Naturally Caspase-Deficient Cells

	References
• Mature red blood cells: noncaspase cysteine proteases, shrinkage, blebbing, phosphatidylserine exposure, phagocytosis	*(40)*
• Yeast: no caspases; most classical apoptotic features; control by ectopic Bcl-2 (N.B. The yeast metacaspase Yor197w/YCA1 might take the role of caspases) (Mol. Cell (2002) 9, 911)	*(38)*
• Leishmania: noncaspase cysteine proteases; phosphatidylserine exposure	*(41)*
• Platelets: calpain-mediated phosphatidylserine exposure	*(42)*
• Plants: no caspases; different PCD mechanisms	*(43)*
• Ced-3 deficient *C. elegans*: CEP-1 overexpression	*(44)*
Mec-4 induced death followed by phagocytosis	*(9)*
• Dictyostelium: vacuolization, phosphatidylserine exposure; AIF-release from mitochondria	*(41,45)*

specific needs, and have diverged widely from those of mammals *(37)*. Because the environmental pressure to provide flexible death responses is very low in this short-lived organism, evolutionary optimization has resulted in a single caspase-dependent apoptosis program. In contrast to mammals, control by mitochondrial proteins may play a minor role, and some degradative enzymes are supplied by the phagocytosing cell rather than by the dying cell itself *(2,3)*. Apoptosis in *C. elegans* is commonly cell-autonomous, i.e., it is not signaled or controlled from outside, and the entire system of death receptors appears to be absent. In accordance with this minimalist program, somatic PCD is not essential for the survival or development in *C. elegans (46)*. Rudimentary remainders of alternative apoptotic programs are, however, still found in the male linker cell, where a possibly Ced-3-independent PCD is triggered from outside *(46)*, and also a role of endonucleotidase derived from mitochondria might be conserved from worm to man *(47,48)*.

The mammalian system of death programs could represent an opposite form of evolutionary direction, where besides the multiple caspases, many other cysteine proteases and mitochondrial factors have taken additional roles in development and life *(3,49)*. The essential nature of some factors (knockout lethality *[36,49]*) combined with redundancy of others (difficulty with interpretation of knockouts *[49]*) has made the study of their specific role in PCD technically challenging. In addition it has remained unclear which mechanisms are essential for commitment to death, and which ones only determine the phenotypic outcome *(19)*.

PCD CAN TAKE MANY FORMS

If one keeps to the strict morphological criteria of apoptosis, including the geometrical shape of chromatin after its condensation (Fig. 2), caspases seem indispensable for apoptosis. However, there are many forms of apoptosis-like PCD where the chromatin condenses to less geometric shapes, and where phagocytosis markers on the plasma membrane are displayed before cell lysis. An array of well-characterized cell-death models occurring in the absence of caspase activation falls into this category *(25,29–34,36,50)* (Fig. 1). Furthermore, an analysis of cell-membrane dynamics in different death models has revealed that an important apoptosis hallmark, zeiosis, can occur independently of caspase activation *(25,51)*. PCD can also occur in the absence of chromatin condensation *(12–15)*. Such necrosis-like PCD is the result of active cellular processes that can be intercepted by, for

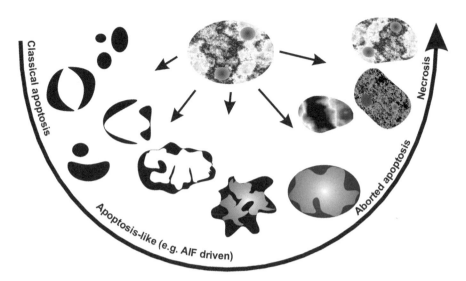

Fig. 2. Nuclear alterations in different forms of PCD. The use of chromatin condensation as a criterion to distinguish apoptosis from apoptosis-like PCD has been inconsistent in the scientific literature, and the potential for overlapping definitions and errors is large. Electron-microscopic examples of classical apoptosis and apoptosis-like PCD *(4)* or the above schematic drawings might provide a general guideline. Control chromatin is speckled showing areas of eu—and heterochromatin, and mostly one—several more condensed micronuclei (top middle). Caspase-dependent chromatin compaction and fragmentation to crescent- or spherical-shaped masses at the nuclear periphery is shown at left. Caspase-independent chromatin margination triggered directly by microinjection of AIF or in a number of models of apoptosis-like death *(4)* is shown at the bottom. Many intermediate forms and also transitions to necrosis are possible. Necrotic morphology is also observed in models where caspases are inhibited before apoptosis is completed (aborted apoptosis).

instance, oxygen-radical scavengers *(12,13,52)*, inhibition of poly(ADP) ribose polymerase (PARP) *(53)*, or mutations in intracellular signaling molecules *(15)*. Further caspase-independent modes of PCD include autophagy, characterized by the formation of large lysosome-derived cytosolic vacuoles *(35,54,55)*, and dark cell death in specialized cells such as chondrocytes *(56)* or neurons *(57)*.

Contrary to earlier expectations, the inhibition of caspase activation does not necessarily protect against cell-death stimuli (Fig. 1). Instead it may reveal, or even enhance, underlying caspase-independent death programs. These programs may resemble apoptosis-like PCD *(25,58–60)*, autophagy *(61)*, or even necrosis *(11,12,13,15,51,62–64)*. In many experimental apoptosis models, including those triggered by death receptors *(12,13,15,64)*, cancer drugs *(65)*, growth-factor deprivation *(61)*, staurosporine *(58)*, anti-CD2 *(58)*, oncogenes *(51)*, colchicine *(60)*, GD3 *(66)*, or expression of Bax-related proteins *(51,67)*, the existence of back-up death pathways has been uncovered following inhibition of caspase activity by pharmaceutical pan-caspase inhibitors. However, several lines of evidence support the relevance of such second-line mechanisms also for normal physiology and pathology. In addition to pharmacological inhibitors, caspase pathways can namely be inactivated by other factors such as mutations *(63)*, energy depletion *(11)*, nitrative/oxidative stress *(17)*, other proteases that are activated simultaneously *(68,69)*, members of the inhibitor of apoptosis protein (IAP) family *(3,70)*, or an array of viral proteins that can silence caspases *(3)*.

Upon caspase inhibition, the alternative death pathways surface also in vivo. They are involved in processes such as the negative selection of lymphocytes *(71,72)*, embryonic removal of interdigital webs *(63)*, tumor necrosis factor (TNF)-mediated liver injury *(73)*, and the death of chondrocytes controlling the longitudinal growth of bones *(56)*. These examples may represent just the tip of the iceberg with regard to the complexity of death signaling in vivo. And the overlapping death pathways

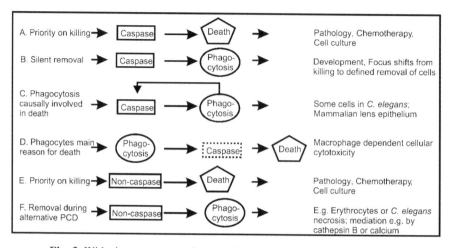

Fig. 3. Widening concept on the role of death mechanisms and phagocytosis.

initiated by a single stimulus seem to be the rule rather than the exception *(15,36)*. The examination of potential crossovers of death pathways that lead eventually to different phenotypical outcomes may open a chance to understand which events really determine commitment to death, and which ones are rather involved in upstream signaling or downstream execution.

THE MEANING OF THE PROGRAM: REMOVAL OF CELLS

Cell biology research has focused considerably on the mechanisms by which caspases kill cells (Fig. 3A), before it became clear that killing might not be the only idea behind PCD. A much less complicated machinery would be required to permeabilize the plasma membrane. The classic caspase-dependent program is in fact optimized to ensure that signals for phagocytosis are displayed well before cellular constituents might be released *(3,74)* (Fig. 3B). In extreme cases, there is in fact a feedback control of phagocytosis on the death program itself to insure that death only occurs when phagocytosis has been initiated *(26,27)* (Fig. 3C,D). Does this also apply to caspase-independent programs? A dominant uptake signal in mammalian cells is the translocation of phosphatidylserine to the outer leaflet of the plasma membrane. Also this "eat-me" indicator is uncoupled from caspase activation in many model systems *(10,14,25,31,38)*, and nonapoptotically dying eukaryotic cells can be efficiently phagocytosed *(10)*. Mechanisms that can lead to the translocation of phospatidylserine and phagocytosis in cells undergoing caspase-independent death include disturbances of cellular calcium homeostasis and protein kinase C (PKC) activation *(10,60)*. Noncaspase cysteine proteases might be involved not only in the alternative death execution (Fig. 3E), but also in alternative phagocytosis signal pathways. For instance, cathepsin B activity is required for the translocation of phospatidylserine in TNF-challenged tumor cells *(25)*. In the apoptosis-like death of platelets, phagocytosis signals are selectively blocked by calpain inhibitors *(42)*. Finally, genetic analysis in *C. elegans* has shown that the same phagocytosis recognition molecules are involved in removing corpses produced by caspase-dependent apoptosis and caspase-independent necrosis *(9)*.

CASPASE-INDEPENDENT PCD: EMERGING MECHANISTIC CLUES

Several molecular mediators of classical caspase-mediated apoptosis pathways were characterized during the last decade *(2,3,75)*, whereas the description of most alternative death routines remained limited to the phenomenological level. Now, recent mechanistic findings have opened a

new era for this field. Like classical apoptosis, alternative death programs can be mediated by proteases and switched on by death receptors or mitochondrial alterations.

Noncaspase Proteases

Caspase-mediated cleavage of specific substrates explains several of the characteristic features of
apoptosis, i.e., cleavage of inhibitor of caspase-activated DNase leads to chromatin changes, lamins
to nuclear shrinkage, cytoskeletal proteins to cytosolic reorganization, and p21-activated kinase-2 or
Rho-activated serine/threonine kinase to blebbing *(2,76)*. The question thus arises: what brings about
the apoptotic features observed in cells dying in caspase-independent manner? The first guess is,
naturally, other proteases (e.g., cathepsins, calpains, serine proteases, and the proteasome complex)
(Fig. 4). Indeed, accumulating data based on activity measurements, protease inhibitors, and/or
genetic depletion support their role as essential co-factors, either upstream or downstream of caspases,
in a number of death paradigms *(20,25,29,77–86)*. Furthermore, many noncaspase proteases can
cleave at least some of the classical caspase substrates, suggesting that they may also mimic the
cellular effects of caspases *(4,5,81,87–90)*.

There is evidence for the ability of other proteases to induce apoptosis-like PCD completely in the
absence of caspase activation; for instance, cathepsin D and B in camptothecin-induced death of liver
cancer cells (29), cathepsin B in TNF-treated fibrosarcoma cells *(25)*, the proteosome in colchicine-
treated neurons *(85)* and calpains in vitamin D-treated breast cancer cells (our own unpublished
observation). The definition of the role of the individual proteases in the complex process of PCD
still requires much careful work. The dependence on certain proteases might be extremely cell type-
and stimulus-specific and depend on the relative expressions, activations, and inactivations of proteases and protease inhibitors (Fig. 5). Genetic approaches need to be combined with meticulous
pharmacological titration of inhibitors *(25)*, because it turns out that *pan*-caspase inhibitors, as well
as many active site inhibitors of other proteases, are highly unspecific at concentrations widely used
to test their role in PCD *(5,25,81,87)*.

Death Receptors as Triggers

The best-studied members of the death-receptor family are TNF receptor 1 (TNF-R1) and Fas
(also known as CD95 or Apo-1) *(see* Chapter 5). Whereas it has long been known that TNF-induced
death can take the shape of either apoptosis or necrosis (91), the ability of Fas to induce necrosis-like
PCD has been described only recently *(11–13,15,17,92)*. In activated primary T lymphocytes, this
caspase-independent necrosis-like PCD seems, at least in some cases, to be the dominant mode of
death *(15)*. This may explain why inhibition of caspase activity in mouse T lymphocytes in vivo does
not induce lymphadenopathy and/or autoimmune disease usually manifested in mice with inactivating mutations in Fas or Fas ligand *(71)*. The complexity of death receptor-induced apoptotic and
necrotic signaling networks exceeds by far that of the simple linear pathway originally suggested by
the discovery of the receptor-triggered caspase cascade. Just as with the different protease families,
the concentration of adapter proteins such as FADD might be one of the switches deciding on the
apoptotic or necrotic pathway triggered by TNF (Fig. 6).

Mitochondria as Integrators

Many models of PCD involve some form of mitochondrial control, and it is useful to consider the
signaling phases up- and downstream of these organelles separately. The pro-apoptotic Bcl-2-related
proteins—such as Bax, Bak, Bid, and Bim—play a dominant role at the mitochondrial stage of PCD
signaling *(2,3)* *(see* Chapters 2 and 3). These proteins translocate to the mitochondria, or change their
conformation and interaction partners on the mitochondria, in response to a variety of death stimuli.
The regulatory counterparts at this level include the anti-apoptotic members of the same family (Bcl-
2, Bcl-x_L and so on). Eventually, the ratio of death and survival signals sensed by the Bcl-2–family
proteins determines whether the cell will live or die *(2,3,19)*.

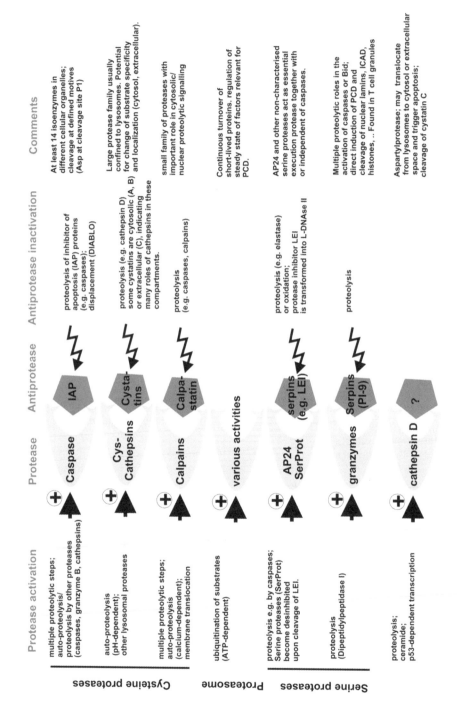

Fig. 4. Death by more than 1000 cuts.

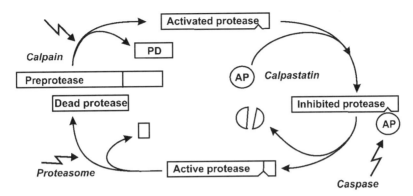

Fig. 5. Interaction of proteases during PCD. Frequently an inactive or weakly active zymogen (preprotease) is activated by cleaving off of a prodomain (PD) (first level of protease family interaction, examples shown in italics). Intracellular protease activity is prevented by specific antiproteases (AP), and ultimate activation requires inactivation of AP (often by proteolysis, second level of protease family interaction). Further proteolysis can lead to inactivation of the active protease and eventually degradation (third level of protease family interaction). The balance of all players in this circle determines which proteases dominate the death process. Pharmacological inhibition of one protease easily shifts the balance to another pathway.

Mitochondria as Triggers

The apoptosome-caspase pathway leading to classical apoptosis *(3,75)* is initiated by the release of cytochrome *c* from the mitochondrial intermembrane space (*see* Chapter 6). Together with other essential factors (e.g., ATP), it triggers assembly of the apoptosome complex, which forms the template for efficient caspase processing. The formation of the apoptosome and the start of the caspase cascade is controlled by caspase-inhibitory factors (IAPs, XIAP). Before the execution caspases can become fully active and produce the typically apoptotic morphology *(2,3)*, these apoptosis-dampening "safety catches" need to be removed by additional proteins (DIABLO/SMAC or Omi/HtrA2) released from mitochondria.

The second mitochondrial death pathway leads to necrotic PCD, without necessarily activating caspases. A prominent example is TNF-induced necrosis-like PCD mediated by mitochondria-derived reactive oxygen species (ROS) *(52)*. Intracellular control of this pathway is indicated by its susceptibility to attenuation by anti-oxidants *(12,13,52)*.

A third distinct pathway from mitochondria is the release of the apoptosis-inducing factor (AIF) from the intermembrane space *(99,102,103)*. Recent genetic evidence indicates that this factor controls PCD in early development—that is, all the hallmarks of early morphogenetic death, including cytochrome *c* release, are prevented by deletion of AIF *(36)*. AIF induces caspase-independent formation of large (50-bp) chromatin fragments, whereas oligonucleosomal DNA fragments are generated only when caspase-activated DNase (CAD) is activated *(3,102)*. This biochemical difference is reflected by slight morphological differences in the shape of the condensed chromatin (Figs. 2 and 8).

Often, more than one pathway seems to be activated simultaneously *(22,75,99,102)*. The cell fate (and death mechanism) is then determined by the relative speed of each process in a given model system, and by the antagonists of the individual pathways differentially expressed in different cell types. AIF, caspases, and ROS can feed back on mitochondria, causing enough structural and functional damage to trigger the release of other death factors, independent of the upstream signals *(3,19,75,102)*.

Defects in any step of the cytochrome *c* or AIF pathways will switch apoptosis to death with a necrotic morphology *(11,51,104)*. This death would still fulfill the criteria of PCD, as it is blocked by the anti-apoptotic oncogenes Bcl-2 or Bcr-Abl *(65,104)* or by the deletion of pro-apoptotic Bax *(105)*.

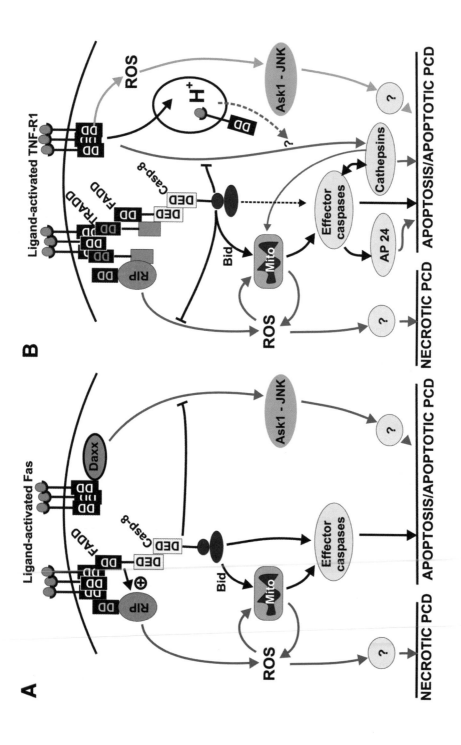

Fig. 6. Multiple death pathways triggered by death receptors. Death-receptor signaling is initiated by ligand-induced receptor trimerization. (**A**) Receptor death domains (DD) of Fas then recruit FADD, RIP, and/or Daxx to the receptor complex. Caspase-8 is activated after recruitment to FADD via death effector domain (DED) interaction, and triggers effector caspases either directly or through a Bid-mediated mitochondrial pathway (*3*) (black arrows). RIP initiates a caspase-independent (red arrows) necrotic pathway mediated by the formation of reactive oxygen species (ROS) (*15*). Daxx activates the ASK1–JNK kinase pathway leading to caspase-independent apoptosis (*93–96*). (**B**) Tumor necrosis factor receptor-1 (TNF-R1) signaling differs from that of Fas on the following

Also, caspase inhibition changes the mode of death, but not its extent, once the signal has arrived at mitochondria *(11,19,51,65,67,104–106)*. Thus it seems that in many models of cell death the master controllers of PCD operate at the mitochondrial level, whereas the decision on the shape of death is taken on the level of caspase activation (19).

There are, however, certain cases where Bcl-2 expression is not protective, and where mitochondria may not have a regulatory role *(34,35,55,107,108)*. Although alternative control mechanisms are not well-characterized, emerging candidates include different chaperone systems, such as heat-shock proteins *(22,34,95)* or ORP150 *(109)*. Organelles that have not received much attention recently, such as the endoplasmic reticulum and lysosomes, might also take an essential role in the control of death *(75,110,111)*.

COMPLEX CONTROL OF TUMOR CELL DEATH

Paradoxically, the cell proliferation induced by enhanced activity of oncoproteins (such as Myc, E1A, E2F, and CDC25) or inactivation of tumor-suppressor proteins (retinoblastoma protein, for example) is often associated with caspase activation and accelerated apoptosis *(112)*. The coupling of cell division to cell death has thus been suggested to act as a barrier that must be circumvented for cancer to occur *(70,112)*. Indeed, high expression of anti-apoptotic proteins (Bcl-2, Bcl-x$_L$, survivin, Bcr-Abl) and/or inactivation of pro-apoptoic tumor-suppressor proteins (p53, p19[ARF], PTEN) controlling caspase-dependent apoptosis pathways are often seen in human tumors *(70,112)*.

Alternative Death Pathways in Cancer

Despite showing severe defects in classical apoptosis pathways, cancer cells have not lost the ability to commit suicide. On contrary, spontaneous apoptosis is common in aggressive tumors, and most of them respond to therapy *(113)*. One explanation may be that defects in the signaling pathways leading to caspase activation may still allow caspase-independent death pathways to execute tumor cell death.

In addition, the alternative death pathways may be enhanced by transformation (Fig. 9). For example, oncogenic Ras can induce caspase- and Bcl-2-independent autophagic death *(55)*, and tumor-associated Src family kinases are involved in caspase-independent cytoplasmic execution of apoptotic programs induced by adenovirus protein E4orf4 *(114)*. Furthermore, a transformation-associated caspase-, p53-, and Bcl-2-independent apoptosis-like death program can be activated in tumor cell lines of different origins by depletion of a 70-kDa heat-shock protein (Hsp70) *(34,115)*. This death is preceded by a translocation of active cysteine cathepsins from lysosomes to cytosol, and inhibitors of their activity partially protect against death. Interestingly, cysteine cathepsins, as well as other noncaspase proteases, are highly expressed in aggressive tumors *(116)*. Thus, expression of protease inhibitors may increase a cancer cell's chances of survival by impairing alternative death routes *(4,25,117)*.

Alternative death pathways can also function at an initial step of tumorigenesis to limit tumor formation. Bin1, a tumor-suppressor protein that is often missing or functionally inactivated in human cancer, can activate a caspase-independent apoptosis-like death process that is blocked by a serine protease inhibitor or simian virus large T antigen, but not by overexpression of Bcl-2 or inactivation

Fig. 6. *(continued)*

steps. 1) Binding of FADD and RIP to the receptor complex requires the adaptor protein TRADD *(3)*. 2) Binding of Daxx to TNF-R1 has not been demonstrated and the Ask1–JNK pathway is activated by ROS *(94,97)* (blue line, caspase involvement unclear). 3) The RIP-mediated necrotic pathway is inhibited by caspase-8 *(15)*. 4) TNF-R1 can initiate a caspase-independent direct cathepsin B-mediated pathway *(25)*. 5) Cathepsin B can enhance the mitochondrial death pathway *(79)*. 6) The final execution of the death—that is, phosphatidylserine exposure, chromatin condensation, and loss of viability—is brought about by effector caspases, the serine protease AP24, or cathepsin B in a cell-type-specific manner *(3,20,25,79)*.

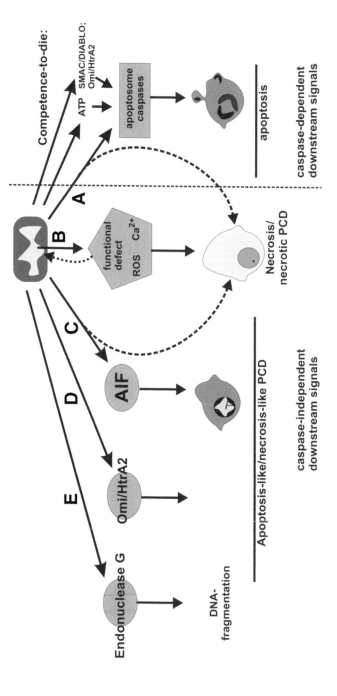

Fig. 7. Mitochondria as integrators in caspase-independent PCD. Death triggers upstream of mitochondria usually do not require caspases (exception: some death receptor signaling). They include a variety of chemotherapeutics, neurotoxins related to 1-methyl-4-phenylpyridinium (MPP) or the retinal cytotoxic pigment A2E, and lipid mediators such as ceramide or the disialoganglioside GD3 (*65,66,98–100*). Most signals can be blocked by anti-apoptotic Bcl-2 family members or survival kinases that act on this level (*3*). Also metabolic factors and drugs can block caspase-independent apoptosis steps at the mitochondrial level (e.g., cyclosporine A), whereas the mitochondrial protein Bar might be involved in caspase scavenging (*101*).

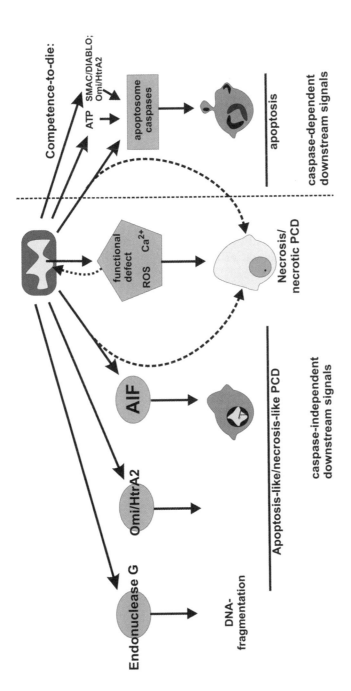

Fig. 8. Mitochondrial as effectors in caspase-independent PCD. Fundamentally different and initially independent signals emanate from mitochondria. (**A**) Cytochrome *c* leads to caspase activation in the apoptosome, and thus triggers classical apoptosis (*2,3*). For full activation of caspases (de-inhibition XIAP of), often the release from mitochondria of Omi/HtrA2 and SMAC/Diablo is also required. (**B**) reactive oxygen species (ROS) and Ca²⁺ induce necrotic PCD (*19,75*). (**C**) The apoptosis inhibitory factor (AIF) is released from mitochondria and triggers apoptotic-like death associated with chromatin condensation and margination, but not advanced chromatin compaction and nuclear fragmentation (*102*). (**D**) Omi/HtrA2 can, apart from displacing XIAP, also act as serine protease triggering caspase-independent cellular apoptotic changes. (**E**) Endonuclease G released from the mitochondrial matrix might play a role in caspase-independent DNA-fragmentation. Some of the released factors (AIF, ROS) may feed back to mitochondria, affecting their function and structure, and therefore trigger one another (*3,75,102*). Lack of essential co-factors for AIF or the proteasome pathway will convert them to necrosis (*4,19*).

of p53 *(35)*. Similarly, promyelocytic leukemia PML/RARA oncoprotein also inhibits caspase-independent PCD induced by the PML tumor-suppressor protein *(118)*. Interestingly, cytoplasmic apoptotic features induced by ectopic expression of PML can even be enhanced by pan-caspase inhibitors *(118)*. It should, however, be noted that PML/RARA can also interfere with caspase activation in some death models *(119)*.

Designing New Therapies Based on Alternative PCD Pathways

Whereas many cancer therapies induce classical apoptosis *(113)*, potential drugs engaging other death routines are emerging (Fig. 9). For instance, the topoisomerase inhibitor camptothecin induces cathepsin D/B-mediated apoptosis-like PCD in hepatocellular carcinoma cells *(29)*; activation of a thrombospondin receptor (CD47) triggers programmed necrosis in B-cell chronic lymphoma cells *(14)*; interferons and arsenite initiate a caspase-independent death pathway, possibly mediated by PML *(118)*; and EB 1089, a synthetic vitamin D analog, kills breast cancer cells through a caspase-independent apoptosis-like PCD *(33)* mediated by calpains (own unpublished observation). Moreover, increased tumor cell death observed in vitro when combining stimuli that activate different death-inducing proteases suggests that therapies activating various PCD pathways simultaneously may also be effective in clinics *(120,121)*.

Experimental gene-therapy approaches also point to alternative death pathways as promising targets for tumor therapy. Expression of Bin1 or adenovirus protein E4orf4 as well as depletion of Hsp70 result in mainly tumor-specific induction of caspase-independent apoptosis-like PCD *(30,34,35,115)*.

ALTERNATIVE CELL DEATH IN THE NERVOUS SYSTEM

Caspase-driven neuronal apoptosis strictly following the classical apoptosome pathway is best documented during development of the nervous system *(49)*, where many superfluous cells are produced and turned over *(122)*, and in vitro cultures of cells derived from developing brain *(75)*. Evidence is scarce for adult neurons, and here caspase-dependent mechanisms may yield to alternative death pathways *(123)*. Notably, a re-evaluation of cell death in caspase knockout mice showed that apoptosis is reduced during development, but cell death in many brain regions proceeds to the same extent in a caspase-independent paraptosis-like fashion *(124)*. Cell suicide in the adult nervous system has serious implications for the whole organism, because turnover is classically very limited. Thus a rapid caspase cascade, which is advantageous for efficient elimination of unwanted or rapidly replaceable cells, is dangerous in the developed brain and must be tightly controlled. For instance, neurons can survive cytochrome *c* release from mitochondria if they do not simultaneously receive a second signal leading to competence to die *(125)*. Neuronally expressed apoptosis inhibitor proteins (IAP, NAIP) buffer the caspase system, and need to be inactivated before classical apoptosis can be executed *(2)*. This buffering capacity may allow for localized caspase activation *(75)* (within synapses or neurites, for example) or sequestration of active caspases *(126)*, without a build-up of the death cascade affecting the entire neuron. Stressed neurons might also acquire a temporary resistance, which allows them to withstand otherwise lethal insults, e.g., by excitotoxins *(127)*. Such circumstances favor activation of slow, caspase-independent elimination routines, where damaged organelles are digested within a stressed cell, and the chance for rescue and reversibility is maintained until the process is complete *(128–130)*.

Although some caspase-dependent apoptosis might occur in adult brain *(75)*, at least part of PCD in chronic neurodegenerative disease follows alternative mechanisms and results in different morphologies *(16,57,105,126,129,131–133)* (Fig. 10) (*see* Chapter 14). Further variation is observed in acute insults such as ischemia or traumatic brain injury (TBI). Here, neurons within one brain region are exposed to different intensities of stress that trigger different death programs. Some of the main excitotoxic processes, such as mitochondrial impairment and dissipation of cell membrane potential,

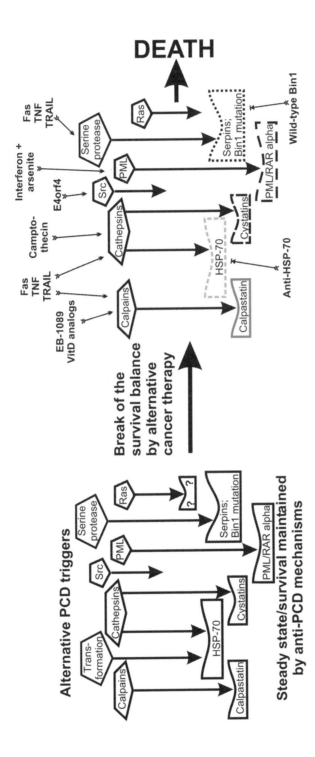

Fig. 9. Alternative death pathways as regulators of tumor cell survival and as putative targets for cancer therapy. (Left) Transformation is associated with upregulation of proteins that sensitize cells to caspase-independent PCD (55,114,116). As a defense line, death-promoting proteins are inactivated or expression of survival proteins is enhanced (4,35,70,118). Analogous changes in proteins regulating caspase-dependent apoptosis have also been demonstrated in cancer (70,112). (Right) Strategies of cancer therapy aimed at facilitating alternative death pathways (14,25,29,30,33–35,114,118).

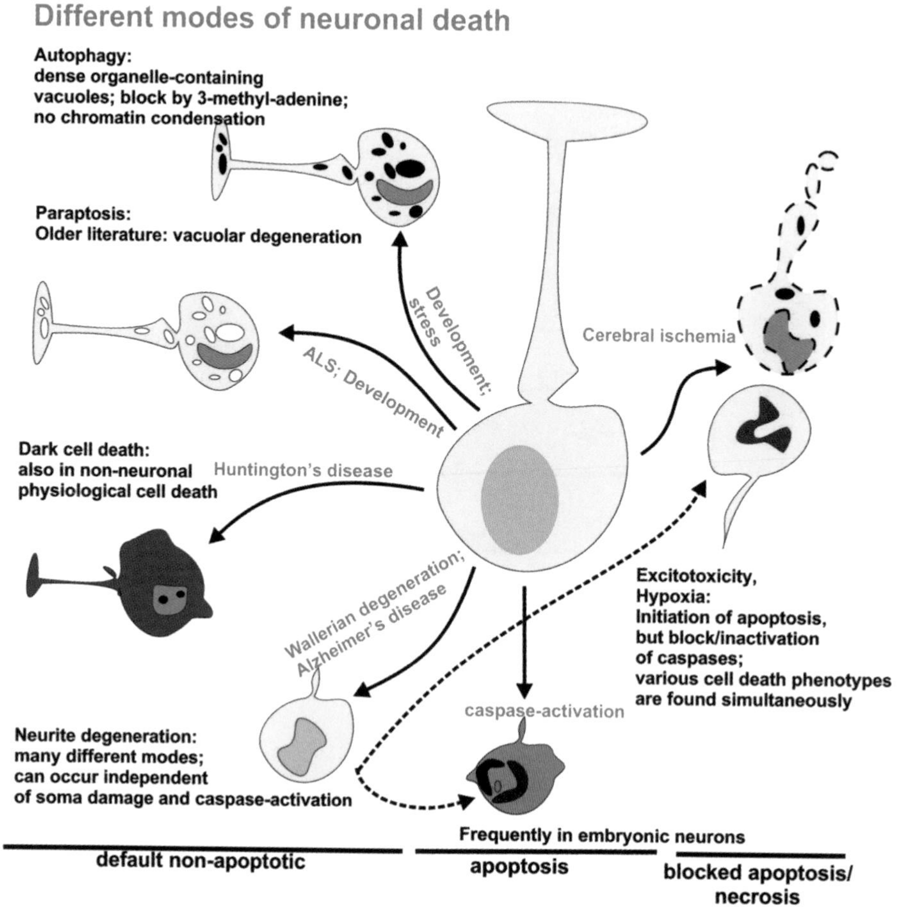

Different modes of neuronal death

Fig. 10. Common modes of PCD. Developmental cell death occurs by caspase-dependent apoptosis *(75)* or morphologically and mechanistically distinct autophagy. In various human diseases or animal models of them, the dominant form of neuronal disease is, for example, dark cell death in a Huntington's disease model *(57)* or paraptosis in a model of amyotrophic lateral sclerosis (ALS) *(16)*. Selective neurite degeneration occurs independently of caspase activation in different situations, and may eventually lead either to caspase-dependent apoptosis of cell bodies or to nonapoptotic death with irregular chromatin condensation *(60,107)*. Excitotoxic death may take many shapes and mechanisms depending on the intensity of insult, the age of the animal, and the brain region affected *(131,132)*. It often results in mixed apoptotic–necrotic features *(19)*, including cellular swelling; blebbing; nuclear pyknosis; display of phosphatidylserine; and some autophagic processes, such as uptake of mitochondria into lysosomes.

differentially impair various secondary routines of PCD *(19,131,132)*. For instance, rapid ATP depletion or disturbance of the intracellular ion composition impair cytochrome c-induced caspase activation, and massive production of nitric oxide (NO) or calpain activation directly inactivate caspases *(19,69)*. Accordingly, cell death has mixed features of apoptosis and necrosis, and might rely on either caspases or calpains as the dominant execution proteases *(85,90)*, or the activation of PARP *(53)* as a controller of programmed necrosis. Another group of proteases implicated as executors of ischemic death are the cysteine cathepsins *(83)*. It is possible that they interact with calpains, and notably there is massive PCD in the brains of mice lacking the cathepsin inhibitor cystatin B *(134)*.

The special shape of neurons (with projections up to 40,000 times longer than their cell bodies) allows degradative processes to be localized to a part of neurons and different death processes to be activated in different subsections of the cell *(19,75)*. For instance, synaptic damage and neurite regression can occur by Bcl-2- and caspase-independent mechanisms *(60,107,135)* and be initially reversible *(128)*, whereas final elimination of cells may depend on caspases or proteasomal activities *(60)*. The role of caspases as enhancers of the final phase of cell degeneration may apply to many common diseases. The longevity of neurons, combined with their dependence on effective intracellular transport, makes them sensitive to a slow form of death associated with the formation of intracellular polypeptide aggregates involving the amyloid-ß precursor protein (APP), ataxins, presenilins, huntingtin, tau, and alpha-synuclein *(75)*. As most of these proteins are caspase targets *(136)* and become more toxic after cleavage, caspases might contribute to the accelerated death of neurons at the end of a caspase-independent degeneration phase, or vice versa, make neurons sensitive to alternative mechanisms without directly participating in death execution *(137)*.

OUTLOOK

The discovery and understanding of alternative death pathways will open new perspectives for the treatment of disease (Fig. 9); one of these therapies (vitamin D analogs) has already advanced into clinical Phase III trials. New options and targets are also emerging for the prevention of death processes in neuro-degenerative disease. Prominent examples that have reached the stage of clinical trials target, for example, PARP in necrosis or calpains in excitotoxicity *(53,90,123)*.

On a more general biological level, the mode of cell death may differentially affect the surrounding tissue *(74)*. The important roles of caspase-independent/alternative death for development of tumor immunity are just emerging (reviewed in ref. *10*). Most recent evidence shows that the mode of cell demise controls the horizontal spread of oncogenic information *(138)* and of infection *(18)*. Because these processes can be favored by caspase-activation, the classical apoptosis pathways can, in fact, be detrimental to the organism. This may explain the need for extremely tight control of caspase-activation by the cellular energy level *(11)*. The apparent paradox that death-bound ATP-depleted cells are not "allowed" to activate caspases may then be explained by the fact that such cells would release activated caspases into the extracellular space upon premature lysis *(139)*. Thus, nonapoptotic death may not only be a passive accidental event, but also in some cases a desirable death option for long-lived organisms having to deal with tumors, infections, and other nonlethal tissue insults throughout their life span.

ACKNOWLEDGMENTS

We acknowledge our colleagues for stimulating discussions and the Danish Cancer Society, the German Research Council, and the Danish Medical Research Council for financial support. We also apologize to those whose work could be cited only indirectly.

REFERENCES

1. Kerr, J. F. R., Wyllie, A. H., and Currie, A. R. (1972) Apoptosis: a basic biological phenomenon with wide-ranging implications in tissue kinetics. *Br. J. Cancer* **26**, 239–257.
2. Hengartner, M. O. (2000) The biochemistry of apoptosis. *Nature* **407(6805)**, 770–776.
3. Strasser, A., O'Connor, L., and Dixit, V. M. (2000) Apoptosis signaling. *Annu. Rev. Biochem.* **69**, 217–245.
4. Leist, M. and Jäättelä, M. (2001) Four deaths and a funeral: from caspases to alternative mechanisms. *Nat. Rev. Mol. Cell Biol.* **2**, 589–598.
5. Johnson, D. E. (2000) Noncaspase proteases in apoptosis. *Leukemia* **14**, 1695–1703.
6. Kitanaka, C. and Kuchino, Y. (1999) Caspase-independent programmed cell death with necrotic morphology. *Cell Death Differ.* **6**, 508–515.
7. Borner, C. and Monney, L. (1999) Apoptosis without caspases: an inefficient molecular guillotine. *Cell Death Differ.* **6**, 497–507.
8. Samali, A., Zhivotovsky, B., Jones, D., Nagata, S., and Orrenius, S. (1999) Apoptosis: cell death defined by caspase activation. *Cell Death Differ.* **6**, 495–496.
9. Chung, S., Gumienny, T. L., Hengartner, M. O., and Driscoll, M. (2000) A common set of engulfment genes mediates removal of both apoptotic and necrotic cell corpses in *C. elegans. Nat. Cell Biol.* **2**, 931–937.

10. Hirt, U. A., Gantner, F., and Leist, M. (2000) Phagocytosis of nonapoptotic cells dying by caspase-independent mechanisms. *J. Immunol.* **164,** 6520–6529.
11. Leist, M., Single, B., Castoldi, A. F., Kuhnle, S., and Nicotera, P. (1997) Intracellular adenosine triphosphate (ATP) concentration: a switch in the decision between apoptosis and necrosis. *J. Exp. Med.* **185,** 1481–1486.
12. Vercammen, D., Beyaert, R., Denecker, G., Goossens, V., Van Loo, G., Declercq, W., et al. (1998) Inhibition of caspases increases the sensitivity of L929 cells to necrosis mediated by tumor necrosis factor. *J. Exp. Med.* **187,** 1477–1485.
13. Vercammen, D., Brouckaert, G., Denecker, G., Van de Craen, M., Declercq, W., Fiers, W., and Vandenabeele, P. (1998) Dual signaling of the Fas receptor: initiation of both apoptotic and necrotic cell death pathways. *J. Exp. Med.* **188,** 919–930.
14. Mateo, V., Lagneaux, L., Bron, D., Biron, G., Armant, M., Delespesse, G., and Sarfati, M. (1999) CD47 ligation induces caspase-independent cell death in chronic lymphocytic leukemia. *Nat. Med.* **5,** 1277–1284.
15. Holler, N., Zaru, R., Micheau, O., Thome, M., Attinger, A., Valitutti, S., et al. (2000) Fas triggers an alternative, caspase-8-independent cell death pathway using the kinase RIP as effector molecule. *Nat. Immunol.* **1,** 489–495.
16. Sperandio, S., de Belle, I., and Bredesen, D. E. (2000) An alternative, nonapoptotic form of programmed cell death. *Proc. Natl. Acad. Sci. USA* **97,** 14376–14381.
17. Leist, M., Single, B., Naumann, H., Fava, E., Simon, B., Kuhnle, S., and Nicotera, P. (1999) Inhibition of mitochondrial ATP generation by nitric oxide switches apoptosis to necrosis. *Exp. Cell Res.* **249,** 396–403.
18. Boise, L. H. and Collins, C. M. (2001) Salmonella-induced cell death: apoptosis, necrosis or programmed cell death? *Trends Microbiol.* **9,** 64–67.
19. Nicotera, P., Leist, M., and Manzo, L. (1999) Neuronal cell death: a demise with different shapes. *Trends Pharmacol. Sci.* **20,** 46–51.
20. Wright, S. C., Schellenberger, U., Wang, H., Kinder, D. H., Talhouk, J. W., and Larrick, J. W. (1997) Activation of CPP32-like proteases is not sufficient to trigger apoptosis: inhibition of apoptosis by agents that suppress activation of AP24, but not CPP32-like activity. *J. Exp. Med.* **186,** 1107–1117.
21. Lacana, E., Ganjei, J. K., Vito, P., and D'Adamio, L. (1997) Dissociation of apoptosis and activation of IL-1beta-converting enzyme/Ced-3 proteases by ALG-2 and the truncated Alzheimer's gene ALG-3. *J. Immunol.* **158,** 5129–5135.
22. Jäättelä, M., Wissing, D., Kokholm, K., Kallunki, T., and Egeblad, M. (1998) Hsp70 exerts its anti-poptotic function downstream of caspase-3-like proteases. *EMBO J.* **17,** 6124–6134.
23. De Maria, R., Zeuner, A., Eramo, A., Domenichelli, C., Bonci, D., Grignani, F., et al. (1999) Negative regulation of erythropoiesis by caspase-mediated cleavage of GATA-1. *Nature* **401,** 489–493.
24. Harvey, K. J., Lukovic, D., and Ucker, D. S. (2000) Caspase-dependent Cdk activity is a requisite effector of apoptotic death events. *J. Cell Biol.* **148,** 59–72.
25. Foghsgaard, L., Wissing, D., Mauch, D., Lademann, U., Bastholm, L., Boes, M., et al. (2001) Cathepsin B acts as a dominant execution protease in tumor cell apoptosis induced by tumor necrosis factor. *J. Cell Biol.* **153,** 999–1009.
26. Hoeppner, D. J., Hengartner, M. O., and Schnabel, R. (2001) Engulfment genes cooperate with ced-3 to promote cell death in *Caenorhabditis elegans*. *Nature* **412,** 202–206.
27. Reddien, P. W., Cameron, S., and Horvitz, H. R. (2001) Phagocytosis promotes programmed cell death in *C. elegans*. *Nature* **412,** 198–202.
28. Los, M., Stroh, C., Janicke, R. U., Engels, I. H., and Schulze-Osthoff, K. (2001) Caspases: more than just killers? *Trends Immunol.* **22,** 31–34.
29. Roberts, L. R., Adjei, P. N., and Gores, G. J. (1999) Cathepsins as effector proteases in hepatocyte apoptosis. *Cell Biochem. Biophys.* **30,** 71–88.
30. Lavoie, J. N., Nguyen, M., Marcellus, R. C., Branton, P. E., and Shore, G. C. (1998) E4orf4, a novel adenovirus death factor that induces p53-independent apoptosis by a pathway that is not inhibited by zVAD-fmk. *J. Cell Biol.* **140,** 637–645.
31. Berndt, C., Mopps, B., Angermuller, S., Gierschik, P., and Krammer, P. H. (1998) CXCR4 and CD4 mediate a rapid CD95-independent cell death in CD4(+) T cells. *Proc. Natl. Acad. Sci. USA* **95,** 12556–12561.
32. Monney, L., Otter, I., Olivier, R., Ozer, H. L., Haas, A. L., Omura, S., and Borner, C. (1998) Defects in the ubiquitin pathway induce caspase-independent apoptosis blocked by Bcl-2. *J. Biol. Chem.* **273,** 6121–6131.
33. Mathiasen, I. S., Lademann, U., and Jäättelä, M. (1999) Apoptosis induced by vitamin D compounds in breast cancer cells is inhibited by Bcl-2 but does not involve known caspases or p53. *Cancer Res.* **59,** 4848–4856.
34. Nylandsted, J., Rohde, M., Brand, K., Bastholm, L., Elling, F., and Jäättelä, M. (2000) Selective depletion of heat shock protein 70 (Hsp70) activates a tumor-specific death program that is independent of caspases and bypasses Bcl-2. *Proc. Natl. Acad. Sci. USA* **97,** 7871–7876.
35. Elliott, K., Ge, K., Du, W., and Prendergast, G. C. (2000) The c-Myc-interacting adaptor protein Bin1 activates a caspase-independent cell death program. *Oncogene* **19,** 4669–4684.
36. Joza, N., Susin, S. A., Daugas, E., Stanford, W. L., Cho, S. K., Li, C. Y., et al. (2001) Essential role of the mitochondrial apoptosis-inducing factor in programmed cell death. *Nature* **410,** 549–554.
37. Aravind, L., Dixit, V. M., and Koonin, E. V. (2001) Apoptotic molecular machinery: vastly increased complexity in vertebrates revealed by genome comparisons. *Science* **291,** 1279–1284.
38. Frohlich, K. U. and Madeo, F. (2000) Apoptosis in yeast: a monocellular organism exhibits altruistic behaviour. *FEBS Lett.* **473,** 6–9.
39. Ameisen, J. C. (1996) The origin of programmed cell death. *Science* **272,** 1278–1279.
40. Bratosin, D., Estaquier, J., Petit, F., Arnoult, D., Quatannens, B., Tissier, J. P., et al. (2001) Programmed cell death in

mature erythrocytes: a model for investigating death effector pathways operating in the absence of mitochondria. *Cell Death Differ.* **8,** 1143–1156.

41. Arnoult, D., Akarid, K., Grodet, A., Petit, P. X., Estaquier, J., and Ameisen, J. C. (2002) On the evolution of programmed cell death: apoptosis of the unicellular eukaryote Leishmania major involves cysteine proteinase activation and mitochondrion permeabilization. *Cell Death Differ.* **9,** 65–81.

42. Wolf, B. B., Goldstein, J. C., Stennicke, H. R., Beere, H., Amarante-Mendes, G. P., Salvesen, G. S., and Green, D. R. (1999) Calpain functions in a caspase-independent manner to promote apoptosis-like events during platelet activation. *Blood* **94,** 1683–1692.

43. Lam, E., Kato, N., and Lawton, M. (2001) Programmed cell death, mitochondria and the plant hypersensitive response. *Nature* **411,** 848–853.

44. Derry, W. B., Putzke, A. P., and Rothman, J. H. (2001) *Caenorhabditis elegans* p53: role in apoptosis, meiosis, and stress resistance. *Science* **294,** 591–595.

45. Wyllie, A. H. and Golstein, P. (2001) More than one way to go. *Proc. Natl. Acad. Sci. USA* **98,** 11–13.

46. Ellis, H. M. and Horvitz, H. R. (1986) Genetic control of programmed cell death in nematode *C. elegans*. *Cell* **44,** 817–829.

47. Parrish, J., Li, L., Klotz, K., Ledwich, D., Wang, X., and Xue, D. (2001) Mitochondrial endonuclease G is important for apoptosis in *C. elegans*. *Nature* **412,** 90–94.

48. Li, L. Y., Luo, X., and Wang, X. (2001) Endonuclease G is an apoptotic DNase when released from mitochondria. *Nature* **412,** 95–99.

49. Los, M., Wesselborg, S., and Schulze-Osthoff, K. (1999) The role of caspases in development, immunity, and apoptotic signal transduction: lessons from knockout mice. *Immunity* **10,** 629–639.

50. Woodle, E. S., Smith, D. M., Bluestone, J. A., Kirkman, W. M., Green, D. R., and Skowronski, E. W. (1997) Anti-human class I MHC antibodies induce apoptosis by a pathway that is distinct from the Fas antigen-mediated pathway. *J. Immunol.* **158,** 2156–2164.

51. McCarthy, N. J., Whyte, M. K. B., Gilbert, C. S., and Evan, G. I. (1997) Inhibition of Ced-3/ICE-related proteases does not prevent cell death induced by oncogenes, DNA damage, or the Bcl-2 homologue Bak. *J. Cell Biol.* **136,** 215–227.

52. Schulze-Osthoff, K., Bakker, A. C., Vanhaesebroeck, B., Beyaert, R., Jacob, W. A., and Fiers, W. (1992) Cytotoxic activity of tumor necrosis factor is mediated by early damage of mitochondrial functions. Evidence for the involvement of mitochondrial radical generation. *J. Biol. Chem.* **267,** 5317–5323.

53. Ha, H. C. and Snyder, S. H. (1999) Poly(ADP-ribose) polymerase is a mediator of necrotic cell death by ATP depletion. *Proc. Natl. Acad. Sci. USA* **96,** 13978–13982.

54. Bursch, W., Ellinger, A., Kienzl, H., Torok, L., Pandey, S., Sikorska, M., Walker, R., and Hermann, R. S. (1996) Active cell death induced by the anti-estrogens tamoxifen and ICI 164 384 in human mammary carcinoma cells (MCF-7) in culture: the role of autophagy. *Carcinogenesis* **17,** 1595–1607.

55. Chi, S., Kitanaka, C., Noguchi, K., Mochizuki, T., Nagashima, Y., Shirouzu, M., et al. (1999) Oncogenic Ras triggers cell suicide through the activation of a caspase-independent cell death program in human cancer cells. *Oncogene* **18,** 2281–2290.

56. Roach, H. I. and Clarke, N. M. (2000) Physiological cell death of chondrocytes in vivo is not confined to apoptosis. New observations on the mammalian growth plate. *J. Bone Joint. Surg. Br.* **82,** 601–613.

57. Turmaine, M., Raza, A., Mahal, A., Mangiarini, L., Bates, G. P., and Davies, S. W. (2000) Nonapoptotic neurodegeneration in a transgenic mouse model of Huntington's disease. *Proc. Natl. Acad. Sci. USA* **97,** 8093–8097.

58. Deas, O., Dumont, C., MacFarlane, M., Rouleau, M., Hebib, C., Harper, F., et al. (1998) Caspase-independent cell death induced by anti-CD2 or staurosporine in activated human peripheral T lymphocytes. *J. Immunol.* **161,** 3375–3383.

59. Luschen, S., Ussat, S., Scherer, G., Kabelitz, D., and Adam-Klages, S. (2000) Sensitization to death receptor cytotoxicity by inhibition of FADD/Caspase signaling: requirement of cell cycle progression. *J. Biol. Chem.* **275,** 24670–24678.

60. Volbracht, C., Leist, M., Kolb, S. A., and Nicotera, P. (2001) Apoptosis in caspase-inhibited neurons. *Mol. Med.* **7,** 36–48.

61. Xue, L., Fletcher, G. C., and Tolkovsky, A. M. (1999) Autophagy is activated by apoptotic signalling in sympathetic neurons: an alternative mechanism of death execution. *Mol. Cell Neurosci.* **14,** 180–198.

62. Khwaja, A. and Tatton, L. (1999) Resistance to the cytotoxic effects of tumor necrosis factor alpha can be overcome by inhibition of a FADD/caspase-dependent signaling pathway. *J. Biol. Chem.* **274,** 36817–36823.

63. Chautan, M., Chazal, G., Cecconi, F., Gruss, P., and Golstein, P. (1999) Interdigital cell death can occur through a necrotic and caspase-independent pathway. *Curr. Biol.* **9,** 967–970.

64. Matsumura, H., Shimizu, Y., Ohsawa, Y., Kawahara, A., Uchiyama, Y., and Nagata, S. (2000) Necrotic death pathway in fas receptor signaling. *J. Cell Biol.* **151,** 1247–1256.

65. Amarante-Mendes, G. P., Finucane, D. M., Martin, S. J., Cotter, T. G., Salvesen, G. S., and Green, D. R. (1998) Anti-apoptotic oncogenes prevent caspase-dependent and independent commitment for cell death. *Cell Death Differ.* **5,** 298–306.

66. Simon, B., Malisan, F., Testi, R., Nicotera, P., and Leist, M. (2001) The disialoganglioside GD3 is released from microglia cells and induces oligodendrocyte apoptosis. *Cell Death Differ.* **9,** 758–767.

67. Xiang, J., Chao, D. T., and Korsmeyer, S. J. (1996) Bax-induced cell death may not require interleukin 1β-converting enzyme-like proteases. *Proc. Natl. Acad. Sci. USA* **93,** 14559–14563.

68. Chua, B. T., Guo, K., and Li, P. (2000) Direct cleavage by the calcium-activated protease calpain can lead to inactivation of caspases. *J. Biol. Chem.* **275,** 5131–5135.

69. Lankiewicz, S., Marc Luetjens, C., Truc Bui, N., Krohn, A. J., Poppe, M., Cole, G. M., et al. (2000) Activation of calpain I converts excitotoxic neuron death into a caspase-independent cell death. *J. Biol. Chem.* **275**, 17064–17071.
70. Jäättelä, M. (1999) Escaping cell death: survival proteins in cancer. *Exp. Cell Res.* **248**, 30–43.
71. Smith, K. G., Strasser, A., and Vaux, D. L. (1996) CrmA expression in T lymphocytes of transgenic mice inhibits CD95 (Fas/APO-1)-transduced apoptosis, but does not cause lymphadenopathy or autoimmune disease. *EMBO J.* **15**, 5167–5176.
72. Doerfler, P., Forbush, K. A., and Perlmutter, R. M. (2000) Caspase enzyme activity is not essential for apoptosis during thymocyte development. *J. Immunol.* **164**, 4071–4079.
73. Kunstle, G., Hentze, H., Germann, P. G., Tiegs, G., Meergans, T., and Wendel, A. (1999) Concanavalin A hepatotoxicity in mice: tumor necrosis factor-mediated organ failure independent of caspase-3-like protease activation. *Hepatology* **30**, 1241–1251.
74. Savill, J. and Fadok, V. (2000) Corpse clearance defines the meaning of cell death. *Nature* **407**, 784–788.
75. Mattson, M. P. (2000) Apoptosis in neurodegenerative disorders. *Nat. Rev. Mol. Cell Biol.* **1**, 120–129.
76. Leverrier, Y. and Ridley, A. J. (2001) Apoptosis: caspases orchestrate the ROCK 'n' bleb. *Nat. Cell Biol.* **3**, E91–E93.
77. Squier, M. K., Miller, A. C., Malkinson, A. M., and Cohen, J. J. (1994) Calpain activation in apoptosis. *J. Cell Physiol.* **159**, 229–237.
78. Deiss, L. P., Galinka, H., Berissi, H., Cohen, O., and Kimchi, A. (1996) Catepsin D protease mediates programmed cell death induced by interferon-γ, Fas/APO-1 and TNF-alpha. *EMBO J.* **15**, 3861–3870.
79. Guicciardi, M. E., Deussing, J., Miyoshi, H., Bronk, S. F., Svingen, P. A., Peters, C., et al. (2000) Cathepsin B contributes to TNF-alpha-mediated hepatocyte apoptosis by promoting mitochondrial release of cytochrome *c*. *J. Clin. Invest.* **106**, 1127–1137.
80. Guicciardi, M. E., Miyoshi, H., Bronk, S. F., and Gores, G. J. (2001) Cathepsin B knockout mice are resistant to tumor necrosis factor-alpha-mediated hepatocyte apoptosis and liver injury: implications for therapeutic applications. *Am. J. Pathol.* **159**, 2045–2054.
81. Waterhouse, N. J., Finucane, D. M., Green, D. R., Elce, J. S., Kumar, S., Alnemri, E. S., et al. (1998) Calpain activation is upstream of caspases in radiation-induced apoptosis. *Cell Death Differ.* **5**, 1051–1061.
82. Vanags, D. M., Porn-Ares, M. I., Coppola, S., Burgess, D. H., and Orrenius, S. (1996) Protease involvement in fodrin cleavage and phosphatidylserine exposure in apoptosis. *J. Biol. Chem.* **271**, 31075–31085.
83. Yamashima, T. (2000) Implication of cysteine proteases calpain, cathepsin and caspase in ischemic neuronal death of primates. *Prog. Neurobiol.* **62**, 273–295.
84. Roberg, K. (2001) Relocalization of cathepsin D and cytochrome c early in apoptosis revealed by immunoelectron microscopy. *Lab. Invest.* **81**, 149–158.
85. Volbracht, C., Fava, E., Leist, M., and Nicotera, P. (2001) Calpain inhibitors prevent nitric oxide-triggered excitotoxic apoptosis. *Neuroreport* **12**, 3645–3648.
86. Kagedal, K., Zhao, M., Svensson, I., and Brunk, U. T. (2001) Sphingosine-induced apoptosis is dependent on lysosomal proteases. *Biochem. J.* **359**, 335–343.
87. Schotte, P., Van Criekinge, W., Van de Craen, M., Van Loo, G., Desmedt, M., Grooten, J., et al. (1998) Cathepsin B-mediated activation of the proinflammatory caspase-11. *Biochem. Biophys. Res. Commun.* **251**, 379–387.
88. Stoka, V., Turk, B., Schendel, S. L., Kim, T. H., Cirman, T., Snipas, S. J., et al. (2001) Lysosomal protease pathways to apoptosis. Cleavage of Bid, not pro-caspases, is the most likely route. *J. Biol. Chem.* **276**, 3149–3157.
89. Gobeil, S., Boucher, C. C., Nadeau, D., and Poirier, G. G. (2001) Characterization of the necrotic cleavage of poly(ADP-ribose) polymerase (PARP-1): implication of lysosomal proteases. *Cell Death Differ.* **8**, 588–594.
90. Wang, K. K. (2000) Calpain and caspase: can you tell the difference? *Trends Neurosci.* **23**, 20–26.
91. Laster, S. M., Wood, J. G., and Gooding, L. R. (1988) Tumor necrosis factor can induce both apoptic and necrotic forms of cell lysis. *J. Immunol.* **141**, 2629–2634.
92. Kawahara, A., Ohsawa, Y., Matsumura, H., Uchiyama, Y., and Nagata, S. (1998) Caspase-independent cell killing by Fas-associated protein with death domain. *J. Cell Biol.* **143**, 1353–1360.
93. Chang, H. Y., Nishitoh, H., Yang, X., Ichijo, H., and Baltimore, D. (1998) Activation of apoptosis signal-regulating kinase 1 (ASK1) by the adapter protein Daxx. *Science* **181**, 1860–1863.
94. Yang, X., Khosravi-Far, R., Chang, H. Y., and Baltimore, D. (1997) Daxx, a novel Fas-binding protein that activates JNK and apoptosis. *Cell* **89**, 1067–1076.
95. Charette, S. J., Lavoie, J. N., Lambert, H., and Landry, J. (2000) Inhibition of Daxx-mediated apoptosis by heat shock protein 27. *Mol. Cell Biol.* **20**, 7602–7612.
96. Charette, S. J., Lambert, H., and Landry, J. (2001) A kinase-independent function of ask1 in caspase-independent cell death. *J. Biol. Chem.* **276**, 36071–36074.
97. Tobiume, K., Matsuzawa, A., Takahashi, T., Nishitoh, H., Morita, K., Takeda, K., et al. (2001) ASK1 is required for sustained activations of JNK/p38 MAP kinases and apoptosis. *EMBO Rep.* **2**, 222–228.
98. De Maria, R., Rippo, M. R., Schuchman, E. H., and Testi, R. (1998) Acidic sphingomyelinase (ASM) is necessary for fas-induced GD3 ganglioside accumulation and efficient apoptosis of lymphoid cells. *J. Exp. Med.* **187**, 897–902.
99. Suter, M., Reme, C., Grimm, C., Wenzel, A., Jäättelä, M., Esser, P., et al. (2000) Age-related macular degeneration. The lipofusion component N-retinyl-N-retinylidene ethanolamine detaches proapoptotic proteins from mitochondria and induces apoptosis in mammalian retinal pigment epithelial cells. *J. Biol. Chem.* **275**, 39625–39630.
100. Leist, M., Volbracht, C., Fava, E., and Nicotera, P. (1998) 1-Methyl-4-phenylpyridinium induces autocrine excitotoxicity, protease activation, and neuronal apoptosis. *Mol. Pharmacol.* **54**, 789–801.
101. Stegh, A. H., Barnhart, B. C., Volkland, J., Algeciras-Schimnich, A., Ke, N., Reed, J. C., and Peter, M. E. (2002)

Inactivation of caspase-8 on mitochondria of Bcl-x$_L$-expressing MCF7-Fas cells. Role for the bifunctional apoptosis regulator protein. *J. Biol. Chem.* **277,** 4351–4360.

102. Susin, S. A., Lorenzo, H. K., Zamzami, N., Marzo, I., Snow, B. E., Brothers, G. M., et al. (1999) Molecular characterization of mitochondrial apoptosis-inducing factor. *Nature* **397,** 441–446.

103. Braun, J. S., Novak, R., Murray, P. J., Eischen, C. M., Susin, S. A., Kroemer, G., et al. (2001) Apoptosis-inducing factor mediates microglial and neuronal apoptosis caused by pneumococcus. *J. Infect. Dis.* **184,** 1300–1309.

104. Daugas, E., Susin, S. A., Zamzami, N., Ferri, K. F., Irinopoulou, T., Larochette, N., et al. (2000) Mitochondrio-nuclear translocation of AIF in apoptosis and necrosis. *FASEB. J.* **14,** 729–739.

105. Miller, T. M., Moulder, K. L., Knudson, C. M., Creedon, D. J., Deshmukh, M., Korsmeyer, S. J., and Johnson, E. M. (1997) Bax deletion further orders the cell death pathway in cerebellar granule cells and suggests a caspase-independent pathway to cell death. *J. Cell Biol.* **139,** 205–217.

106. Hirsch, T., Marchetti, P., Susin, S. A., Dallaporta, B., Zamzami, N., Marzo, I., et al. (1997) The apoptosis-necrosis paradox. Apoptogenic proteases activated after mitochondrial permeability transition determine the mode of cell death. *Oncogene* **15,** 1573–1581.

107. Finn, J. T., Weil, M., Archer, F., Siman, R., Srinivasan, A., and Raff, M. C. (2000) Evidence that Wallerian degeneration and localized axon degeneration induced by local neurotrophin deprivation do not involve caspases. *J. Neurosci.* **20,** 1333–1341.

108. Schierle, G. S., Leist, M., Martinou, J. C., Widner, H., Nicotera, P., and Brundin, P. (1999) Differential effects of Bcl-2 overexpression on fibre outgrowth and survival of embryonic dopaminergic neurons in intracerebral transplants. *Eur. J. Neurosci.* **11,** 3073–3081.

109. Tamatani, M., Matsuyama, T., Yamaguchi, A., Mitsuda, N., Tsukamoto, Y., Taniguchi, M., et al. (2001) ORP150 protects against hypoxia/ischemia-induced neuronal death. *Nat. Med.* **7,** 317–323.

110. Leist, M. and Jäättelä, M. (2001) Triggering of apoptosis by cathepsins. *Cell Death Differ.* **8,** 324–326.

111. Ferri, K. R. and Kroemer, G. (2001) Organelle-specific initiation of cell death pathways. *Nat. Cell Biol.* **3,** E255–E263.

112. Schmitt, C. A. and Lowe, S. W. (1999) Apoptosis and therapy. *J. Pathol.* **187,** 127–137.

113. Kerr, J. F. R., Winterford, C. M., and Harmon, B. V. (1994) Apoptosis. Its significance in cancer and cancer therapy. *Cancer* **73,** 2013–2026.

114. Lavoie, J. N., Champagne, C., Gingras, M. C., and Robert, A. (2000) Adenovirus E4 open reading frame 4-induced apoptosis involves dysregulation of Src family kinases. *J. Cell Biol.* **150,** 1037–1056.

115. Nylandsted, J., Brand, K., and Jäättelä, M. (2000) Heat shock protein 70 is required for the survival of cancer cells. *Ann. NY Acad. Sci.* **926,** 122–125.

116. Duffy, M. J. (1996) Proteases as prognostic markers in cancer. *Clin. Cancer Res.* **2,** 613–618.

117. Alexander, C. M., Howard, E. W., Bissell, M. J., and Werb, Z. (1996) Rescue of mammary epithelial cell apoptosis and entactin degradation by a tissue inhibitor of metalloproteinases-1 transgene. *J. Cell Biol.* **135,** 1669–1677.

118. Quignon, F., De Bels, F., Koken, M., Feunteun, J., Ameisen, J. C., and de The, H. (1998) PML induces a novel caspase-independent death process. *Nat. Genet.* **20,** 259–265.

119. Wang, Z. G., Ruggero, D., Ronchetti, S., Zhong, S., Gaboli, M., Rivi, R., and Pandolfi, P. P. (1998) PML is essential for multiple apoptotic pathways. *Nat. Genet.* **20,** 266–272.

120. Wang, Q., Yang, W., Uytingco, M. S., Christakos, S., and Wieder, R. (2000) 1,25-Dihydroxyvitamin D3 and all-trans-retinoic acid sensitize breast cancer cells to chemotherapy-induced cell death. *Cancer Res.* **60,** 2040–2048.

121. Mathiasen, I. S., Hansen, C. M., Foghsgaard, L., and Jäättelä, M. (2001) Sensitization to TNF-induced apoptosis by 1,25-dihydroxy vitamin D(3) involves up-regulation of the TNF receptor 1 and cathepsin B. *Int. J. Cancer* **93,** 224–231.

122. Raff, M. C. (1992) Social controls on cell survival and cell death. *Nature* **356,** 397–400.

123. Johnson, M. D., Kinoshita, Y., Xiang, H., Ghatan, S., and Morrison, R. S. (1999) Contribution of p53-dependent caspase activation to neuronal cell death declines with neuronal maturation. *J. Neurosci.* **19,** 2996–3006.

124. Oppenheim, R. W., Flavell, R. A., Vinsant, S., Prevette, D., Kuan, C. Y., and Rakic, P. (2001) Programmed cell death of developing mammalian neurons after genetic deletion of caspases. *J. Neurosci.* **21,** 4752–4760.

125. Deshmukh, M., Kuida, K., and Johnson, E. M., Jr. (2000) Caspase inhibition extends the commitment to neuronal death beyond cytochrome c release to the point of mitochondrial depolarization. *J. Cell Biol.* **150,** 131–143.

126. Stadelmann, C., Deckwerth, T. L., Srinivasan, A., Bancher, C., Bruck, W., Jellinger, K., and Lassmann, H. (1999) Activation of caspase-3 in single neurons and autophagic granules of granulovacuolar degeneration in Alzheimer's disease. Evidence for apoptotic cell death. *Am. J. Pathol.* **155,** 1459–1466.

127. Hansson, O., Petersen, A., Leist, M., Nicotera, P., Castilho, R. F., and Brundin, P. (1999) Transgenic mice expressing a Huntington's disease mutation are resistant to quinolinic acid-induced striatal excitotoxicity. *Proc. Natl. Acad. Sci. USA* **96,** 8727–8732.

128. Yamamoto, A., Lucas, J. J., and Hen, R. (2000) Reversal of neuropathology and motor dysfunction in a conditional model of Huntington's disease. *Cell* **101,** 57–66.

129. Jellinger, K. A. and Stadelmann, C. H. (2000) The enigma of cell death in neurodegenerative disorders. *J. Neural. Transm. Suppl.* 21–36.

130. Xue, L., Fletcher, G. C., and Tolkovsky, A. M. (2001) Mitochondria are selectively eliminated from eukaryotic cells after blockade of caspases during apoptosis. *Curr. Biol.* **11,** 361–365.

131. Roy, M. and Sapolsky, R. (1999) Neuronal apoptosis in acute necrotic insults: why is this subject such a mess? *Trends Neurosci.* **22,** 419–422.

132. Fujikawa, D. G. (2000) Confusion between neuronal apoptosis and activation of programmed cell death mechanisms in acute necrotic insults. *Trends Neurosci.* **23,** 410–411.
133. Colbourne, F., Sutherland, G. R., and Auer, R. N. (1999) Electron microscopic evidence against apoptosis as the mechanism of neuronal death in global ischemia. *J. Neurosci.* **19,** 4200–4210.
134. Pennacchio, L. A., Bouley, D. M., Higgins, K. M., Scott, M. P., Noebels, J. L., and Myers, R. M. (1998) Progressive ataxia, myoclonic epilepsy and cerebellar apoptosis in cystatin B-deficient mice. *Nat. Genet.* **20,** 251–258.
135. Sagot, Y., Dubois-Dauphin, M., Tan, S. A., de Bilbao, F., Aebischer, P., Martinou, J. C., and Kato, A. C. (1995) Bcl-2 overexpression prevents motoneuron cell body loss but not axonal degeneration in a mouse model of a neurodegenerative disease. *J. Neurosci.* **15,** 7727–7733.
136. Wellington, C. L. and Hayden, M. R. (2000) Caspases and neurodegeneration: on the cutting edge of new therapeutic approaches. *Clin. Genet.* **57,** 1–10.
137. Zhang, Y., Goodyer, C., and LeBlanc, A. (2000) Selective and protracted apoptosis in human primary neurons microinjected with active caspase-3, -6, -7, and -8. *J. Neurosci.* **20,** 8384–8389.
138. Bergsmedh, A., Szeles, A., Henriksson, M., Bratt, A., Folkman, M. J., Spetz, A. L., and Holmgren, L. (2001) Horizontal transfer of oncogenes by uptake of apoptotic bodies. *Proc. Natl. Acad. Sci. USA* **98,** 6407–6411.
139. Hentze, H., Schwoebel, F., Lund, S., Kehl, M., Ertel, W., Wendel, A., Jäättelä, M., and Leist, M. (2001) In vivo and in vitro evidence for extracellular caspase activity released from apoptotic cells. *Biochem. Biophys. Res. Commun.* **283,** 1111–1117.

II
Apoptosis in Action

8

Apoptosis in Plants, Yeast, and Bacteria

Ron Mittler and Vladimir Shulaev

INTRODUCTION

Apoptosis is a highly organized and genetically controlled mode of programmed cell death (PCD) in animals, crucial for the development and homeostasis of metazoan organisms. For a long time, apoptosis or PCD were mainly thought to occur in multi-cellular organisms. Recently, apoptosis-like cell death was described in both prokaryotic and eukaryotic unicellular organisms including bacteria, yeast, slime molds, kinetoplastid parasites, ciliate, and dinoflagellate, suggesting the existence of an ancient, genetically controlled, cell-death machinery conserved throughout the biological kingdom. In addition, recent studies suggest that PCD in plants may be very similar to apoptosis in animals.

PCD IN BACTERIA

It has been postulated that, similar to multicellular organisms, many developmental processes in bacteria involve PCD (1). Processes such as lysis of mother cells during sporulation of *Bacillus subtilus*, lysis of vegetative cells during fruit body formation in myxobacteria, and death of damaged cells, include PCD as part of a genetically controlled program.

One of the most studied examples of PCD in bacteria includes death of mother cells during sporulation of *Bacillus subtilus* (1). During sporulation in *B. subtilus* the mother cell is actively lysed by the autolysins CwlB, CwlC, and CwlH, produced prior to spore release. Lysis can be induced by various environmental stimuli including nutritional status, cell density, heat-shock, and cell-cycle signals. These activate a signal-transduction cascade involving the SpoA transcription factor, which in turn activates a cascade of interdependent sporulation-specific sigma factors. The expression of the autolysins CwlC and CwlH by sporulation-specific $E\sigma^K$ RNA polymerase is controlled by a terminal σ^K factor and requires the coat protein transcriptional activator GerE. This process is tightly controlled genetically and several mutants affecting autolysis have been identified and characterized.

An additional example of PCD in bacteria is autolysis during fruit body formation in *Myxococcus* bacteria (1). Autolysis is induced by "autocides," compounds that are bactericidal to the producing bacteria but have no effect on other bacteria. Several classes of "autocides," including fatty acids and glucosamine, can trigger autolysis.

Similar to apoptotic death of damaged metazoan cells, cell death in bacteria also occurs in response to many damaging agents, such as penicillin, cephalosporin, and glycopeptides. Simple genetic screens designed to identify genes preventing cell lysis in the presence of these agents have identified

From: *Essentials of Apoptosis: A Guide for Basic and Clinical Research*
Edited by: X-M. Yin and Z. Dong © Humana Press Inc., Totowa, NJ

several genes that control cell death and survival, and have provided intriguing data on the molecular mechanisms of PCD in bacteria. In general, the binding of a bactericidal antibiotic to its target significantly inhibits bacterial growth but does not directly cause cell death. A complex signal-transduction cascade within the bacterial cell is required to activate lytic enzymes such as LytA in *Streptococcus pneumoniae* that subsequently dissolve the cell wall during autolysis.

Using *S. pneumoniae* as a model, Novak et al. *(2)* showed that binding of penicillin to its target initiates an intrinsic two-component sensor-regulator system, VncR/S, that triggers multiple cell-death pathways, and proposed a model for the control of bacterial cell death. A screen of transposon-mutagenized *S. pneumonia* collection led to the identification of 10 candidate genes responsible for penicillin tolerance. A sensor gene, named VncS, was found to encode a histidine kinase and is homologous to VanS, a kinase regulating vancomycin resistance in *Enterococcus faecalis*. A mutation in this sensor gene led to bacterial survival in the presence of multiple antibiotics with unrelated mode of action (involving cell wall, ribosome, and DNA gyrase inhibitions). VncS regulates the expression of another protein, named VncR, which activates autolysin LytA responsible for lysis of bacterial cell walls. The signal that triggers the VncR/S system was recently identified as Pep^{27}, a secreted peptide containing 27 amino acids that initiate the cell-death program *(2)*. Pep^{27} is constantly produced by bacteria and is secreted by a dedicated ABC-transporter, but it takes a critical concentration threshold for this peptide to activate VanS and trigger PCD. Most likely, antibiotics like penicillin and vancomycin increase the production of a Pep^{27} peptide leading to the activation of PCD.

PCD can also be controlled in bacteria via "addiction modules," a unique genetic system consisting of two genes *(3)*. One gene encodes a stable toxin whereas the other encodes a labile antitoxin. Addiction modules were originally discovered on extrachromosomal elements of *Escherichia coli* and were thought to be a very precise mechanism for preventing plasmid loss. Two classes of addiction modules were identified in *E. coli* depending on whether both toxin and antitoxin are proteins, or toxin is a protein and antitoxin is an unstable antisense RNA molecule. Additional homologous modules were discovered on the bacterial chromosome, suggesting that PCD is inherent in bacteria. Among the best-characterized addiction modules are extrachromosomal elements *ccdAB* of plasmid F, *pemI/K* of plasmid R100, *kis/kid* of plasmid R1, *parDE* of RK2/RP4, and those located on the bacterial chromosome *mazEF* system *(3)*. The mechanism by which these systems execute cell death is very similar. Usually addiction module consist of two adjacent genes (e.g., MazE and MazF genes of the *chpA* locus) that are co-expressed. These encode a stable toxic component and liable antitoxic component. When the two genes are expressed and both toxin and antitoxin products are synthesized, they form a complex and prevent cell killing by the toxin. When the continuous expression of the liable antitoxin is inhibited, the toxin remains uninhibited and causes cell death. Addiction modules can be triggered by various stimuli that induce PCD in bacteria. It was shown that the *E. coli mazEF* module can be triggered by several antibiotics (rifampicin, chloramphenicol, and spectinomycin) that are general inhibitors of transcription and/or translation, implicating PCD as the possible mode of action for this group of antibiotics *(4)*.

The evolutionary significance of PCD in bacteria and other unicellular organisms is being widely discussed *(1)*. The existence of a genetic program for PCD in unicellular prokaryotic organisms like bacteria may be advantageous on the population level, considering the fact that many bacterial species live in highly differentiated bacterial communities. Multicellularity is now generally accepted as a common bacterial trait *(5)*. There is a significant body of evidence suggesting that different cells in bacterial communities are able to communicate via sophisticated transducing mechanisms involving diverse classes of signal molecules. The beneficial role of PCD may include the elimination of damaged or mutated cells, limiting the replication of viruses and phages, or reducing competition for nutrients. PCD can therefore be an altruistic event serving to benefit the survival of a population as a whole.

PCD IN YEAST

Initial database searches indicated that the yeast genome lacks genes with obvious homology to many critical components of metazoan apoptotic death machinery, including Bax/Bcl-2 family, Apaf-1/Ced-4, p53, or caspases. This led to a conclusion that PCD does not exist in yeast. However, new evidence has accumulated recently implying that yeast can undergo PCD and that many components of the yeast PCD and the metazoan apoptosis may be conserved. In addition, more detailed analysis of the yeast genome has shown that several yeast proteins share homology with metazoan apoptotic proteins and may be members of an evolutionary conserved family of cell-death proteins.

Evidence suggesting that at least part of the apoptotic machinery may be present in yeast comes from studies on the overexpression of mammalian apoptotic genes in yeast. Owing to the absence of obvious homologs of major genes involved in apoptosis, the yeast system was used extensively to study the role of individual molecular components in apoptosis. Surprisingly, expression of several proteins involved in metazoan apoptosis induced cell death in both the budding yeast *Saccharomyces cerevisiae* and the fission yeast *Schizosaccharomyces pombe (6)*. Heterologous expression of the pro-apoptotic Bcl-2 family proteins Bax and Bak, caspases, CED-4/ApaF-1, or p53 in yeast triggered apoptosis-like cell death. Furthermore, cell death induced by proapoptotic genes was suppressed by anti-apoptotic Bcl-2 family proteins. Co-expression of several members of Bcl-2 family that are anti-apoptotic in animals, including Bcl-X_L, Bcl-2, Mcl-1, and Al, suppressed yeast lethality induced by Bax and Bak.

The identification of specific mutation (S565G) in the cell-division cycle gene CDC48 *(7)* that can trigger cell death further indicated the presence of genetically controlled PCD machinery in yeast. Cell death in the CDC48 mutant resembled many features typical to apoptosis: exposure of phosphatidylserine at the outer layer of the cytoplasmic membrane, DNA fragmentation, and chromatin condensation and fragmentation. PCD has also been implicated as a mechanism of eliminating of senescing yeast mother cell *(8)*.

Reactive oxygen intermediates (ROI) are important mediators of cell death in many organisms. It has recently been suggested that ROI play a central role in the induction and execution of PCD in yeast *(6,7)*. Cell death can be triggered in yeast by various treatments that result in ROI production such as low external doses of hydrogen peroxide or depletion of glutathione, lipid hydroperoxide or exposure to acetic acid. Cell death induced by these ROI-generating stimuli can be prevented by oxygen-radical depletion or hypoxia. Moreover, inhibition of translation by cyclohexamide inhibited cell death induced by H_2O_2, suggesting the active participation of a cellular machinery in this cell-death process *(7)*. Additionally, the apoptotic phenotype of CDC48[S565G] is due to the accumulation of ROI *(7)*. The human Bcl-2 protein can also prevent cell death in some mutant yeast caused by oxidative stress, further implying that some components of PCD may be evolutionary conserved.

ROI may play an important role in the yeast aging process as well. A recent study indicated that deletions of genes coding for superoxide dismutase (SOD) or catalase, as well as changes in atmospheric oxygen concentration, resulted in ROI accumulation in yeast mother cells and significantly shortened their life span *(8)*. The addition of the antioxidant glutathione, on the other hand, increased the life span of yeast cells. Accumulation of ROI in the mitochondria of aging mother cells was accompanied by phenotypic markers of yeast apoptosis.

Mitochondria play a crucial role in apoptosis in animals *(9)*. Disruption of mitochondrial membranes releases apoptogenic factors such as cytochrome *c* and apoptosis-inducing factor (AIF) activating either caspases-dependent or caspase-independent cell-death pathways. Evidence is accumulating in support of a central role for mitochondria in yeast PCD. This points to the presence of a mechanism of mitochondrial membrane disruption in yeast, similar to that of animals. The recent discovery of a novel mammalian ion channel, named mitochondrial apoptosis-induced channel, whose activity correlates with the onset of apoptosis, and the finding of a similar channel activity in mitochondrial outer membranes of yeast expressing human Bax, may indicate that release of cell death-inducing factors from mitochondria occurs via a similar mechanism in mammals and yeast *(10)*.

Although it was generally accepted that there are no homologs of metazoan apoptotic proteins present in yeast, recent data indicate that some yeast proteins have regions with similarity to mammalian proteins involved in cell death. The Pbh1p protein from *S. pombe (11)* and the Bir1p protein from *S. cerevisiae (12)* are some of the proteins that share homology with members of the evolutionary conserved cell death protein family. These proteins contain BIR (Baculovirus IAP repeat) domains characteristic to a family of inhibitor of apoptotis proteins (IAPs). Another protein that shares similarity with human apoptotic proteins was recently identified in *S. pombe*. It is a homolog of human Rad9 protein, SpRad9 *(13)*. SpRad9 contains a group of amino acids with similarity to the Bcl-2 homology 3 domain, and may be a member of the BH3-only family of apoptosis promoting proteins. SpRad9 can induce apoptosis in human cells by interacting with human Bcl-2 through the region similar to BH3 domain. An ancient family of caspase-like proteins, namely metacaspases, was also identified in the *S. cerevisae* genome using the caspase-like domain of the *Dictyostelium* sequence as a query for BLAST search *(14)*. Of note, a recent study has identified a yeast protein with structural homology to mammalian caspases *(15)*. More importantly, this protein named Yeast Caspase-1 exhibits caspase enzymatic activity, mediating not only apoptosis induced by hydrogen peroxide in yeast but also the death process of overaged cultures *(15)*.

PCD IN PLANTS

Plants are simple multicellular organisms that vary considerably from animals. On the cellular level, plant cells contain rigid cell walls composed mainly of cellulose, a large vacuole that participates in many molecular and biochemical processes, and a specialized organelle responsible for conducting photosynthesis, i.e., the chloroplast. Plants do not have an immune system similar to animals, but are capable of controlling and coordinating a large array of defenses when attacked by pathogens. In addition, the development of plants is solely determined by cell division and not by a combination of cell division and cell migration. Despite these differences, plants, much like animals, rely heavily on PCD for their proper development and response to different environmental insults *(16,17)*. Plants may therefore represent an interesting example of how a simple multicellular organism adapts and uses PCD for its specific evolutionary needs.

Although similar in its conceptual and functional definition, in many characteristic and mechanistic aspects PCD in plants is different from apoptosis in animals. The lack of an immune system, for example, lowers the need for the formation of apoptotic bodies and the "clean" elimination of cells, because an inflammatory response is not likely to occur when the content of a plant cell is spilled. In addition, in most examples of PCD in plants, the cell wall remains intact, making the trafficking of apoptotic bodies from one cell to the other, or the engulfment of a dying cell by neighboring cells, impossible. Despite their relative simplicity, plants appear to encode and execute a large number of different pathways for PCD.

Types of PCD in Plants

There are three major types of PCD in plants: developmental PCD, pathogen-induced PCD, and environmental (abiotic) PCD.

Developmental PCD

Plants use PCD throughout their development and reproduction cycles. Examples of PCD range from death of root-tip cells, through the death of differentiating xylem vessels, leaf cells, stem cells, and the different processes of PCD in floral organs. These include PCD during the abortion of floral organs, the abortion of megaspores, tapetum degradation, synergid cell death, and PCD involved in pollen growth and self-incompatibility. There are also examples of PCD during embryogenesis, such as the death of the suspensor, and the death of storage cells *(18–20)*. Senescence, a defined developmental program in plants, is also considered to be a PCD process *(18)*.

The most characterized example of developmental PCD in plants is the death of xylem cells *(21)*. These cells compose the water-conducting vessels of the plant, and transmit water from the roots to the leaves and the reproductive organs. Cells of the tracheary elements (TE), the principal component of this tissue, differentiate from progenitor cambium and procambium cells. Their differentiation involves a complex process of cell expansion, thickening and patterning of cell walls, dismantling of the wall plates that connect each TE with the upper and lower part of the tube, and PCD of the protoplast. This process is controlled by the plant hormones auxin and cytokinin and involves the activation of a genetic program that dismantles the dying cells, in order not to leave a cell corpse clogging the xylem vessel. Although the latter sounds like a true apoptotic process, TE undergo PCD in a unique manner. They first accumulate a large number of hydrolytic enzymes such as proteases, nucleases, and lipases in their vacuole. When they reach the developmental point of PCD, the vacuole swells, ruptures, and releases all of its deadly hydrolytic content into the cytosol. The plasma membrane remains intact and the cell undergoes a process of near-complete autolysis. Following the digestion of most cellular components, the plasma membrane is also ruptured and the degraded content of the cell is released and possibly uptaked as nutrients by neighboring cells. Thus, although determined by a genetic PCD program, this process is very different from apoptosis in animal cells *(21)*.

Pathogen-Induced PCD

In general, there appear to be two types of pathogen-induced PCD in plants: PCD activated by the plant as part of an active defense mechanism against pathogen attack *(16)*, and PCD induced in plant cells by certain pathogens that feed on dead plant cells (necrotrophic pathogens; ref. *22*).

The interaction of plants with pathogens is dependent on a set of plant-encoded genes (resistance genes), and a set of pathogen-encoded genes (virulance genes). This interaction, termed "gene-for-gene," determines whether the plant will recognize the pathogen and activate its defenses, and thereby become resistant, or will fail to recognize the pathogen, will not activate defense mechanisms, and thereby become infected *(16)*. Interestingly, some of these gene-for-gene genes were found to be similar to animal genes controlling bacterial pathogenicity *(23)*. One of the key defense mechanisms activated by plants upon pathogen recognition is a type of PCD called the hypersensitive response (HR). This response is activated in cells that directly come into contact with the pathogen, as well as in neighboring cells that surround the primary infected cells. These cells will undergo a rapid process of PCD that will result in the formation of a lesion composed of dead cells (Fig. 1A). It is thought that this rapid cell-death process slows down the propagation of the invading pathogen and allows the plant to activate other defenses such as pathogenesis-related proteins and other chemicals and compounds with antimicrobial activity *(24)*. Because most of the pathogens that activate this response are obligate parasites, they cannot replicate within dead cells and their growth is inhibited. Recently, a number of plant genes that control this response were cloned and characterized *(17)*. Mutations in these genes resulted in uncontrollable activation of PCD and the formation of HR lesions on plants in the absence of pathogens. A similar phenotype, that is, the spontaneous development of lesions in the absence of pathogens, can also appear on plants following the expression of foreign genes that alter plant metabolism *(24)*. This finding suggests that many different signals in plants can result in the activation of the HR-PCD pathway.

Major players involved in the activation of PCD during the HR are ROI, nitric oxide (NO), calcium and proton pumps, mitogen-activated protein kinase (MAPK), and salicylic acid (SA). It is believed that the initial recognition of the pathogen via the gene-for-gene response activates a signal-transduction pathway that involves the translocation of calcium and protons across the plasma membrane into the cytosol, protein phosphorylation/dephosporylation events, the activation of enzymes that generate ROI such as NADPH-oxidase and peroxidases, and the accumulation of NO and SA.

Fig. 1. Induction of PCD in plants. (**A**) Hypersensitive response (HR) lesions induced by plants in response to infection by tobacco mosaic virus. These lesions (1–2 mm diameter) result from the activation of a PCD pathway in infected cells and in cells that surrounds the infected region. (**B**) Induction of HR-PCD in transgenic tobacco plants expressing a bacterial proton pump. The induction of a proton flux across the plasma membrane of plant cells by the pump mimics an early signal-transduction event that occurs during the HR and activates the HR-PCD pathway. This response, termed a "lesion mimic response," results in the formation of HR lesions on leaves, in the absence of a pathogen.

Manipulations of proton and calcium homeostasis pharmacologically or genetically were shown to activate the HR (ref. *24*; Fig. 1B). A combination of enhanced production of ROI and NO was also shown to activate the HR in the absence of a pathogen, and SA was shown to facilitate the formation of ROI as well as to inhibit catalase and ascorbate peroxidase, two of the key ROI removal enzymes in plants (*18,25*). These processes are thought to be orchestrated by a cascade of MAPKs (*26*).

The morphological and biochemical characteristics of HR-PCD appear to depend on the nature of the plant pathogen interaction. Thus, some types of HR-PCD are accompanied by the formation of DNA ladders and apoptotic bodies, and some are not. In all cases, however, the corpse of the dead cell, composed mainly of cell walls and cell debris, is left behind and is not completely eliminated.

Plant-pathogen interactions that result in the successful infection of plants and the development of disease symptoms are sometimes accompanied by the activation of PCD in plant cells by the pathogen. For many years, it was thought that disease symptoms that included host cell death were caused by toxins produced by the pathogens. However, it was recently discovered that similar symptoms could be induced spontaneously in mutants in the absence of a pathogen. Moreover, it was recently reported that animal genes that inhibit apoptosis can inhibit the formation of these symptoms (described below). Thus, it appears as if some necrotrophic pathogens that feed on dead plant tissues actually activate a plant PCD pathway to induce cell death in plant tissues. As opposed to the type of cell death described earlier for the HR, little is known about the processes involved in this response. However, it is thought that ROI play an important role in this response as well (*27*).

Environmental PCD

A number of abiotic stress conditions were found to induce PCD in plants. These include anaerobiosis, ultraviolet (UV) and ionizing radiation, and oxidative stress. Anaerobiosis that results from flooding of fields was shown to induce in plants a process of PCD in stem and root cells. This process creates a new tissue called aeronchyma, composed mainly of empty spaces. Long tunnels are formed within stems and connect the upper part of the plant with its root system. These are not visible from the outside but they enhance the rate of oxygen diffusion to the submerged roots. The plant hormone ethylene is thought to regulate this process *(18)*.

Oxidative stress induced by different compounds such as paraquat, a herbicide that enhances the rate of superoxide generation in cells, or by ionizing radiation, induce cell death in plants. Recently it was reported that the animal anti-apoptotic genes Bcl-X_L and Ced-9 can inhibit paraquat- and radiation-induced cell death in plants, suggesting that this cell death process may be a PCD *(28)*. Other environmental stresses such as high salt, UV radiation, and certain toxins were shown to induce cell death that was accompanied by internucleosomal DNA cleavage, suggesting that it may involve some processes similar to apoptosis in animals.

Molecular Mechanisms of PCD in Plants

Many of the molecular switches that control PCD in plants are unknown. The sequencing of the entire genome of the flowering plant *Arabidopsis thaliana* failed to identify homologs of key regulators of apoptosis in animals such as caspases and Bcl-2 or Bax. Plants were, however, found to contain a group of proteases named metacaspases, and homologs of BI-1 (Bax inhibitor-1), and DAD-1 *(24,29)*. There are two types of metacaspases in plants: Type I, which contains a predicted caspase-like proteolytic domain but lacks the death effector domain; and Type II, which contains, in addition to the caspase-like domain, an N-terminal zinc finger and proline-rich domains, also found in LSD-1, a protein involved in the control of PCD during the HR. Supporting the involvement of caspases in PCD in plants (possibly mediated by the metacaspase family) are studies that demonstrated the suppression of HR-PCD by synthetic peptides that act as inhibitors of caspase activity *(29)*. In addition, caspase-like protease activity was demonstrated in plant cells undergoing HR-PCD *(29)*. Additional players that may be similar to some of those controlling PCD in animals are small GTP-binding proteins of the Ras class and cystein-sensitive proteases.

Although Bax homologs were not found in plant cells, expression of Bax in plants induces PCD *(30)*. Moreover, this induction requires the proper oligomerization and cellular localization of Bax in plants. The BI-1 protein was recently identified by a screen of human genes that inhibit Bax toxicity in yeast. Homologs of this protein were found in plants. Co-expression of Bax and the plant homolog of BI-1 resulted in the inhibition of Bax-induced cell death in plants *(30)*. These results suggest an active role for the mitochondria in plant PCD processes. In support of this hypothesis, overexpression of animal anti-apoptotic genes such as Bcl-2, Bcl-X_L, and Ced-9 was found to inhibit pathogen-induced PCD and oxidative stress-induced PCD in plants *(28)*. A recent homology search using BI-1 sequences revealed the possible existence of a functional homolog of Bcl-2 in plants (called ABR proteins; ref. *29*). However, further experimental work is required to support this possibility. Interestingly, the expression of the different animal anti-apoptotic genes in plants did not seem to affect certain developmental processes that require PCD such as xylem formation, suggesting that different PCD pathways in plants may be controlled by different mechanisms.

CONCLUSIONS

The finding of PCD in prokaryotes, unicellular euokaryotes, and simple multicellular organism such as plants suggest that PCD is a basic biological process that may be essential for the survival of almost all living organisms. Interestingly, at least some mechanisms of PCD appear to be conserved between yeast, plants, and animals. These appear to involve the mitochondrion as a central player.

Because the mitochondrion is believed to have originated from an endosymbiotic event between a primitive eukaryotic cell and a proteobacteria, the finding of different PCD processes in bacteria may extend the link between yeast, plants, and animals to include bacteria and mechanisms of PCD in bacteria that involve the rupture of bacterial cells. These may resemble the leakage of proteins from mitochondria during apoptosis. Because PCD appears to be a key mechanism for the survival of almost all known organisms, deeper understanding of the evolutionary path of PCD in different organisms will considerably enhance our perception of the balance between life and death, and the different driving forces that affect and shape all living organisms.

REFERENCES

1. Lewis, K. (2000) Programmed death in bacteria. *Microbiol. Mol. Biol. Rev.* **64,** 503–514.
2. Novak, R., Charpentier, E., Braun, J. S., and Tuomanen, E. (2000) Signal transduction by a death signal peptide: uncovering the mechanism of bacterial killing by penicillin. *Mol. Cell* **5,** 49–57.
3. Engelberg-Kulka, H. and Glaser, G. (1999) Addiction modules and programmed cell death and antideath in bacterial cultures. *Annu. Rev. Microbiol.* **53,** 43–70.
4. Sat, B., Hazan, R., Fisher, T., Khaner, H., Glaser, G., and Engelberg-Kulka, H. (2001) Programmed cell death in *Escherichia coli*: some antibiotics can trigger *mazEF* lethality. *J. Bacteriol.* **183,** 2041–2045.
5. Shapiro, J. A. (1998) Thinking about bacterial populations as multicellular organisms. *Annu. Rev. Microbiol.* **52,** 81–104.
6. Frohlich, K. U. and Madeo, F. (2000) Apoptosis in yeast: a monocellular organism exhibits altruistic behaviour. *FEBS Lett.* **473,** 6–9.
7. Madeo, F., Frohlich, E., Ligr, M., Grey, M., Sigrist, S. J., Wolf, D. H., and Frohlich, K. U. (1999) Oxygen stress: a regulator of apoptosis in yeast. *J. Cell Biol.* **145,** 757–767.
8. Laun, P., Pichova, A., Madeo, F., Fuchs, J., Ellinger, A., Kohlwein, S., et al. (2001) Aged mother cells of *Saccharomyces cerevisiae* show markers of oxidative stress and apoptosis. *Mol. Microbiol.* **39,** 1166–1173.
9. Ranganath, R. M. and Nagashree, N. R. (2001) Role of programmed cell death in development. *Int. Rev. Cytol.* **202,** 159–242.
10. Pavlov, E. V., Priault, M., Pietkiewicz, D., Cheng, E. H. Y., Antonsson, B., Manon, S., et al. (2001) A novel, high conductance channel of mitochondria linked to apoptosis in mammalian cells and Bax expression in yeast. *J. Cell Biol.* **155,** 725–731.
11. Rajagopalan, S. and Balasubramanian, M. K. (1999) *S. pombe* Pbh1p: an inhibitor of apoptosis domain containing protein is essential for chromosome segregation. *FEBS Lett.* **460,** 187–190.
12. Yoon, H. J. and Carbon, J. (1999) Participation of Bir1p, a member of the inhibitor of apoptosis family, in yeast chromosome segregation events. *Proc. Natl. Acad. Sci. USA* **96,** 13208–13213.
13. Komatsu, K., Hopkins, K. M., Lieberman, H. B., and Wang, H. G. (2000) *Schizosaccharomyces pombe* Rad9 contains a BH3-like region and interacts with the anti-apoptotic protein Bcl-2. *FEBS Lett.* **481,** 122–126.
14. Uren, A. G., O'Rourke, K., Aravind, L., Pisabarro, M. T., Seshagiri, S., Koonin, E. V., and Dixit, V. M. (2000) Identification of paracaspases and metacaspases: Two ancient families of caspase-like proteins, one of which plays a key role in MALT lymphoma. *Mol. Cell* **6,** 961–967.
15. Madeo, F., Herker, E., Maldener, C., Wissing, S., Lachelt, S., Herlan, M., et al. (2002) A caspase-related protease regulates apoptosis in yeast. *Mol. Cell* **9,** 911–917.
16. Dangl, J. L., Dietrich, R. A., and Richberg, M. H. (1996) Death don't have no mercy: cell death programs in plant-microbe interactions. *Plant Cell* **8,** 1793–1807.
17. Beers, E. P. and McDowell, J. M. (2001) Regulation and execution of programmed cell death in response to pathogens, stress and developmental cues. *Curr. Opin. Plant Biol.* **4,** 561–567.
18. Mittler, R. (1998) Cell death in plants, in *When Cells Die: A Comprehensive Evaluation of Apoptosis and Programmed Cell Death*, Zakeri, Z., Tilly, J., Lockshin, R. A., eds., New York, NY: John Wiley & Sons, pp. 147–174.
19. Greenberg, J. T. (1996) Programmed cell death: a way of life for plants. *Proc. Natl. Acad. Sci. USA* **93,** 12094–12097.
20. Lam, E., Fukuda, H., and Greenberg, J. T., eds. (2000) Programmed cell death in higher plants. *Plant Mol. Biol.* **44,** 245–318.
21. Fukuda, H. (1996) Xylogenesis: initiation, progression, and cell death. *Ann. Rev. Plant Physiol. Plant Mol. Biol.* **47,** 299–325.
22. Dickman, M. B., Park, Y. K., Oltersdorf, T., Li, W., Clemente, T., and French, R. (2001) Abrogation of disease development in plants expressing animal antiapoptotic genes. *Proc. Natl. Acad. Sci. USA* **98,** 6957–6962.
23. Rahme, L. G., Tan, M. W., Le, L., Wong, S. M., Tompkins, R. G., Calderwood, S. B., and Ausubel, F. M. (1997) Use of model plant hosts to identify Pseudomonas aeruginosa virulence factors. *Proc. Natl. Acad. Sci. USA* **94,** 13245–13250.
24. Mittler, R. and Lam, E. (1996) Sacrifice in the face of foes: pathogen-induced programmed cell death in higher plants. *Trends Microbiol.* **4,** 10–15.
25. Delledonne, M., Zeier, J., Marocco, A., and Lamb, C. (2001) Signal interactions between nitric oxide and reactive oxygen intermediates in the plant hypersensitive disease resistance response. *Proc. Natl. Acad. Sci. USA* **98,** 13454–13459.

26. Zhang, S. and Klessig, D. F. (2001) MAPK cascades in plant defense signaling. *Trends Plant Sci.* **6**, 520–527.
27. Govrin, E. M. and Levine, A. (2000) The hypersensitive response facilitates plant infection by the necrotrophic pathogen Botrytis cinerea. *Curr. Biol.* **10**, 751–757.
28. Mitsuhara, I., Malik, K. A., Miura, M., and Ohashi, Y. (1999) Animal cell-death suppressors Bcl-x(L) and Ced-9 inhibit cell death in tobacco plants. *Curr. Biol.* **9**, 775–778.
29. Lam, E., Kato, N., and Lawton, M. (2001) Programmed cell death, mitochondria and the plant hypersensitive response. *Nature* **411**, 848–853.
30. Kawai-Yamada, M., Jin, L., Yoshinaga, K., Hirata, A., and Uchimiya, H. (2001) Mammalian Bax-induced plant cell death can be down-regulated by overexpression of Arabidopsis Bax Inhibitor-1 (AtBI-1). *Proc. Natl. Acad. Sci. USA* **98**, 12295–12300.

9

Programmed Cell Death in *C. elegans*

Yi-Chun Wu and Ding Xue

INTRODUCTION

Studies in the nematode *Caenorhabditis elegans* established that programmed cell death is a normal, genetically determined part of development and is controlled by a number of specific genes. Genetic analyses have ordered these genes into a pathway. This cell death pathway is evolutionarily conserved and provides a basis for understanding programmed cell death in more complex organisms, including humans.

C. ELEGANS AS A MODEL ORGANISM FOR STUDIES OF PROGRAMMED CELL DEATH

C. elegans is a small (adult animals are approx 1 mm in length), free-living worm with a short generation time (3 d at 20°C) (Fig. 1). It can feed on bacteria *Escherichia coli* and is cultivated on Petri dishes in the laboratory *(1)*. Because *C. elegans* is small and transparent, its internal structures can be visualized with the light microscope. Furthermore, using high-magnification Nomarski differential interference contrast optics, cell divisions and cell deaths can be observed and followed in living animals *(2–4)*. For these reasons, *C. elegans* has been an excellent model organism for experimental analyses and has proven to be exceptionally well-suited for the study of programmed cell death.

During the development of the *C. elegans* adult hermaphrodite from the fertilized egg, 1090 somatic nuclei are generated by essentially invariant patterns of cell divisions. Of these 1090 somatic cells, 131 undergo programmed cell death *(2–4)*. When observed with Nomarski optics, the dying cell adopts a refractile and raised, flattened button-like appearance (Fig. 2). When viewed using an electron microscope, the cells that undergo programmed cell death in *C. elegans* display characteristic features of apoptosis observed in mammals, including cell shrinkage, chromatin aggregation, and phagocytosis of cell corpse *(5)*. The entire process of cell death from the birth to the disappearance of the cell by phagocytosis occurs within approximately an hour *(2,6)*.

THE GENETIC PATHWAY FOR PROGRAMMED CELL DEATH

Genetic approaches have been taken to identify genes that are involved in controlling different aspects of *C. elegans* programmed cell death. Genetic analyses have placed these genes in five

From: *Essentials of Apoptosis: A Guide for Basic and Clinical Research*
Edited by: X-M. Yin and Z. Dong © Humana Press Inc., Totowa, NJ

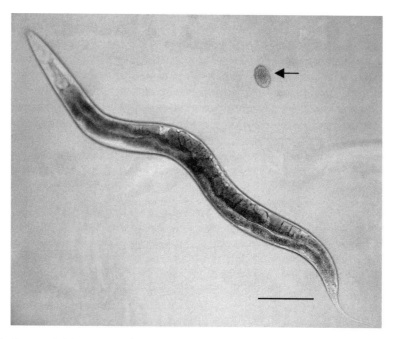

Fig. 1. A *C. elegans* adult hermaphrodite and an embryo viewed using bright-field microscopy. The embryo is indicated by an arrow. Dorsal is up and anterior is to the left for the adult. The bar represents 0.1 mm.

sequential and distinct steps: 1) the decision of individual cells whether to live or die; 2) the activation of the cell-killing machinery in a cell that is committed to die; 3) the execution of the cell-killing process; 4) the phagocytosis of the dying cell by its neighboring cell; and 5) the degradation of the dead cell *(7)* (Fig. 3). Genes acting in the first step affect only specific cells. By contrast, genes that function in the later four steps appear to affect all somatic cell deaths and are likely act in a pathway common to all cell deaths.

In the following sections, we first review our current understanding of genes that function as global cell-death regulators during the cell-death activation and then discuss how these regulators may be controlled by cell-fate specification genes to commit the life vs death fates of specific cells. Last, we focus on genes that function in the cell death execution and in the engulfment and the degradation of cell corpses.

ACTIVATION OF CELL DEATH

The Genetic Pathway for Cell-Killing

Three death-promoting genes *egl-1*, *ced-3*, and *ced-4* are required for most, if not all, programmed cell death in *C. elegans*. Strong loss-of-function (lf) mutations in any of these genes lead to the survival of essentially all cells that normally undergo programmed cell death during the development of wild-type animals *(8,9)*. Genetic mosaic analyses were carried out to determine whether these death-promoting genes act in cells that are doomed to die to mediate or promote a cell suicide process or instead function in adjacent cells to promote the death of dying cells as "murders." In these experiments, mosaic animals that contain both genotypically wild-type cells and genotypically mutant *ced-3* (or *ced-4*) cells were generated *(10)*. It was found that in these mosaic animals cells that are genotypically wild-type are capable of undergoing programmed cell death and those cells that are genotypically mutant for the *ced-3* (or the *ced-4*) gene fail to die. These findings indicate that both

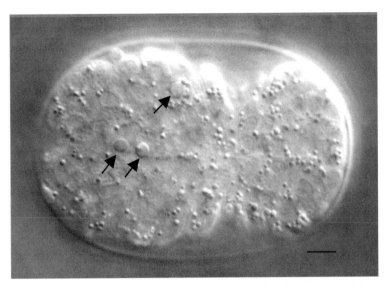

Fig. 2. A Nomarski photomicrograph of an embryo with apoptotic cells. Three cells indicated by arrows underwent programmed cell death in a bean/comma stage embryo and exhibited a refractile raised-button-like appearance. The bar represents 5 μm.

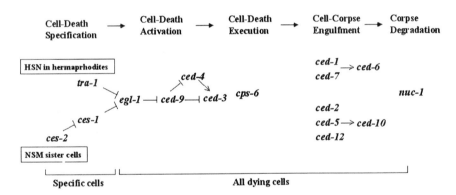

Fig. 3. Genetic pathway of programmed cell death in *C. elegans*. Five sequential steps of programmed cell death are indicated. In the cell-death specification step, genes involved in regulating the death fates of two specific cell types (HSN neurons and sister cells of NSM neurons) are shown. There are two partially redundant pathways (*ced-1, -6, -7* and *ced-2, -5, -10, -12,* respectively) that mediate the engulfment of cell corpses.

ced-3 and *ced-4* genes function within dying cells to cause cell death and provide the first genetic evidence that cells undergoing programmed cell death are killed by an intrinsic suicide mechanism.

In contrast to those death-promoting genes, the activity of the *ced-9* gene protects a majority of cells from undergoing programmed cell death during *C. elegans* development *(11)*. Loss-of-function mutations in *ced-9* cause embryonic lethality, as a consequence of massive ectopic deaths of cells that normally live *(11)*.

To understand how *egl-1, ced-3, ced-4,* and *ced-9* coordinate to regulate programmed cell death, two approaches have been taken to order their functions in a pathway leading to cell death. First, genetic epistasis analyses have been performed to define the relationships of the killer genes *egl-1, ced-3,* and *ced-4* with respect to the protector gene *ced-9,* taking advantage that the phenotype of mutants defective in any of these killer genes is opposite to the Ced-9(lf) phenotype. Specifically, double-mutant combinations such as *ced-9;ced-3*, *ced-4;ced-9*, and *ced-9;egl-1* were generated, and the phenotypes of double mutants were compared with those of respective single mutants. For example, *ced-9;ced-3* and *ced-4;ced-9* double mutants are viable and have extra surviving cells as observed in *ced-3* and *ced-4* single mutants *(11)*. This result indicates that the ectopic cell deaths and lethality caused by loss-of-function mutations in *ced-9* are suppressed by loss-of-function mutations in *ced-3* or *ced-4* and are dependent on the activities of *ced-3* and *ced-4*. Therefore, *ced-9* likely acts upstream of *ced-3* and *ced-4* to negatively regulate the activities of these two death-promoting genes. By contrast, *ced-9;egl-1* double mutants are lethal and exhibit a large amount of ectopic cell deaths as observed in *ced-9(lf)* single mutants *(9)*, indicating that loss of *egl-1* function does not suppress ectopic deaths caused by loss of *ced-9* function *(9)*. Therefore, *egl-1* likely acts upstream to negatively regulate *ced-9*.

Transcriptional overexpression experiments have been used to help order the functions of death-promoting genes *egl-1*, *ced-3*, and *ced-4*. Overexpression of *egl-1*, *ced-3*, or *ced-4* in C. elegans can induce deaths of cells in which one of these genes is ectopically expressed *(12)*. The resulting killing effects vary in different genetic backgrounds. For example, the cell killing caused by overexpression of *egl-1* is greatly reduced in *ced-3(lf)* or *ced-4(lf)* mutants *(9)*, consistent with the model that *ced-3* and *ced-4* act genetically downstream of *egl-1*. Similarly, the cell killing caused by overexpression of *ced-4* is greatly reduced in *ced-3(lf)* mutants *(12)*. By contrast, cell killing mediated by overexpression of *ced-3* does not seem to be affected by the absence of the *ced-4* activity *(12)*. These observations indicate that *ced-3* likely acts genetically downstream of *ced-4*.

Molecular Identities of egl-1, ced-3, ced-4, and ced-9

CED-9 is similar to the product of human proto-oncogene *bcl-2 (13)*, which plays a similar role in preventing apoptosis in mammals *(14–18)*. *ced-9* and *bcl-2* are two members of an expanding gene family that play important roles in regulating apoptosis in diverse organisms (for review, *see* refs. *19* and *20*). All members of this gene family contain at least one Bcl-2 homology region (BH domain), and some members, such as CED-9 and Bcl-2, consist of up to four BH domains *(19)*. The C. elegans killer gene *egl-1* encodes a relatively small protein of 91 amino acids with a potential BH3 motif that has been found in all death-promoting Bcl-2/CED-9 family members *(9)*.

CED-3 belongs to a family of cysteine proteases called caspases (<u>c</u>ysteine <u>asp</u>artate-specific pro-te<u>ase</u>), which cleave their substrates exclusively after an aspartate amino acid *(21)*. CED-3, like many other caspases, is first synthesized as a 56 kDa proenzyme and can be proteolytically activated to generate an active cysteine protease that contains two protease subunits derived from the cleavage products *(21,22*; Fig. 4D). Protease activity assay of several mutant CED-3 proteins showed that the extents of reduction of CED-3 in vitro protease activities correlate directly with the extents of reduction of *ced-3* in vivo killing activities *(22)*. These findings indicate that the CED-3 protease activity is important for *ced-3* to cause programmed cell death in C. elegans. The *ced-4* gene encodes a protein similar to mammalian Apaf-1, which is an activator of caspases during apoptosis in mammals *(23–26)*.

Molecular Model for the Killing Process

Molecular studies of C. elegans killer and protector genes not only reveal molecular identities of these genes, but also facilitate further biochemical and immunocytochemical analyses of these genes. These studies have provided important insights into how EGL-1, CED-3, CED-4, and CED-9 proteins function in a protein interaction cascade leading to the activation of programmed cell death.

CED-4 has been shown to interact physically with CED-9 in vitro *(27)*. Endogenous CED-4 and CED-9 proteins are co-localized at mitochondria in C. elegans embryos as detected using antibodies against CED-4 and CED-9 proteins *(28)*. In embryos in which cells had been induced to die by

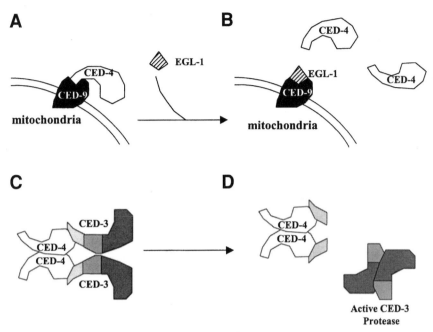

Fig. 4. The molecular model for the activation of programmed cell death. (**A**) In living cells CED-4 and CED-9 form a complex, which is tethered to mitochondria through CED-9. CED-3 may associate with CED-4 at the mitochondria or exist elsewhere without binding to CED-4. However, in either case, CED-3 remains an inactive proenzyme in living cells. (**B**) In the cells that are doomed to die, the death-initiator protein EGL-1 binds to CED-9 and triggers the release of CED-4 (or the CED-4/CED-3 complex) from CED-9. (**C**) Released CED-4 proteins undergo oligomerization, which brings two CED-3 proenzymes to close proximity and (**D**) leads to CED-3 autoproteolytic activation.

overexpression of *egl-1*, CED-4 assumed a perinuclear instead of mitochondrial localization *(28)*. This translocalization of CED-4 is mediated through interaction between EGL-1 and CED-9, as biochemical assays show that EGL-1 can bind CED-9 *(9,29)* and this binding induces release of CED-4 from CED-9/CED-4 complex *(30)* (Fig. 4A,B). The *ced-9* gain-of-function mutation *n1950*, which substitutes glycine 169 with glutamate, impairs the binding of EGL-1 to CED-9 but does not affect association of CED-9 with CED-4. As a result, this mutation inhibits the EGL-1-induced translocation of CED-4 and results in inhibition of programmed cell death *(30)*.

In addition to interacting with CED-9, CED-4 has been shown to interact with CED-3 *(31,32)*. The binding of CED-3 and CED-9 to CED-4 is not mutually exclusive *(31,32)*. This observation leads to the hypothesis that in living cells CED-3, CED-4, and CED-9 may co-exist as a ternary protein complex in which CED-3 remains an inactive proenzyme. However, the CED-3 subcellular localization has not yet been determined. It is also possible that CED-3 does not associate with CED-4/CED-9 complex and exists as inactive monomer elsewhere in the cell. In either case, EGL-1-induced dissociation of CED-4 from CED-9 may allow the formation of CED-3/CED-4 complex (Fig. 4). In addition to translocating to the perinuclear region, CED-4 may also undergo self-oligomerization, which appears to be important for CED-3 activation, as mutant CED-4 proteins that cannot self-oligomerize fail to activate CED-3 killing activity in mammalian cells *(31)*. These findings lead to the hypothesis that CED-4 oligomerization may bring CED-3 zymogens to close proximity and thus facilitate intermolecular proteolytic cleavage between CED-3 zymogens to activate the CED-3 protease (Fig. 4C,D). Activated CED-3 proteases may cause cell death by cleaving key substrates and hence lead to systematic cell disassembly and the eventual recognition and phagocytosis of the cell corpse by its neighboring cell.

In addition to CED-4, CED-3 can interact with CED-9. Moreover, CED-9 is an excellent substrate of the CED-3 protease in vitro. Mutations that destroy both CED-3 cleavage sites in CED-9 markedly reduce the death-protective activity of CED-9 in vivo, suggesting that CED-3 cleavage sites are important for CED-9 death-protective function *(33)*. Cleavage of CED-9 by CED-3 generates a carboxyl-terminal product that resembles Bcl-2 in sequence and is sufficient to mediate interaction with CED-4 *(33)*. These results suggest that CED-9 may inhibit cell death in *C. elegans* by two distinct mechanisms. First, CED-9 may directly inhibit the CED-3 protease activity through its CED-3 cleavage sites, probably acting as a competitive inhibitor. Second, CED-9 may indirectly inhibit the activation of CED-3 by forming a complex with CED-4 through its carboxyl-terminal Bcl-2 homology regions.

SPECIFICATION OF THE LIFE VS DEATH FATES

Genetic studies in *C. elegans* suggest that the life vs death decisions of individual cells are controlled by cell-type specific regulatory genes *(34)*. These regulatory genes appear to control cell deaths by regulating the expression or activities of key components in the central cell-killing pathway. The control of the life vs death fates of two specific cell types, hermaphrodite-specific neurons (HSNs) and neurosecretory motor (NSM) sister cells, will be discussed below.

HSN Neurons

The hermaphrodite-specific neurons (HSN) of *C. elegans*, which control the egg-laying behavior in hermaphrodites, are generated embryonically in both hermaphrodites and males but undergo programmed cell death specifically in males, because they are not needed in males *(4)*. In these sexually dimorphic HSNs, the cell-killer gene *egl-1* is under the direct transcriptional control of the *C. elegans* sex-determination pathway. TRA-1A, a zinc-finger protein, is the terminal global regulator of somatic sexual fate and binds to the *egl-1* gene in vitro *(35)*. Specific mutations in a *cis*-regulatory element of the *egl-1* gene that disrupt the binding of the TRA-1A to the *egl-1* gene result in transcriptional activation of *egl-1* in the HSNs and subsequent deaths of HSNs not only in males but also in hermaphrodites *(35)*. Therefore, *tra-1* helps to prevent the cell-death fate of the HSNs in hermaphrodites by transcriptionally repressing the expression of the *egl-1* gene (Fig. 3).

Sister Cells of NSM Neurons

Two genes, *ces-1* and *ces-2* (cell death specification), are important in determining the life vs death fates of the sister cells of the two neurosecretory motor (NSM) neurons in *C. elegans* pharynx *(34)*. Either the loss-of-function mutation in *ces-2* or gain-of-function mutations in *ces-1* can prevent these two cells from adopting their normal apoptotic cell fates. Interestingly, these cells undergo programmed cell death normally in *ces-1(lf)* mutants and *ces-1(lf);ces-2(lf)* double mutants. These observations suggest that the activity of *ces-1* normally is suppressed to allow NSM sister cells to die and that *ces-2* likely acts upstream of *ces-1* to suppress the activity of *ces-1*.

The *ces-1* gene encodes a Snail family zinc-finger protein *(36)*. CES-2, a putative bZIP (basic leucine-zipper) transcription factor *(37)*, can bind to the *ces-1* gene in vitro and may thus directly repress *ces-1* transcription *(36)*. These findings suggest that a transcriptional regulatory cascade may control the deaths of NSM sister cells in *C. elegans*.

The relationship of the *ces* genes with the death-promoting genes has mainly been inferred from the phenotype of *ces-1;egl-1* double mutants *(9)*. In *ces-1(lf);egl-1(lf)* mutants, NSM sister cells survive as they do in the *egl-1(lf)* mutants. This observation indicates that *egl-1* likely acts downstream of *ces-1* to cause programmed death and *ces-1* may negatively regulate the activity of *egl-1* in these cells. Therefore, *ces-2, ces-1*, and *egl-1* function in a negative regulatory chain to regulate the death fate of NSM sister cells (Fig. 3).

ENGULFMENT OF CELL CORPSES

Once a cell undergoes programmed death, the cell corpse is rapidly engulfed by one of its neighboring cells *(2,6)*. Genetic analyses have identified at least seven genes, *ced-1, ced-2, ced-5, ced-6, ced-7, ced-10,* and *ced-12,* which function in the cell-corpse engulfment process in *C. elegans (6,38–41)*. Mutations in any of these genes block the engulfment of many dying cells and lead to persistence of cell corpses. Genetic analysis of double-mutant combinations suggests that these seven genes fall into two classes: *ced-1, ced-6, ced-7* in one and *ced-2, ced-5, ced-10, ced-12* in the other. Single mutants or double mutants within the same class show relatively weak engulfment defects, whereas double mutants between two classes show much stronger engulfment defects. This finding suggests that dying cells likely present at least two different engulfment-inducing signals, which can be recognized by distinct molecules from engulfing cells, and both signaling events are required for efficient and complete phagocytosis.

The engulfment process consists of three sequential steps: 1) the recognition of a cell corpse by an engulfing cell, 2) transduction of the engulfing signal to the cellular machinery in the engulfing cell, and 3) the phagocytosis of the cell corpse by the engulfing cell. The genes *ced-1, ced-6,* and *ced-7,* which define one engulfment pathway, appear to encode components of a signaling pathway involved in cell-corpse recognition. CED-1 is similar to the human scavenger receptor SREC and may function as a corpse-recognizing phagocytic receptor because CED-1 protein was found to cluster around dying cells *(42)*. CED-7 is similar to ABC (ATP-binding cassette) transporters *(42,43)* and may play a role in promoting or mediating cell-corpse recognition by CED-1 as CED-1 receptors fail to cluster around dying cells in mutants defective in the *ced-7* gene *(42,43)*. The CED-6 protein contains a PTB (phosphotyrosine-binding) domain *(44)* and may act as a signaling adaptor downstream of CED-1 and CED-7 (Fig. 5).

The *ced-2, ced-5, ced-10,* and *ced-12* genes, which define the other engulfment pathway, also control the migration of specific somatic cells, the gonadal distal tip cells (DTC) of *C. elegans*. These four genes encode conserved components of the Rac GTPase signaling pathway involved in regulating actin cytoskeleton rearrangement essential for cell-corpse phagocytosis and cell migration. CED-2 is a CrkII-like adaptor, consisting of one SH2 and two SH3 domains *(45)*. CED-5 is similar to human DOCK180 that physically interacts with human CrkII *(46)*. CED-10 is a *C. elegans* homolog of mammalian Rac GTPase *(45)*, which controls cytoskeletal dynamics and cell shape change (for review, *see* ref. *47*). CED-12 contains a potential PH (pleckstrin-homology) domain and an SH3-binding motif *(39–41)*. Transcriptional overexpression studies suggest that *ced-2, ced-5,* and *ced-12* function at the same step upstream of *ced-10* during the engulfment process. Biochemical analysis indicates that CED-2, CED-5, and CED-12 form a ternary complex in vitro and so do their human homologs *(39–41)*. Based on these findings, it has been postulated that the engulfing signal induces the formation and translocation of CED-2, CED-5, and CED-12 ternary complex to the plasma membrane of engulfing cells and the subsequent activation of CED-10 GTPase, leading to extension of membrane processes around cell corpses (Fig. 5).

DEGRADATION OF CHROMOSOMAL DNA IN DYING CELLS

Degradation of chromosomal DNA has been thought to be a crucial step and a hallmark of apoptosis. Apoptotic DNA degradation in *C. elegans* has been studied with the aid of DNA-staining techniques. For example, DAPI or Feulgen dye has been used to visualize DNA for *in situ* staining, and the terminal deoxynuleotidyl transferase mediated dUTP-biotin nick end labeling (TUNEL) technique, which was initially developed to specifically label dying cells because DNA degradation causes these cells to have more free DNA ends than do viable cells *(48)*, has been applied to detect the DNA intermediates with 3'-hydroxyl ends during the degradation process.

When stained with TUNEL, only a small subset of cells that undergo programmed cell death in wild-type *C. elegans* embryos is TUNEL-positive *(49)*, suggesting that DNA degradation is a rapid

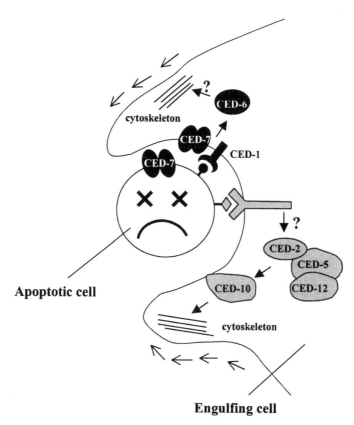

Fig. 5. The molecular model for the cell corpse-engulfment process. The engulfment process is mediated by two partially redundant pathways. Molecules in the CED-1, CED-6, and CED-7 pathway are labeled in black. CED-1 and CED-7 act on the surface of engulfing cells to mediate cell-corpse recognition and to transduce the engulfing signal through CED-6 to the cellular machinery of the engulfing cells for engulfment. CED-7 also acts in dying cells. Molecules in the CED-2, CED-5, CED-10, and CED-12 pathway are labeled in gray. The CED-2/CED-5/CED-12 ternary complex mediates the signaling event from unidentified engulfing signal(s) and receptor(s) to activate CED-10 during phagocytosis.

process and TUNEL only labels apoptotic cells during a transient intermediate stage. Interestingly, mutant embryos defective in *nuc-1*, which encodes a mammalian DNAse II homolog, have many more TUNEL-reactive nuclei than do wild-type embryos *(49)*. This finding indicates that mutations in *nuc-1* allow the generation of TUNEL-reactive DNA breaks, but block the subsequent conversion of these TUNEL-reactive DNA ends to TUNEL-unreactive ones. Like *nuc-1* mutations, the mutation in the *cps-6* gene (CED-3 protease suppressor), which encodes a homolog of mammalian mitochondrial endonuclease G, also increases the number of TUNEL-reactive nuclei *(50)*. Interestingly, *cps-6*: *nuc-1* double mutants have more TUNEL-reactive nuclei than *cps-6* or *nuc-1* single mutants, suggesting that *cps-6* and *nuc-1* likely function in a partially redundant fashion to destroy TUNEL-reactive DNA ends. However, *cps-6* and *nuc-1* appear to play different roles during apoptosis. A loss-of-function mutation in the *cps-6* gene not only delays the appearance of embryonic cell corpses during development, but also can block the deaths of some cells if the activity of other cell-death components is compromised *(50)*, suggesting that *cps-6* is important for the normal progression and execution of apoptosis. In contrast, mutations in *nuc-1* do not appear to affect either the execution of cell death or the engulfment of cell corpses. Furthermore, *nuc-1* mutants are also defective in the degradation of DNA from ingested bacteria in the intestinal lumens. These observations indicate that *cps-6* is a more specific cell-death nuclease and may function at an earlier stage in the apoptotic DNA degradation process than *nuc-1* does. CPS-6 is the first mitochondrial protein that is shown to be important for apoptosis in invertebrates, underscoring the conserved role of mitochondria in regulating apoptosis.

SUMMARY

Genetic studies in *C. elegans* have identified more than a dozen genes that function in different aspects of programmed cell death. These genes have defined a programmed cell death pathway that is evolutionarily conserved. In *C. elegans* the life vs death fates of cells appear to be controlled at the level of transcription. Once the decision to die has been made, a protein interaction cascade involving EGL-1, CED-3, CED-4, and CED-9 is responsible for the activation of the cell-death machinery which initiates various cell disassembly processes such as the apoptotic DNA degradation process involving CPS-6 and NUC-1. The dying cell presents at least two distinct engulfing signals to their neighboring cells for phagocytosis. These two signals are mediated by two partially redundant pathways: CED-1, CED-6, and CED-7 in one and CED-2, CED-5, CED-10, and CED-12 in the other. Finally, the dying cell is completely degraded in the engulfing cell.

REFERENCES

1. Brenner, S. (1974) The genetics of *Caenorhabditis elegans*. *Genetics* **77**, 71–94.
2. Sulston, J. E. and Horvitz, H. R. (1977) Post-embryonic cell lineages of the nematode, *Caenorhabditis elegans*. *Dev. Biol.* **56**, 110–156.
3. Kimble, J. and Hirsh, D. (1979) The postembryonic cell lineages of the hermaphrodite and male gonads in *Caenorhabditis elegans*. *Dev. Biol.* **70**, 396–417.
4. Sulston, J. E., Schierenberg, E., White, J. G., and Thomson, J. N. (1983) The embryonic cell lineage of the nematode *Caenorhabditis elegans*. *Dev. Biol.* **100**, 64–119.
5. Robertson, A. G. and Thomson, J. N. (1982) Morphology of programmed cell death in the ventral nerve cord of *Caenorhabditis elegans* larvae. *J. Embryol. Exp. Morph.* **67**, 89–100.
6. Hedgecock, E., Sulston, J., and Thomson, J. N. (1983) Mutations affecting programmed cell deaths in the nematode *Caenorhabditis elegans*. *Science* **220**, 1277–1279.
7. Ellis, R. E., Yuan, J., and Horvitz, H. R. (1991) Mechanisms and functions of cell death. *Annu. Rev. Cell Biol.* **7**, 663–698.
8. Ellis, H. M. and Horvitz, H. R. (1986) Genetic control of programmed cell death in the nematode *C. elegans*. *Cell* **44**, 817–829.
9. Conradt, B. and Horvitz, H. R. (1998) The *C. elegans* protein EGL-1 is required for programmed cell death and interacts with the Bcl-2-like protein CED-9. *Cell* **93**, 519–529.
10. Yuan, J. Y. and Horvitz, H. R. (1990) The *Caenorhabditis elegans* genes *ced-3* and *ced-4* act cell autonomously to cause programmed cell death. *Dev. Biol.* **138**, 33–41.
11. Hengartner, M. O., Ellis, R. E., and Horvitz, H. R. (1992) *Caenorhabditis elegans* gene *ced-9* protects cells from programmed cell death. *Nature* **356**, 494–499.
12. Shaham, S. and Horvitz, H. R. (1996) Developing *Caenorhabditis elegans* neurons may contain both cell-death protective and killer activities. *Genes Dev.* **10**, 578–591.
13. Hengartner, M. O. and Horvitz, H. R. (1994) *C. elegans* cell survival gene *ced-9* encodes a functional homolog of the mammalian proto-oncogene bcl-2. *Cell* **76**, 665–676.
14. Tsujimoto, Y. and Croce, C. M. (1986) Analysis of the structure, transcripts, and protein products of bcl-2, the gene involved in human follicular lymphoma. *Proc. Natl. Acad. Sci. USA* **83**, 5214–5218.
15. Seto, M., Jaeger, U., Hockett, R. D., Graninger, W., Bennett, S., Goldman, P. (1988) Alternative promoters and exons, somatic mutation and deregulation of the Bcl-2-Ig fusion gene in lymphoma. *EMBO J.* **7**, 123–131.
16. Cleary, M. L., Smith, S. D., and Sklar, J. (1986) Cloning and structural analysis of cDNAs for bcl-2 and a hybrid bcl-2/immunoglobulin transcript resulting from the t(14;18) translocation. *Cell* **47**, 19–28.
17. Vaux, D. L., Cory, S., and Adams, J. M. (1988) Bcl-2 gene promotes haemopoietic cell survival and cooperates with c-myc to immortalize pre-B cells. *Nature* **335**, 440–442.
18. Nunez, G., London, L., Hockenbery, D., Alexander, M., McKearn, J. P., and Korsmeyer, S. J. (1990) Deregulated Bcl-2 gene expression selectively prolongs survival of growth factor-deprived hemopoietic cell lines. *J. Immunol.* **144**, 3602–3610.
19. Reed, J. C. (1998) Bcl-2 family proteins. *Oncogene* **17**, 3225–3236.
20. Adams, J. M. and Cory, S. (2001) Life-or-death decisions by the Bcl-2 protein family. *Trends Biochem. Sci.* **26**, 61–66.
21. Yuan, J., Shaham, S., Ledoux, S., Ellis, H. M., and Horvitz, H. R. The *C. elegans* cell death gene *ced-3* encodes a protein similar to mammalian interleukin-1B-converting enzyme. *Cell* **75**, 641–652.

22. Xue, D., Shaham, S., and Horvitz, H. R. (1996) The *Caenorhabditis elegans* cell-death protein Ced-3 is a cysteine protease with substrate specificities similar to those of the human CPP32 protease. *Genes Dev.* **10**, 1073–1083.

23. Yuan, J. and Horvitz, H. R. (1992) The *Caenorhabditis elegans* cell death gene *ced-4* encodes a novel protein and is expressed during the period of extensive programmed cell death. *Development* **116**, 309–320.

24. Zou, H., Henzel, W. J., Liu, X., Lutschg, A., and Wang, X. (1997) Apaf-1, a human protein homologous to *C. elegans* CED-4, participates in cytochrome c-dependent activation of caspase-3. *Cell* **90**, 405–413.

25. Cecconi, F., Alvarez-Bolado, G., Meyer, B. I., Roth, K. A., and Gruss, P. (1998) Apaf1 (CED-4 homolog) regulates programmed cell death in mammalian development. *Cell* **94**, 727–737.

26. Yoshida, H., et al. (1998) Apaf1 is required for mitochondrial pathways of apoptosis and brain development. *Cell* **94**, 739–750.

27. Spector, M. S., Desnoyers, S., Hoeppner, D. J., and Hengartner, M. O. (1997) Interaction between the *C. elegans* cell-death regulators CED-9 and CED-4. *Nature* **385**, 653–656.

28. Chen, F., et al. (2000) Translocation of *C. elegans* CED-4 to nuclear membranes during programmed cell death. *Science* **287**, 1485–1489.

29. del Peso, L., Gonzalez, V. M., and Nunez, G. (1998) *Caenorhabditis elegans* EGL-1 disrupts the interaction of CED-9 with CED-4 and promotes CED-3 activation. *J. Biol. Chem.* **273**, 33495–33500.

30. Parrish, J., Metters, H., Chen, L., and Xue, D. (2000) Demonstration of the in vivo interaction of key cell death regulators by structure-based design of second-site suppressors. *Proc. Natl. Acad. Sci. USA* **97**, 11916–11921.

31. Yang, X., Chang, H. Y., and Baltimore, D. (1998) Essential role of CED-4 oligomerization in CED-3 activation and apoptosis. *Science* **281**, 1355–1357.

32. Chinnaiyan, A. M., O'Rourke, K., Lane, B. R., and Dixit, V. M. (1997) Interaction of CED-4 with CED-3 and CED-9: a molecular framework for cell death. *Science* **275**, 1122–1126.

33. Xue, D. and Horvitz, H. R. (1997) *Caenorhabditis elegans* CED-9 protein is a bifunctional cell-death inhibitor. *Nature* **390**, 305–308.

34. Ellis, R. E. and Horvitz, H. R. (1991) Two *C. elegans* genes control the programmed deaths of specific cells in the pharynx. *Development* **112**, 591–603.

35. Conradt, B. and Horvitz, H. R. (1999) The TRA-1A sex determination protein of *C. elegans* regulates sexually dimorphic cell deaths by repressing the *egl-1* cell death activator gene. *Cell* **98**, 317–327.

36. Metzstein, M. M. and Horvitz, H. R. (1999) The *C. elegans* cell death specification gene *ces-1* encodes a snail family zinc finger protein. *Mol. Cell* **4**, 309–319.

37. Metzstein, M. M., Hengartner, M. O., Tsung, N., Ellis, R. E., and Horvitz, H. R. (1996) Transcriptional regulator of programmed cell death encoded by *Caenorhabditis elegans* gene *ces-2*. *Nature* **382**, 545–547.

38. Ellis, R., Jacobson, D. M., and Horvitz, H. R. (1991) Genes required for the engulfment of cell corpses during programmed cell death in *C. elegans*. *Genetics* **129**, 79–94.

39. Wu, Y. C., Tsai, M. C., Cheng, L. C., Chou, C. J., and Weng, N. Y. (2001) *C. elegans* CED-12 acts in the conserved crkII/dock180/rac pathway to control cell migration and cell corpse engulfment. *Dev. Cell* **1**, 491–502.

40. Zheng, Z., Caron, E., Harwieg, E., Hall, A., and Horvitz, H. R. (2001) The *C. elegans* PH domain protein CED-12 regulates cytoskeletal reorganization via a rho/rac GTPase signaling pathway. *Dev. Cell* **1**, 477–489.

41. Gumienny, T. L., Brugnera, E., Tosello-Trampont, A. C., Kinchen, J. M., Haney, L. B., Nishiwaki, K., et al. (2001) CED-12/elmo, a novel member of the crkii/dock180/rac pathway, is required for phagocytosis and cell migration. *Cell* **107**, 27–41.

42. Zhou, Z., Hartwieg, E., and Horvitz, H. R. (2001) CED-1 is a transmembrane receptor that mediates cell corpse engulfment in *C. elegans*. *Cell* **104**, 43–56.

43. Wu, Y. C. and Horvitz, H. R. (1998) The *C. elegans* cell corpse engulfment gene *ced-7* encodes a protein similar to ABC transporters. *Cell* **93**, 951–960.

44. Liu, Q. A. and Hengartner, M. O. (1998) Candidate adaptor protein CED-6 promotes the engulfment of apoptotic cells in *C. elegans*. *Cell* **93**, 961–972.

45. Reddien, P. W. and Horvitz, H. R. (2000) CED-2/CrkII and CED-10/Rac control phagocytosis and cell migration in *Caenorhabditis elegans*. *Nat. Cell Biol.* **2**, 131–136.

46. Wu, Y. C. and Horvitz, H. R. (1998) *C. elegans* phagocytosis and cell-migration protein CED-5 is similar to human DOCK180. *Nature* **392**, 501–504.

47. Hall, A. (1992) Ras-related GTPases and the cytoskeleton. *Mol. Biol. Cell* **3**, 475–479.

48. Gavrieli, Y., Sherman, Y., and Ben-Sasson, S. A. (1992) Identification of programmed cell death *in situ* via specific labeling of nuclear DNA fragmentation. *J. Cell Biol.* **119**, 493–501.

49. Wu, Y. C., Stanfield, G. M., and Horvitz, H. R. (2000) NUC-1, a *Caenorhabditis elegans* DNase II homolog, functions in an intermediate step of DNA degradation during apoptosis. *Genes Dev.* **14**, 536–548.

50. Parrish, J., Li, L., Klotz, K., Ledwich, D., Wang, X., Xue, D. (2001) Mitochondrial endonuclease G is important for apoptosis in *C. elegans*. *Nature* **412**, 90–94.

10

Cell Death in *Drosophila*

Sujin Bao and Ross L. Cagan

INTRODUCTION

Only relatively recently has programmed cell death (PCD) been fully appreciated as a central component of development and disease. Removal of damaged cells represents one of the most important defenses our body has to prevent pathology, and abnormal removal of cells has been observed in a number of degenerative diseases. In this review, we emphasize the importance of cell death during development, with a brief excursion at the end to examine neurodegeneration. The fruitfly *Drosophila* has been at the center of a number of important discoveries regarding development and signal transduction, so it is perhaps surprising that it has been a relatively late entrant into the world of programmed cell death. However, flies have been making up for lost time.

Why produce cells during development that are then removed by programmed cell death? The details of cell death often seem specific to the tissue using it, but some basic principles have emerged. During cell division, some determinants are partitioned selectively to one daughter cell; for example, the developing *Caenhorhabditis elegans* embryo presents several examples of segregating away unneeded determinants to permit one type of neuron to survive while the sister cell is removed by programmed cell death (reviewed in ref. *1*). A more common use of cell death in higher organisms is during patterning, an issue we will discuss in some detail below. After establishing a crudely-patterned structure, spatially selective cell death then "sculpts" the tissue towards a final structure. Examples abound and include the hollowing out of certain neural tubes *(2)*, the removal of intervening tissue to form digits *(3,4)*, the shaping of kidney glomeruli (for a review *see* ref. *5*), and the tightening of rhombomere boundaries in the hindbrain *(6)*. These examples all share an important issue that is not at all understood: how are some cells removed in a spatially restricted fashion while others in the region are left alone?

A BRIEF PRIMER ON CELL DEATH

Programmed cell death refers to the process by which particular cells are removed by design. Cells are most commonly removed by apoptosis, a process that includes membrane alteration and blebbing, organelle dissolution, coherent DNA fractionation, and eventual engulfment by neighboring cells or dedicated macrophages. This process has been seen in all metazoans examined to date, and the process of apoptosis is remarkably well-conserved. Several excellent reviews have explored our knowledge of this process (e.g., *see* refs. *7–12*). This section serves as a brief review of those aspects of cell death that are relevant to work in *Drosophila*.

From: *Essentials of Apoptosis: A Guide for Basic and Clinical Research*
Edited by: X-M. Yin and Z. Dong © Humana Press Inc., Totowa, NJ

Table 1
Conserved Intrinsic Death Machinery

	Mammalian	*Drosophila*	*C. elegans*
Death activator	Smac/Diablo, HtrA2	Rpr, Hid, Grim	Not known
Inhibitor of apoptosis protein	NIAP, cIAP1, CIAP2 XIAP, Survivin	DIAP1, DIAP2	IAP-1, IAP-2
Caspases			
Initiator caspases	Caspases-1, -2, -4, -5, -8,-9, -10, etc.	DRONC, DREDD	Ced-3
Effector caspases	Caspases-3, -6, -7, -14	DRICE, DCP-1, DECAY	Not known
Apoptosis activating factor	Apaf-1	DARK	Ced-4
Bcl-2 family			
Pro-apoptotic ligand with BH3 domain only	Bad, Bod, Bid, Bik, Blk, Hrk, Nip3, Nix	Drob/DBORG-1 DBORG-2	Egl-1
Pro-apoptotic channel members	Bax, Bak, Bok, Diva	Not known	Not known
Anti-apoptotic channel members	Bcl-2, $Bclx_L$, Bcl_W Mcl-1, Bfl-1/A1	Not known	Ced-9

Cell Intrinsic Machinery

Although a caspase-independent pathway has been suggested to mediate cell death involving Bax group proteins (for a review *see* ref. *13*), the cysteine aspartases known as caspases appear to be the major final mediators of cell death. Caspases exist in two categories: upstream caspases or "initiator caspases" that are targeted by upstream cell death effectors to trigger the cell death pathway, and downstream or "effector caspases." In flies, the upstream caspases include Dredd and Dronc; the downstream caspases are DCP-1, DRICE, and DECAY (Table 1). Although they are not fully characterized in flies, activated effector caspases target a number of death substrates in mammals, including poly (ADP) ribose polymerase (PARP), nucleases, gelsolin, actin, lamin, fodrin, and so on. Together, these factors complete the task of destroying the cell. Regulation of effector caspases, therefore, represents a central step in regulating the cell-death process.

One mechanism by which effector caspases are regulated is through the "inhibitor of apoptosis protein" family or IAPs (Table 1). IAPs complex directly with effector caspases, covering their site of self-cleavage and activation (for reviews *see* refs. *9, 14*). IAPs are ubiquitously expressed; their presence prevents most cells from undergoing apoptotic cell death. Regulation of IAPs is an area of intense study. In flies, the novel cytoplasmic proteins Reaper (Rpr), Grim, or Hid can physically interact with IAP and antagonize its anti-apoptotic properties: binding by Rpr, Grim, or Hid to IAPs disrupts the caspase/IAP complex to induce apoptosis *(15,16)*. Complete removal of these three genes through an overlapping deficiency results in the loss of most apoptotic cell death in the embryo *(17)*. Although clear orthologs of Reaper, Grim, or Hid have not been reported in mammals, they share an "RGH" domain with the death activator Smac/Diablo and HtrA2 *(18–20)*, which also target IAPs. Recently, Morgue, a ubiquitin-conjugating enzyme-related protein, has been shown to bind IAP directly and promote its degradation in the fly retina *(21,22)*.

Another mechanism by which effector caspases are activated is through inclusion within a multimeric complex that includes Apaf-1 (the fly ortholog is Dark/Hac-1) and cytochrome *c* (for reviews *see* refs. *23, 24*). This "apoptosome" complex is formed when cytochrome *c* is released from mitochondria to form a cytochrome c/Apaf-1/caspase complex. The *Drosophila* Apaf-1 ortholog Dark was shown to interact with initiator Caspases Dredd and Dronc *(25)*, facilitating their enzymatic activation; importantly, Dark contains a consensus cytochrome c binding site and presumably forms a similar apoptosome complex. However, the precise role played by cytochrome c in *Drosophila* is not clear *(10)*. Mutations in Dark suppress the normal programmed cell death and ectopic killing induced by Rpr, Hid, and Grim *(25–27)*.

In mammals, release of cytochrome c and other mitochondrial factors is regulated by members of the Bcl-2 family (Table 1). The Bcl-2 family includes members that protect from apoptotic cell death (e.g., Bcl-2, Bcl-X_L) and those that promote cell death (e.g., Bax, Bok, Bik, Bad, and Bid). Although the details of how Bcl-2 family members regulate cell death is still not fully understood, the balance between death-protective and death-promoting members appears to determine whether a cell will respond to death stimuli by surviving or by dying (for a review *see* ref. *28*). At least some data indicates that death-promoting factors such as Bax act by promoting release of mitochondrial factors such as cytochrome c. The sole *Drosophila* Bcl-2 family members identified to date are Drob/Debcl/Dborg-1 and Dborg-2 *(29–31)*. Based on sequence and some biochemistry, both appear to promote cell death, suggesting they are most similar to Bax or Bok. The *Drosophila* genome does not appear to contain a clear candidate for a true "pro-life" Bcl-2 ortholog, a surprising omission (Table 1).

Cell Extrinsic Machinery

While the molecules and mechanisms involved in triggering, executing, and regulating apoptotic cell death within a cell are becoming well-understood, only now are we beginning to understand the connections between this process and cell signaling, either extracellular or intracellular. Importantly, until recently most of our understanding of cell death has come from studies in the immune system or in tissue-culture cells. For example, in the immune system, regulation through the tumor necrosis factor (TNF) receptor and Fas mediate death signaling through activation of initiator caspases such as caspase-8, which in turn cleave and activate downstream effector caspases (for a review *see* ref. *32*). This work has given us important information about the basal death machinery, but we are beginning to look at developing epithelia in an effort to understand its spatial regulation. For example, simply releasing death-promoting factors across an epithelium is an efficient method to ensure the entire tissue goes away.

An important example of invoking cell death in response to external stimuli involves signaling through the Ras signal-transduction pathway. Following stimuli from growth factors and extracellular matrix, activated Ras signaling can regulate apoptosis, either positively or negatively, by controlling the activity of multiple effectors. In fibroblasts and lymphocytes, Ras can promote apoptosis through either accumulating p53, p16, or p21, upregulating Fas ligand expression or activating JNK pathway *(33–35)*. In flies, activation of Ras signaling can block cell death (see below); interestingly, it can also lead to apoptotic cell death in surrounding cells that have not received ectopic Ras activation (i.e., nonautonomously; *36*), suggesting that a death-promoting factor is being released to neighbors.

In contrast, activation of Ras in epithelial cells can inhibit apoptosis induced by detachment from extracellular matrix through activation of the PKB/Akt pathway *(37)* and Ras can protect neurons from apoptosis induced by neurotrophic factor withdrawal through the Raf/MAP kinase pathway *(35)*. Similarly, in flies the EGF receptor ortholog dEGFR acts through dRas1 signaling to promote survival. During Drosophila embryogenesis (see below), activation of dEGFR leads to phosphorylation of the death activator Hid and suppression of its proapoptotic activity *(38)*. Additionally, the Ras/Raf/MAPK signaling pathway has been shown to promote survival both by downregulating Hid expression and by inactivating Hid activity through phosphorylation *(39,40)*.

Finally, a few words about an important cell-death regulator that reacts to internal signals. Genomic instability and the excess accumulation of free oxygen radicals can also trigger apoptotic cell death

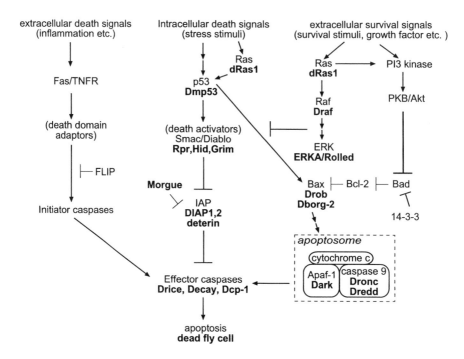

Fig. 1. Outline of cell-death machinery. Although the signals to trigger death or survival can be diverse and dynamic, they converge in the regulation of caspases through the intrinsic cell-death machinery, which is conserved from worms to flies to mammals (*see* also Table 1) Some special fly orthologs are noted in bold. For details see text.

through the tumor-suppressor protein p53. Increased levels of p53 due to cell-cycle imbalance or DNA damage induce the expression of Bax to promote apoptosis *(41)*. In mammals, p53 appears to regulate both cell-cycle progression and cell death. *Drosophila* Dmp53 appears to target exclusively cell death; one reason is its ability to directly target the death activator gene *reaper (42–44)*.

Figure 1 presents many of the players in the cell-death game.

EARLY USE OF CELL DEATH IN OOGENESIS AND EMBRYOGENESIS

Cell Death in Drosophila Oogenesis

Drosophila oogenesis can be divided into 14 stages during its maturation from a germarium stem cell to a mature oocyte *(45)*. During early stages (S1–S8) of oogenesis, the *Drosophila* egg chamber encloses a cluster of 16 interconnected germline cells surrounded by a monolayer of follicle cells. One of these 16 germline cells will become the oocyte and the other 15 will develop into nurse cells which support the oocyte by transporting proteins and RNAs into the oocyte through cytoplasmic bridges. Programmed cell death plays an important role in a number of steps. For example, defective egg chambers will be removed by apoptosis between stages 7 and 8 of oogenesis *(46)*. After dumping their contents, the remaining depleted nurse cells are removed from the egg chambers during stages 12 and 13, again by apoptotic cell death (Fig. 2). The dying nurse cells are phagocytosed by overlying follicle cells.

Most components of the intrinsic cell death machinery in *Drosophila* are found in nurse cells, including death-activator genes *reaper*, *hid*, and *grim*; anti-apoptotic genes *diap1* and *diap2*; caspase orthologs *dredd*, *dronc*, *dcp-1*, and *decay*; and the Bcl-2 family member *drob-1/debcl/dborg-1 (29–31; 47–51)*. At stages 10–11, nurse cells undergo a dramatic rearrangement of filamentous actin to

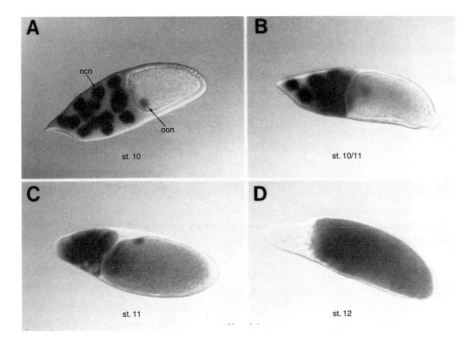

Fig. 2. Programmed cell death removes nurse cells. Nuclei are marked by an enhancer trap line: **(A)** stage 10, **(B)** stage 10B/11; **(C)** stage 11; **(D)** stage 12. ncn, nurse cell nuclei; oon, oocyte nucleus. Adapted with permission from ref. *(123).*

form actin bundles that, due to myosin-based contraction, transport cytoplasm through the ring canal (reviewed in ref. *52*). Interestingly, *dcp-1* egg chambers display delayed actin-bundle formation and defects in nurse cell nuclear-envelope perforation and contraction; the result is a "dumpless" phenotype. This indicates that *dcp-1*-mediated cell death is required to initiate rapid cytoplasm transport as well as the removal of nurse cells *(49)*. Surprisingly, oogenesis proceeds normally in germline clones homozygous for the H99 deficiency that removes *reaper, grim*, and *hid (47)*. These death activators are widely utilized in other developmental processes, suggesting that oogenesis uses at least some unique cell death regulators.

Cell-cell signaling also plays an important role in regulating nurse cell death. The BMP2/4 ortholog Dpp is expressed in the follicle cells while its receptors are expressed in both the nurse cells and follicle cells. Several experiments suggest that Dpp signaling is central to regulating cell death during oogenesis. Reducing the activity of either Dpp or its receptor Saxophone in egg chambers produces defective anterior eggshell structures. In addition, germline cells mutant for *saxophone* display defects in actin-bundle formation and failure of the nurse cells to complete cytoplasm transfer *(53)*, suggesting Dpp signaling orchestrates programmed cell death in the germline. The circulating steroid hormone ecdysone also provides a death-inducing signal during *Drosophila* oogenesis. In females injected with ecdysone, stage 9 egg chambers are almost completely eliminated by apoptosis *(54)*. However, little is understood to date about the link between upstream signals and activation of caspases during oogenesis.

Cell Death in the Embryonic Epidermis

Cell death is utilized throughout embryonic development, and the embryo has provided a number of important insights into the role of programmed cell death during normal epithelial patterning. An

excellent example is the establishment of segments. Cell division and cell movement create 14 segments that represent the fundamental segmental units of the embryo. Each segment expresses
hedgehog and *engrailed* at its posterior margin, whereas *wingless* is expressed within its anterior.
wingless- and *hedgehog*-expressing cells signal each other to stabilize segment boundaries (for a
review *see* ref. *55*). In late-stage wild-type embryos, about 10 diverse epidermal cell types are generated within each segment; each cell type contributes to a different part of the cuticular covering that
serves to mark positional information.

Time-lapse microscopic studies using acridine orange (AO)-injected embryos found that approx
40–45 dying cells were detected in the ectoderm of each segment during stages 12–14 *(56)*. Although
the number of dying cells varies from embryo to embryo, the pattern of cell death within each embryo
is conserved. Approximately three-fourths of the dying cells were located in or immediately adjacent
to the *engrailed* posterior stripe. Some dying cells form clusters at specific locations along the dorsal-
ventral axis and the expression pattern of Reaper correlates closely with this distribution (Fig. 3).
Dead cells are eventually engulfed by their neighbors *(57)*.

Segment polarity genes *wingless*, *hedgehog*, and their signal-pathway components play an essential role in patterning the embryonic epidermis *(58–60)*. Loss of *wingless* signaling leads to abnormally small embryos with a number of patterning defects *(61–63)*. These defects are likely due to
inappropriate cell death: disruption of the *wingless* signal components *dishevelled* or *armadillo* leads
to both patterning defects and increased cell death within the embryonic epidermis *(62,63)*. Using
time-lapse imaging studies, Pazdera et al. *(56)* showed that removal of *wingless* signaling resulted in
a fivefold increase in the number of dying cells within the anterior of each segment; this death
occurred in cells approximately six rows away from the *wingless* secreting cells. Conversely, activating the *wingless* pathway by reducing *naked* function led to a sixfold increase in apoptosis in the
posterior region *(56)*.

This regional increase in cell death appears to be a common phenomenon of mutations that affect
axis formation. The best evidence to date suggests that *wingless* and *hedgehog* do not directly activate the apoptotic pathway. Instead, cell death is a response common to mutations in a number of
signaling pathways, suggesting that embryos monitor the precision with which segments are made.
Surprisingly, correcting axis defects appears to use only cell death and not, for example, cell cycle. In
an elegant set of experiments, Namba et al. *(64)* demonstrated that an overall increase in the gene
dosage of the anterior determinant *bicoid* leads initially to an enlarged head region; however, the
embryo compensates by increasing cell death in its anterior regions and decreasing it in the posterior
(64). A decrease in *bicoid* led to the complementary response. Cell-cycle levels were unaffected.
These experiments point to the central role of selective cell death to clean up the normal errors that
occur during development. Cell death acts as a buffer against errors, and provides precision to the
emerging embryo. As we shall see later in this review, other tissues such as the eye utilize death in
much the same way.

Death in the CNS Midline

As tissues involute during gastrulation, two stripes of mesectoderm in each side of the ventral
midline are brought together and eventually generate a distinct set of 6–8 central nervous system
(CNS) midline nerve-cell precursors (CMPs) per segment. These cells later differentiate into approx
25 neurons and glia. During this process, two-thirds of the developing midline glia die and are quickly
phagocytosed by migrating macrophages. In contrast, all of the associated ventral unpaired median
(VUM) neurons survive *(65,66)*. As expected, cell death in midline cells is caspase-dependent: mid-
line-targeted expression of the caspase inhibiting protein p35 blocks normal cell death as well as
death due to ectopically expressed *reaper* and *hid (66)*.

Expression of *reaper*, *hid*, and *grim* can be detected in dying midline cells and these death-activator
genes are essential for patterning CNS midline cell death *(65–68)*. Analysis of mutations that remove
different combinations of *reaper*, *hid*, and *grim* demonstrated that these three death-activator genes

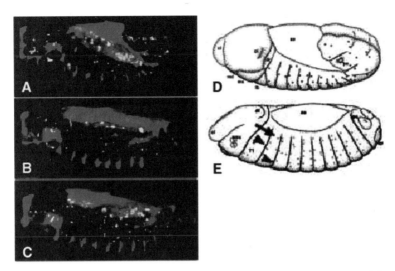

Fig. 3. Cell-death pattern in the embryonic epidermis. Apoptotic cells detected by Acridine Orange (light gray) are shown relative to each segment border (stripes) in an embryo at stages 12–14 (**A–C**). A map of cell death in the abdominal epidermis is illustrated during stages 11–12 (**D**) and stages 13–14 (**E**). The arrow and arrowheads indicate three clusters of dying cells along the dorsal-ventral axis. Adapted with permission from ref. *(56)*.

normally act synergistically to promote cell death at the midline *(66)*. Also, ectopic expression of *reaper* or *hid* alone failed to induce ectopic cell death in midline cells, whereas co-expression of *reaper* and *hid* results in rapid midline cell death. *grim* can also act cooperatively with *reaper* or *hid* to induce a higher level of midline cell death than either gene alone.

Interestingly, *grim* presents some unique features. Ectopic expression of *grim* alone was sufficient to induce ectopic midline cell death *(69)*. Ectopic Diap2, which blocks both *reaper*- and *hid*-induced cell death, fails to block *grim*-induced cell death of midline cells *(69)*.

Recently, a connection between surface signaling and the downstream cell-death effectors has been identified. The majority of midline cells are kept alive through activation of the dEGFR pathway. Secretion of the dEGFR ligand Spitz from neurons promotes activation of downstream effector ERKA/Rolled, which in turn phosphorylates and inactivates Hid *(38)*. By calibrating this signaling, presumably, the correct number of cells can be retained. How this calibration is achieved, however, is not understood.

Embryonic Head Development

The emerging embryonic head undergoes axial induction during a process of "head involution," in which the head is "swallowed" inward at the mouth regions. Based on the expression of *reaper* as a marker for regions of death, normal head morphogenetic movements utilize widespread cell death during retraction of the clypeolabrum during early head involution, formation of the dorsal ridge, fusion and involution of dorsal structures of the pharynx and mouth cavity, and segregation of progenitor cells of the brain *(70)*. Although the early stage of head development, such as formation of dorsal ridge and pharynx, proceeds almost normally without apoptotic death, programmed cell death is required for later stages of head development. Embryos homozygous for Df(3L)H99, a deletion that uncovers the death-activator genes *reaper*, *hid*, and *grim,* fail to demonstrate normal morphogenesis associated with head involution including retraction of the clypeolabrum, formation of dorsal pouch, and fusion of gnathal lobes *(70)*. This suggests that shaping and reducing the size of the head are all important aspects of normal involution.

Another important role for programmed cell death is to correct misplaced cells. This can be seen in mutations that alter the distribution of cells in the head. Early expression of *wingless* is essential for the proper development of the anterior brain region *(71)*. During embryonic stages 9–10 when *wingless*-expressing protocephalic neuroblasts begin to delaminate, loss of *wingless* activity leads to incorrect cell-fate determination within *wingless*-expressing cells and their neighbors. In these mutants, increased cell death is observed at the anterior end of the embryonic protocerebrum: approx one-half of the protocerebrum is deleted by programmed cell death at late embryonic stages *(71)*. Again, removal of incorrectly specified cells is a recurring theme in development; how mispositioned cells are recognized and removed represents an enduring mystery in the cell-death field.

CELL DEATH DURING METAMORPHOSIS

Metamorphosis, the wonderful process that turns a larva into an adult fly, is regulated by changes in the titer of the steroid hormone ecdysone. A sharp rise in ecdysone at the end of larval development triggers puparium formation and the onset of prepupal development. A second major ecdysone pulse 10 h later initiates the prepupal-pupal transition. Pupal development then proceeds for most of the next 4 d until ecdysone titers return to basal levels and the adult fly emerges or "ecloses." During prepupal and early pupal development, the larval tissues such as the larval midgut and salivary glands undergo histolysis *(72)*. Programmed cell death also plays an integral part in the development of imaginal discs, islands of tissue that will give rise to most of the adult structures. Precisely patterned cell death has been reported in the mature larval and young pupal wing and eye imaginal discs. Between 50 and 80% of metamorphosis, virtually all of the midline glia in the CNS of *Drosophila* undergo programmed cell death *(73)*. Below we focus on cell death during metamorphosis of larval tissues and in the developing wing. The special case of the developing eye deserves its own section.

Midgut and Salivary Gland

During metamorphosis, the larval midgut and salivary gland undergo histolysis and are replaced by related adult tissue. Dramatic morphological changes in the midgut occur between 2–4 h after puparium formation (APF), marked by contraction of the midgut body, disappearance of the gastric caeca, and reduction of the proventriculus *(74)*. Acridine orange and TUNEL staining in these structures confirm widespread apoptosis *(74)*. High-level expression of the initiator caspase Dronc was also observed in midgut and salivary-gland cells from late third instar larvae, preceding apoptosis in these tissues *(50)*. In support of the essential role of caspases during destruction of these tissues, ectopic expression of the anti-apoptotic Baculovirus protein p35 inhibited cell death in these tissues *(74)*. The dying midgut cells are replaced by elongating adult midgut cells, and are eventually expelled as a "meconium" after eclosion. In contrast, programmed cell death in the salivary gland occurs at 10 h APF, correlating to the second ecdysone pulse. The cells in the gland detach from the basement membrane surrounding them and are degraded rapidly by 14.5 h APF.

Ecdysone signaling is a key trigger for programmed cell death in larval tissues. In cultured tissues, expression of *reaper* and *dronc* was upregulated while the caspase inhibitor *diap2* was downregulated after addition of ecdysone *(50,74)*. Transcription factors within the Broad-Complex (BR-C) and E74A, two early targets of ecdysone, can function together with the ecdysone receptor to induce expression of cell-death activators *reaper* and *hid*; the transcription factor E75B, also an early response target of ecdysone, can downregulate *diap2* by repressing the ß-FTZ-F1 orphan nuclear receptor *(75)*.

The difference in the timing of death between these two ecdysone-dependent events—midgut and salivary gland—may reside in their distinct expression of death activators. *reaper* and *hid* were detected in the prepupal midgut (0 h APF) but at 12 h APF in the salivary gland, consistent with the timing of cell death reported in these tissues. Also, the caspase inhibitor *diap2* is induced in the mid-prepupal salivary gland and repressed during later stages.

Cell Death in the Developing Wing

Adult wings develop from wing imaginal discs. These discs invaginate from the embryonic ectoderm as simple pouches of epithelium during embryogenesis and remain as such until metamorphosis, when they differentiate and fold into their final form. Based on Hoechst and TUNEL staining, a steady average of 1.4% of wing cells are apoptotic throughout the entire second and early third instar larval wing disc *(76)*. The location of these apoptotic cells is not random: dying cells localize preferentially at the wing/notum border in the late larval period; in the early pupal wing (0–12 h APF), the majority of dying cells are found at the wing hinge; at 12–20 h APF they appear in the pupal wing blade, and at 20–24 h APF dying cells are distributed along the periphery of the wing blade and at the wing margin. After 24 h APF, there is no detectable cell death in the pupal wing *(76)*.

Activation of the *Drosophila* c-Jun amino-terminal kinase (DJNK) pathway is sufficient to initiate cell death within the wing primordium in the late third instar larva: expressing a constitutively active form of the DJNK kinase Hemipterous led to massive cell death in both the proximal and distal wing primordia *(77)*. What regulates this evolving pattern of programmed cell death in the wing? One clear candidate is Dpp, which acts as a morphogen to establish the antero-posterior and proximo-distal axes. In the distal wing, reduced activity of Dpp activates DJNK-dependent cell death, an effect that is further enhanced by reduced Wingless activity *(77)*. Conversely, ectopic expression of Dpp can induce DJNK-dependent apoptosis in proximal wing cells. Thus, similar to the embryo, programmed cell death appears to correct mispatterning along the proximo-distal axis when normal signals are distorted.

CELL DEATH IN THE DEVELOPING EYE

The *Drosophila* compound eye is composed of approx 750 identical units or ommatidia. Each adult ommatidium contains eight photoreceptor neurons, four cone cells, and two primary pigment cells. Between ommatidia lies an interweaving hexagonal lattice composed of secondary/tertiary pigment cells and mechanosensory bristles (Fig. 4). In our humble opinion, the compound eye represents one of the most beautiful structures in nature.

It is also an excellent amplifier of mutation-induced defects: misplacement of just a few ommatidia can make the entire eye seem jumbled, similar to throwing a stone in a pond. For this reason, a number of mutations that affect cell fate and cell death have been discovered or studied in the eye. Also, the retina offers the opportunity to study cell death at the level of single cells and within the context of an emerging, patterning neuroepithelium.

The eye has been particularly popular as a tool for overexpression studies, in part because the fly can survive without its eye (at least in a vial). For example, ectopic expression of the death activators *reaper, hid,* and *grim* can induce extensive cell death in the eye, leading to a small or ablated eye *(68,78–80)*.

Cell Death in the Larval Eye

Drosophila retinal development occurs within a special retinal epithelium, the "eye disc." This retinal epithelium is specified by the combined efforts of a number of genes, including eyeless, twin of *eyeless, sine oculus, eyes absent,* and *dachsund.* Loss of, for example, *eyes absent* activity will result in loss of the eye progenitor cells by programmed cell death, leading to complete loss of the adult eye *(81)*.

After the imaginal tissue is specified as eye primordia, cells within the eye simply proliferate without commitment to a specific retinal cell fate until the final larval stage—the third instar—when retinal pattern formation begins in a wave of morphogenesis that sweeps across the eye disc from posterior to anterior *(82)*. At the front of this wave is the "morphogenetic furrow," the region in which cells prepare for differentiation by arresting in G_1 of the cell cycle. A low level of cell death can be observed anterior to and 10–12 rows posterior to the morphogenetic furrow *(83)*. Failure to

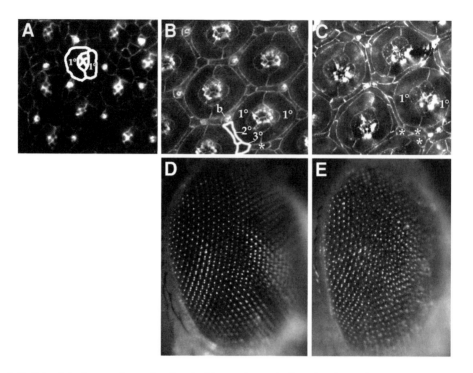

Fig. 4. Cell death in the pupal eye visualized with a probe to the junctional protein Armadillo. During early *Drosophila* eye development, an excess of lattice cells are generated (**A**, 22 h APF); one ommatidial core is outlined. Approximately one-third of these cells are then eliminated by programmed cell death in the pupa, yielding an exquisitely precise hexagonal array of ommatidia (**B**, 42 h APF) and a smoothly patterned adult eye (**D**). *IrreC-rst* mutations block lattice cell death (**C**, 42 h APF). Compare the region of three 2°-like cells (asterisks) to the single 2° found in the same position in panel B. The result is a "rough" adult eye, which contains disordered rows of ommatidial lenses (**E**). Labeled cell types in (B): c, cone cell; 1°, primary pigment cell; 2°, 3° lattice of secondary, and tertiary pigment cells; b, bristle.

propagate the morphogenetic furrow (e.g., in *hedgehog* mutants), leads to ectopic cell death (for a review *see* ref. *84*).

Abnormal regulation of the cell cycle can also lead to removal of cells by apoptosis. Within the morphogenetic furrow, the entry of cells into G_1 arrest requires several genes, including the *cyclin* genes, *string*, *dEGFR*, *dpp*, and *roughex*. Cells with loss-of-function *roughex* mutations fail to enter G_1 arrest, leading to errors in pattern formation and cell-fate determination. These cells—which attempt to become neurons while still dividing—undergo apoptosis *(85)*.

Within the morphogenetic furrow, a patterned array of proneural clusters is established and cell fate of the first neuron, photoreceptor R8, is determined. The cell number in each proneural cluster and spacing between the clusters are controlled by coordinated action of a number of genes, including *scabrous, Notch, atonal, dEGFR, spitz, argos,* and *Star*. These factors continue to play a prominent role in subsequent differentiation of the remaining photoreceptors (R1–R7) and the supporting glial-like cone and primary pigment cells (for a review *see* refs. *86, 87*). In most cases, mutation or ectopic expression of these genes will result in global mispatterning, leading to ectopic cell death (for a review *see* ref. *88*).

In a set of elegant studies, Baker and Yu demonstrated that dEGFR is required for cell-cycle progression from G_2 phase to M phase in cells between emerging ommatidia. Ommatidia produce the dEGFR ligand Spitz to ensure survival of most interommatidial cells; those cells furthest from ommatidia often fail to survive *(89)*. This is the first step in ensuring the correct number of interommatidial cells.

Cell Death in the Pupal Eye

Starting at 22 h APF, secondary and tertiary pigment cells (2°/3°s for simplicity) and the bristle complex (composed of four cells) begin to rearrange themselves with the goal of forming an invariant hexagonal lattice between the ommatidial array (Fig. 4). In the end, each ommatidium will share six secondaries, three tertiaries, and three bristles with its neighbors. This process involves extensive cell movement and, of course, extensive programmed cell death: about 2000 excess interommatidial precursor cells (IPCs) in the pupal retina will be removed *(83,90)*. Our laboratory has been working to understand the link between morphogenesis and cell death, and how these work together to create a proper hexagon.

Many of the cell-death regulatory factors that we discussed earlier are active in the pupal retina as well. Blocking caspase activity, either by ectopic expression of the caspase inhibitor P35 or by overexpression of DIAP1 or DIAP2, leads to ectopic 2°/3° cells within the lattice *(68,78–80,91)*. Furthermore, mutation of initiator caspase *dredd* suppresses *rpr*- and *grim*-induced cell death in the eye *(48)*.

Prior to cell death, IPCs switch their arrangement from lying in multiple layers ("side-by-side") between ommatidia to lying in single file ("end-to-end"; Fig. 4). Work with the *irregular chiasm C-roughest (irreC-rst)* locus indicates that this morphogenesis may represent an important first step in the cell-death process *(83,92)*. The *irreC-rst* locus encodes a single pass transmembrane protein of the immunoglobulin superfamily that preferentially accumulates at the border between primary pigment cells and IPCs in the wild-type pupal retina *(92,93)*. Mutations that reduce *irreC-rst* activity block cell death in the pupal eye, giving rise to doubled or tripled secondary pigment cells lying side-by-side (Fig. 4). By contrast, *echinus* mutations also result in a partial block of lattice cell death, but the additional cells are aligned end-to-end *(83)*. These subtle differences in the organization of IPCs can be used to determine which step each mutant acts at: for example, the failure of *irreC-rst* IPCs to rearrange indicates that IrreC-rst acts at an earlier stage than *echinus*.

Recent work is beginning to shed light on the cell extrinsic factors that mediate cell death across the pupal retina. Two transmembrane proteins have been implicated in providing a cell death signal, IrreC-rst and Notch. Notch, a type I transmembrane receptor, is expressed in the IPCs during the stage of pupal programmed cell death *(94)*. Reduction of *Notch* activity in the early pupal eye blocks programmed cell death in IPCs *(83,95)*. Interestingly, the subcellular localization of IrreC-rst protein is altered in *Notch* mutants *(92,96)*, although the link between these two interesting transmembrane proteins is not understood.

Cell-ablation studies revealed that primary pigment cells oppose this death signaling by providing a "life" signal to the IPCs *(97)*. A number of studies have implicated signaling through the *Drosophila* EGF receptor (dEGFR). The dEGFR is expressed primarily in the IPCs and its ligand is expressed in the neighboring cone cells and primary pigment cells *(97)*. Importantly, loss of *dEGFR* activity leads to a loss of lattice cells *(98,99)*, whereas ectopic activation of dEGFR partially blocked programmed cell death *(97,98)*. This "life" signal appears to be mediated through the dRas1 signal-transduction pathway, counteracting a Notch-imposed "death" signal *(97,100–102)*.

Together, these data suggest a model in which ommatidia signal IPCs through the dEGFR, blocking cell death in most cells; excess cells are then removed by Notch signaling. Unfortunately, this model cannot account for the spatial precision of the interommatidial lattice: how are just the correct number of cells removed in the correct positions?

Finally, a special word about the periphery of the retina. The outer rim of the retina contains a row of stunted ommatidia; these perimeter "ommatidia" probably exist to aid normal pathfinding and perhaps ensure straight ommatidial rows near the retina's edge. They are removed by apoptotic cell death between 60 and 70 h APF at 20°C' *(83)*. Surprisingly, this cell death was left intact in *irreC-rst* mutations, suggesting that the edges of the eye disc require other, separate cell-death signals.

CELL DEATH IN THE NEWLY EMERGING ADULT

At the end of metamorphosis (about 4 d after pupariation), pupae push through the pupal case (eclosion) and the adult fly emerges with unexpanded wings and long, thin, and relatively unpigmented bodies. Cell death is observed in muscles and neurons of the newly emerged adult. Many of the abdominal muscles used for eclosion and wing-spreading behavior die by 12 h after eclosion *(103)*. Most of the ptilinal muscles, a special set of muscles involved in retracting the ptilinum into the head capsule, die, and are absorbed within 24 h after eclosion *(104)*. Ligation experiments demonstrated that muscle degeneration is triggered by a signal from the anterior region that occurs about 1 h before eclosion.

Most neuronal death occurred in the dorsal and lateral regions of the abdominal and metathoracic neuromeres *(103)*. In contrast to muscular cell death, neural cell death is triggered after eclosion. The type II neurons, a group of approx 300 neurons that express 10-fold higher levels of the A isoform of the ecdysone receptor (EcR-A) than other central neurons, start to die by 4 h after adult emergence *(105)*. Dying neurons reach their maximal abundance at 6–8 h and decrease to none by 24 h after eclosion *(103)*.

Ecdysone also plays an essential role in regulating programmed cell death in the muscle breakdown and neuronal death that occurs in the emerging adult *(103,105,106)*. A decline in the levels of Ecdysone at the end of metamorphosis is required for death of ventral CNS type II neurons; these neurons in turn express high levels of EcR-A *(105)*. Treatment of flies with 20-hydroxyecdysone 3 h before the onset of neuronal degeneration prevents the death of these neurons. Also, *reaper*, *hid*, and *grim* are expressed in all dying n4 cells, a subset of type II neurons *(106)*. Interestingly, hormone injection or decapitation of newly emerged adults demonstrated that *reaper* transcript accumulation is regulated by both steroid hormone titer and a signal from the head of the emerging adult *(106)*. Currently, little is known about the genetic control of programmed cell death in muscle and neuron after adult emergence. Mutations in two different loci delay cell death in the ptilinal head muscle *(104)* but the genes involved are still not clear.

DEGENERATION IN ADULTS

A chapter on cell death in *Drosophila* would not be complete without at least a few words about degeneration of adult structures. Adult fly cells, similar to our own, retain the capacity to undergo apoptosis. Many cells, particularly in the nervous system, require continual trophic support for their survival. This has been perhaps best demonstrated with the mutual requirement for connections between photoreceptor neurons in the eye and their target neurons in the optic lobes of the brain. Failure to establish and maintain proper connections leads to loss of the optic lobes (e.g., ref. *107*) and, conversely, eventual degeneration of the photoreceptor neurons themselves *(108–110)*. Recently, a beautiful set of experiments has begun to define mutations that lead to degeneration of the nervous system, particularly the brain. Seymour Benzer and colleagues have initiated studies that have identified a number of interesting loci with names such as *spongecake*, *eggroll*, *drop dead*, and *bubblegum* *(111–113)*.

The mechanisms that regulate cell death is an increasingly hot topic in studies of phototransduction mutants; these studies have particular relevance to a number of human diseases leading to blindness such as *retinitis pigmentosa*. Mutations in several components of the rhodopsin signaling pathway can lead to apoptotic cell death; these mutations mirror mutations in human genes associated with blindness to a remarkable degree (reviewed in ref. *114, 115*). Significantly, the degeneration observed in these mutations can be rescued by expressing the caspase inhibitor P35 in the eye; vision appears to be fully restored *(116,117)*. The links between mutations that alter light-mediated signaling and the triggering of the apoptotic machinery remain mysterious.

Although *Drosophila* is often thought of as a great model for studying development, these mutations herald a growing interest in the cell death field: the use of flies to study issues of adult human disease. The ability to target genes—including disease forms of genes—to specific fly tissues allows us to take advantage of the myriad of fly tools to better dissect the underlying basis of a pathogenesis. These include Parkinson's disease *(118)* and polyglutamine repeat-based diseases such as Spinocerebellar ataxia type 1 and type 3 and Huntington's disease *(119–121)*. Once established, these models can be subjected to genetic screening, and an increasing wealth of new factors are being discovered that participate in the cell-death process (reviewed in ref. *122*).

SUMMARY

Drosophila has firmly placed itself as an important contributor to the cell-death field. The ability to use genetics to identify functional components of the cell-death machinery has made *Drosophila* an important tool both for identifying new death factors and as a testing site to determine if interactions observed in vitro or in tissue-culture cells can be identified in normal developing tissues. Recently, cell-death work in *Drosophila* has been extended to include a more direct assault on specific questions of human disease. These studies represent the next logical step in utilizing *Drosophila*'s remarkable features.

REFERENCES

1. Metzstein, M. M., Stanfield, G. M., and Horvitz, H. R. (1998) Genetics of programmed cell death in C. elegans: past, present and future. *Trends Genet.* **14,** 410–416.
2. Coucouvanis, E. and Martin, G. R. (1995) Signals for death and survival: a two-step mechanism for cavitation in the vertebrate embryo. *Cell* **83,** 279–287.
3. Saunders, J. W., Jr. and Fallon, J. F. (1966) Cell death in morphogenesis, in *Major Problems of Developmental Biology* (Locke, M., ed.), New York: Academic Press, pp. 289–314.
4. Mori, C., Nakamura, N., Kimura, S., Irie, H., Takigawa, T., and Shiota, K. (1995) Programmed cell death in the interdigital tissue of the fetal mouse limb is apoptosis with DNA fragmentation. *Anat. Rec.* **242,** 103–110.
5. Conlon, I. and Raff, M. (1999) Size control in animal development. *Cell* **96,** 235–244.
6. Graham, A., Heyman, I., and Lumsden, A. (1993) Even-numbered rhombomeres control the apoptotic elimination of neural crest cells from odd-numbered rhombomeres in the chick hindbrain. *Development* **119,** 233–245.
7. Jacobson, M. D., Weil, M., and Raff, M. C. (1997) Programmed cell death in animal development. *Cell* **88,** 347–354.
8. Reed, J. C. and Kroemer, G. (2000) Mechanisms of mitochondrial membrane permeabilization. *Cell Death Differ.* **7,** 1145.
9. Song, Z. and Steller, H. (1999) Death by design: mechanism and control of apoptosis. *Trends Cell Biol.* **9,** M49–M52.
10. Abrams, J. M. (1999) An emerging blueprint for apoptosis in *Drosophila*. *Trends Cell Biol.* **9,** 435–440.
11. Bangs, P. and White, K. (2000) Regulation and execution of apoptosis during *Drosophila* development. *Dev. Dyn.* **218,** 68–79.
12. Vernooy, S. Y., Copeland, J., Ghaboosi, N., Griffin, E. E., Yoo, S. J., and Hay, B. A. (2000) Cell death regulation in *Drosophila*: conservation of mechanism and unique insights. *J. Cell Biol.* **150,** F69–F76.
13. Adams, J. M. and Cory, S. (1998) The Bcl-2 protein family: arbiters of cell survival. *Science* **281,** 1322–1326.
14. Stennicke, H. R., Ryan, C. A., and Salvesen, G. S. (2002) Reprieval from execution: the molecular basis of caspase inhibition. *Trends Biochem. Sci.* **27,** 94–101.
15. Wang, S. L., Hawkins, C. J., Yoo, S. J., Muller, H. A., and Hay, B. A. (1999) The *Drosophila* caspase inhibitor DIAP1 is essential for cell survival and is negatively regulated by HID. *Cell* **98,** 453–463.
16. Goyal, L., McCall, K., Agapite, J., Hartwieg, E., and Steller, H. (2000) Induction of apoptosis by *Drosophila* reaper, hid and grim through inhibition of IAP function. *EMBO J.* **19,** 589–697.
17. White, K., Grether, M. E., Abrams, J. M., Young, L., Farrell, K., and Steller, H. (1994) Genetic control of programmed cell death in *Drosophila*. *Science* **264,** 677–683.
18. Du, C., Fang, M., Li, Y., Li, L., and Wang, X. (2000) Smac, a mitochondrial protein that promotes cytochrome c-dependent caspase activation by eliminating IAP inhibition. *Cell* **102,** 33–42.
19. Verhagen, A. M., Ekert, P. G., Pakusch, M., Silke, J., Connolly, L. M., Reid, G. E., et al. (2000) Identification of DIABLO, a mammalian protein that promotes apoptosis by binding to and antagonizing IAP proteins. *Cell* **102,** 43–53.
20. Verhagen, A. M., Silke, J., Ekert, P. G., Pakusch, M., Kaufmann, H., Connolly, L. M., et al. (2002) HtrA2 promotes cell

death through its serine protease activity and its ability to antagonize inhibitor of apoptosis proteins. *J. Biol. Chem.* **277**, 445–454.

21. Hays, R., Wickline, L., and Cagan, R. L. (2002) Morgue mediates apoptosis in the *Drosophila* retina by promoting degradation of DIAP1. *Nat. Cell Biol.* **4**, 425–431.

22. Wing, J. P., Schreader, B. A., Yokokura, T., Wang, Y., Andrews, P. S., Huseinovic, N., et al. (2002) *Drosophila* Morgue is a novel F box/ubiquitin conjugase domain protein important for grim-reaper mediated apoptosis. *Nat. Cell Biol.* **4**, 451–456.

23. Cryns, V. and Yuan, J. (1998) Proteases to die for. *Genes Dev.* **12**, 1551–1570.

24. Newton, K. and Strasser, A. (1998) The Bcl-2 family and cell death regulation. *Curr. Opin. Genet. Dev.* **8**, 68–75.

25. Rodriguez, A., Oliver, H., Zou, H., Chen, P., Wang, X., and Abrams, J. M. (1999) Dark is a *Drosophila* homologue of Apaf-1/CED-4 and functions in an evolutionarily conserved death pathway. *Nat. Cell Biol.* **1**, 272–279.

26. Kanuka, H., Sawamoto, K., Inohara, N., Matsuno, K., Okano, H., and Miura, M. (1999) Control of the cell death pathway by Dapaf-1, a *Drosophila* Apaf-1/CED-4-related caspase activator. *Mol. Cell* **4**, 757–769.

27. Zhou, L., Song, Z., Tittel, J., and Steller, H. (1999) HAC-1, a *Drosophila* homolog of APAF-1 and CED-4 functions in developmental and radiation-induced apoptosis. *Mol. Cell* **4**, 745–755.

28. Oltvai, Z. N. and Korsmeyer, S. J. (1994) Checkpoints of dueling dimers foil death wishes. *Cell* **79**, 189–192.

29. Brachmann, C. B., Jassim, O. W., Wachsmuth, B. D., and Cagan, R. L. (2000) The *Drosophila* bcl-2 family member dBorg-1 functions in the apoptotic response to UV-irradiation. *Curr. Biol.* **10**, 547–550.

30. Igaki, T., Kanuka, H., Inohara, N., Sawamoto, K., Nunez, G., Okano, H., and Miura, M. (2000) Drob-1, a *Drosophila* member of the Bcl-2/CED-9 family that promotes cell death. *Proc. Natl. Acad. Sci. USA* **97**, 662–667.

31. Colussi, P. A., Quinn, L. M., Huang, D. C., Coombe, M., Read, S. H., Richardson, H., and Kumar, S. (2000) Debcl, a proapoptotic Bcl-2 homologue, is a component of the *Drosophila* melanogaster cell death machinery. *J. Cell Biol.* **148**, 703–714.

32. Ashkenazi, A. and Dixit, V. M. (1998) Death receptors: signaling and modulation. *Science* **281**, 1305–1308.

33. Serrano, M., Lin, A. W., McCurrach, M. E., Beach, D., and Lowe, S. W. (1997) Oncogenic ras provokes premature cell senescence associated with accumulation of p53 and p16INK4a. *Cell* **88**, 593–602.

34. Latinis, K. M., Carr, L. L., Peterson, E. J., Norian, L. A., Eliason, S. L., and Koretzky, G. A. (1997) Regulation of CD95 (Fas) ligand expression by TCR-mediated signaling events. *J. Immunol.* **158**, 4602–4611.

35. Xia, Z., Dickens, M., Raingeaud, J., Davis, R. J., and Greenberg, M. E. (1995) Opposing effects of ERK and JNK-p38 MAP kinases on apoptosis. *Science* **270**, 1326–1331.

36. Karim, F. D. and Rubin, G. M. (1998) Ectopic expression of activated Ras1 induces hyperplastic growth and increased cell death in *Drosophila* imaginal tissues. *Development* **125**, 1–9.

37. Khwaja, A., Rodriguez-Viciana, P., Wennstrom, S., Warne, P. H., and Downward, J. (1997) Matrix adhesion and Ras transformation both activate a phosphoinositide 3-OH kinase and protein kinase B/Akt cellular survival pathway. *EMBO J.* **16**, 2783–2793.

38. Bergmann, A., Tugentman, M., Shilo, B. Z., and Steller, H. (2002) Regulation of cell number by MAPK-dependent control of apoptosis: a mechanism for trophic survival signaling. *Dev. Cell* **2**, 159–170.

39. Bergmann, A., Agapite, J., McCall, K., and Steller, H. (1998) The *Drosophila* gene hid is a direct molecular target of Ras-dependent survival signaling. *Cell* **95**, 331–341.

40. Kurada, P. and White, K. (1998) Ras promotes cell survival in *Drosophila* by downregulating hid expression. *Cell* **95**, 319–329.

41. Miyashita, T. and Reed, J. C. (1995) Tumor suppressor p53 is a direct transcriptional activator of the human bax gene. *Cell* **80**, 293–299.

42. Ollmann, M., Young, L. M., Di Como, C. J., Karim, F., Belvin, M., Robertson, S., et al. (2000) *Drosophila* p53 is a structural and functional homolog of the tumor suppressor p53. *Cell* **101**, 91–101.

43. Jin, S., Martinek, S., Joo, W. S., Wortman, J. R., Mirkovic, N., Sali, A., et al. (2000) Identification and characterization of a p53 homologue in *Drosophila* melanogaster. *Proc. Natl. Acad. Sci. USA* **97**, 7301–7306.

44. Brodsky, M. H., Nordstrom, W., Tsang, G., Kwan, E., Rubin, G. M., and Abrams, J. M. (2000) *Drosophila* p53 binds a damage response element at the reaper locus. *Cell* **101**, 103–113.

45. King, R. C. (1970) *Ovarian Development in* Drosophila Melanogaster. New York: Academic Press.

46. Giorgi, F. and Deri, P. (1976) Cell death in ovarian chambers of *Drosophila* melanogaster. *J. Embryol. Exp. Morphol.* **35**, 521–533.

47. Foley, K. and Cooley, L. (1998) Apoptosis in late stage *Drosophila* nurse cells does not require genes within the H99 deficiency. *Development* **125**, 1075–1082.

48. Chen, P., Rodriguez, A., Erskine, R., Thach, T., and Abrams, J. M. (1998) Dredd, a novel effector of the apoptosis activators reaper, grim, and hid in *Drosophila*. *Dev. Biol.* **201**, 202–216.

49. McCall, K. and Steller, H. (1998) Requirement for DCP-1 caspase during *Drosophila* oogenesis. *Science* **279**, 230–234.

50. Dorstyn, L., Colussi, P. A., Quinn, L. M., Richardson, H., and Kumar, S. (1999) DRONC, an ecdysone-inducible *Drosophila* caspase. *Proc. Natl. Acad. Sci. USA* **96**, 4307–4312.

51. Dorstyn, L., Read, S. H., Quinn, L. M., Richardson, H., and Kumar, S. (1999) DECAY, a novel *Drosophila* caspase related to mammalian caspase-3 and caspase-7. *J. Biol. Chem.* **274**, 30778–30783.

52. Mahajan-Miklos, S. and Cooley, L. (1994) Intercellular cytoplasm transport during *Drosophila* oogenesis. *Dev. Biol.* **165**, 336–351.

53. Twombly, V., Blackman, R. K., Jin, H., Graff, J. M., Padgett, R. W., and Gelbart, W. M. (1996) The TGF-beta signaling pathway is essential for *Drosophila* oogenesis. *Development* **122**, 1555–1565.

54. Soller, M., Bownes, M., and Kubli, E. (1999) Control of oocyte maturation in sexually mature *Drosophila* females. *Dev. Biol.* **208,** 337–351.
55. DiNardo, S., Heemskerk, J., Dougan, S., and O'Farrell, P. H. (1994) The making of a maggot: patterning the *Drosophila* embryonic epidermis. *Curr. Opin. Genet. Dev.* **4,** 529–534.
56. Pazdera, T. M., Janardhan, P., and Minden, J. S. (1998) Patterned epidermal cell death in wild-type and segment polarity mutant *Drosophila* embryos. *Development* **125,** 3427–3436.
57. Tepass, U., Fessler, L. I., Aziz, A., and Hartenstein, V. (1994) Embryonic origin of hemocytes and their relationship to cell death in *Drosophila*. *Development* **120,** 1829–1837.
58. Jürgens, G., Wieschaus, E., Nüsslein-Volhard, C., and Kluding, H. (1984) Mutations affecting the pattern of the larval cuticle in *Drosophila* melanogaster. II. Zygotic loci on the third chromosome. *Roux Arch. Dev. Biol.* **193,** 283–295.
59. Nüsslein-Volhard, C., Wieschaus, E., and Kluding, H. (1984) Mutations affecting the pattern of the larval cuticle in *Drosophila* melanogaster. I. Zygotic loci on the second chromosome. *Roux Arch. Dev. Biol.* **193,** 267–282.
60. Wieschaus, E., Nüsslein-Volhard, C., and Jürgens, G. (1984) Mutations affecting the pattern of the larval cuticle in *Drosophila* melanogaster. III. Zygotic loci on the X chromosome and the fourth chromosome. *Roux Arch. Dev. Biol.* **193,** 296–307.
61. Bejsovec, A. and Wieschaus, E. (1993) Segment polarity gene interactions modulate epidermal patterning in *Drosophila* embryos. *Development* **119,** 501–517.
62. Perrimon, N. and Mahowald, A. P. (1987) Multiple functions of segment polarity genes in *Drosophila*. *Dev. Biol.* **119,** 587–600.
63. Klingensmith, J., Noll, E., and Perrimon, N. (1989) The segment polarity phenotype of *Drosophila* involves differential tendencies toward transformation and cell death. *Dev. Biol.* **134,** 130–145.
64. Namba, R., Pazdera, T. M., Cerrone, R. L., and Minden, J. S. (1997) *Drosophila* embryonic pattern repair: how embryos respond to bicoid dosage alteration. *Development* **124,** 1393–1403.
65. Sonnenfeld, M. J. and Jacobs, J. R. (1995) Apoptosis of the midline glia during *Drosophila* embryogenesis: a correlation with axon contact. *Development* **121,** 569–578.
66. Zhou, L., Schnitzler, A., Agapite, J., Schwartz, L. M., Steller, H., and Nambu, J. R. (1997) Cooperative functions of the reaper and head involution defective genes in the programmed cell death of *Drosophila* central nervous system midline cells. *Proc. Natl. Acad. Sci. USA* **94,** 5131–5136.
67. Zhou, L., Hashimi, H., Schwartz, L. M., and Nambu, J. R. (1995) Programmed cell death in the *Drosophila* central nervous system midline. *Curr. Biol.* **5,** 784–790.
68. Chen, P., Nordstrom, W., Gish, B., and Abrams, J. M. (1996) grim, a novel cell death gene in *Drosophila*. *Genes Dev.* **10,** 1773–1782.
69. Wing, J. P., Zhou, L., Schwartz, L. M., and Nambu, J. R. (1998) Distinct cell killing properties of the *Drosophila* reaper, head involution defective, and grim genes. *Cell Death Differ.* **5,** 930–939.
70. Nassif, C., Daniel, A., Lengyel, J. A., and Hartenstein, V. (1998) The role of morphogenetic cell death during *Drosophila* embryonic head development. *Dev. Biol.* **197,** 170–286.
71. Richter, S., Hartmann, B., and Reichert, H. (1998) The wingless gene is required for embryonic brain development in *Drosophila*. *Dev. Genes Evol.* **208,** 37–45.
72. Bodenstein, D. (1965) The postembryonic development of *Drosophila*, in *Biology of Drosophila* (Demerec, M., ed.), New York: Hafner Publishing Co, pp. 275–367.
73. Awad, T. A. and Truman, J. W. (1997) Postembryonic development of the midline glia in the CNS of *Drosophila*: proliferation, programmed cell death, and endocrine regulation. *Dev. Biol.* **187,** 283–297.
74. Jiang, C., Baehrecke, E. H., and Thummel, C. S. (1997) Steroid regulated programmed cell death during *Drosophila* metamorphosis. *Development* **124,** 4673–4683.
75. Jiang, C., Lamblin, A. F., Steller, H., and Thummel, C. S. (2000) A steroid-triggered transcriptional hierarchy controls salivary gland cell death during *Drosophila* metamorphosis. *Mol. Cell* **5,** 445–455.
76. Milan, M., Campuzano, S., and Garcia-Bellido, A. (1997) Developmental parameters of cell death in the wing disc of *Drosophila*. *Proc. Natl. Acad. Sci. USA* **94,** 5691–5696.
77. Adachi-Yamada, T., Fujimura-Kamada, K., Nishida, Y., and Matsumoto, K. (1999) Distortion of proximodistal information causes JNK-dependent apoptosis in *Drosophila* wing. *Nature* **400,** 166–169.
78. Grether, M. E., Abrams, J. M., Agapite, J., White, K., and Steller, H. (1995) The head involution defective gene of *Drosophila* melanogaster functions in programmed cell death. *Genes Dev.* **9,** 1694–1708.
79. Hay, B. A., Wassarman, D. A., and Rubin, G. M. (1995) *Drosophila* homologs of baculovirus inhibitor of apoptosis proteins function to block cell death. *Cell* **83,** 1253–1262.
80. White, K., Tahaoglu, E., and Steller, H. (1996) Cell killing by the *Drosophila* gene reaper. *Science* **271,** 805–807.
81. Bonini, N. M., Leiserson, W. M., and Benzer, S. (1993) The eyes absent gene: genetic control of cell survival and differentiation in the developing *Drosophila* eye. *Cell* **72,** 379–395.
82. Ready, D. F., Hanson, T. E., and Benzer, S. (1976) Development of the *Drosophila* retina, a neurocrystalline lattice. *Dev. Biol.* **53,** 217–240.
83. Wolff, T. and Ready, D. F. (1991) Cell death in normal and rough eye mutants of *Drosophila*. *Development* **113,** 825–839.
84. Heberlein, U. and Moses, K. (1995) Mechanisms of *Drosophila* retinal morphogenesis: the virtues of being progressive. *Cell* **81,** 987–990.
85. Thomas, B. J., Gunning, D. A., Cho, J., and Zipursky, L. (1994) Cell cycle progression in the developing *Drosophila* eye: roughex encodes a novel protein required for the establishment of G1. *Cell* **77,** 1003–1014.

86. Zipursky, S. L. and Rubin, G. M. (1994) Determination of neuronal cell fate: lessons from the R7 neuron of *Drosophila. Annu. Rev. Neurosci.* **17,** 373–397.

87. Rusconi, J. C., Hays, R., and Cagan, R. L. (2000) Programmed cell death and patterning in *Drosophila. Cell Death Differ.* **7,** 1063–1070.

88. Bonini, N. M. and Fortini, M. E. (1999) Surviving *Drosophila* eye development: integrating cell death with differentiation during formation of a neural structure. *Bioessays* **21,** 991–1003.

89. Baker, N. E. and Yu, S. Y. (2001) The EGF receptor defines domains of cell cycle progression and survival to regulate cell number in the developing *Drosophila* eye. *Cell* **104,** 699–708.

90. Cagan, R. L. and Ready, D. F. (1989) The emergence of order in the *Drosophila* pupal retina. *Dev. Biol.* **136,** 346–362.

91. Hay, B. A., Wolff, T., and Rubin, G. M. (1994) Expression of baculovirus P35 prevents cell death in *Drosophila. Development* **120,** 2121–2129.

92. Reiter, C., Schimansky, T., Nie, Z., and Fischbach, K. F. (1996) Reorganization of membrane contacts prior to apoptosis in the *Drosophila* retina: the role of the IrreC-rst protein. *Development* **122,** 1931–1940.

93. Ramos, R. G., Igloi, G. L., Lichte, B., Baumann, U., Maier, D., Schneider, T., et al. (1993) The irregular chiasm C-roughest locus of *Drosophila,* which affects axonal projections and programmed cell death, encodes a novel immunoglobulin-like protein. *Genes Dev.* **7,** 2533–2547.

94. Kooh, P. J., Fehon, R. G., and Muskavitch, M. A. (1993) Implications of dynamic patterns of Delta and Notch expression for cellular interactions during *Drosophila* development. *Development* **117,** 493–507.

95. Cagan, R. L. and Ready, D. F. (1989) Notch is required for successive cell decisions in the developing *Drosophila* retina. *Genes Dev.* **3,** 1099–1112.

96. Gorski, S. M., Brachmann, C. B., Tanenbaum, S. B., and Cagan, R. L. (2000) Delta and notch promote correct localization of irreC-rst. *Cell Death Differ.* **7,** 1011–1013.

97. Miller, D. T. and Cagan, R. L. (1998) Local induction of patterning and programmed cell death in the developing *Drosophila* retina. *Development* **125,** 2327–2335.

98. Freeman, M. (1996) Reiterative use of the EGF receptor triggers differentiation of all cell types in the *Drosophila* eye. *Cell* **87,** 651–660.

99. Sawamoto, K., Okano, H., Kobayakawa, Y., Hayashi, S., Mikoshiba, K., and Tanimura, T. (1994) The function of argos in regulating cell fate decisions during *Drosophila* eye and wing vein development. *Dev. Biol.* **164,** 267–276.

100. Baker, N. E. and Rubin, G. M. (1992) Ellipse mutations in the *Drosophila* homologue of the EGF receptor affect pattern formation, cell division, and cell death in eye imaginal discs. *Dev. Biol.* **150,** 381–396.

101. Miyamoto, H., Nihonmatsu, I., Kondo, S., Ueda, R., Togashi, S., Hirata, S., et al. (1995) canoe encodes a novel protein containing a GLGF/DHR motif and functions with Notch and scabrous in common developmental pathways in *Drosophila. Genes Dev.* **9,** 612–625.

102. Matsuo, T., Takahashi, K., Kondo, S., Kaibuchi, K., and Yamamoto, D. (1997) Regulation of cone cell formation by Canoe and Ras in the developing *Drosophila* eye. *Development* **124,** 2671–2680.

103. Kimura, K. I. and Truman, J. W. (1990) Postmetamorphic cell death in the nervous and muscular systems of *Drosophila* melanogaster. *J. Neurosci.* **10,** 403–401.

104. Kimura, K. and Tanimura, T. (1992) Mutants with delayed cell death of the ptilinal head muscles in *Drosophila. J. Neurogenet.* **8,** 57–69.

105. Robinow, S., Talbot, W. S., Hogness, D. S., and Truman, J. W. (1993) Programmed cell death in the *Drosophila* CNS is ecdysone-regulated and coupled with a specific ecdysone receptor isoform. *Development* **119,** 1251–1259.

106. Robinow, S., Draizen, T. A., and Truman, J. W. (1997) Genes that induce apoptosis: transcriptional regulation in identified, doomed neurons of the *Drosophila* CNS. *Dev. Biol.* **190,** 206–213.

107. Xiong, W. C. and Montell, C. (1995) Defective glia induce neuronal apoptosis in the repo visual system of *Drosophila. Neuron* **14,** 581–590.

108. Steller, H., Fischbach, K. F., and Rubin, G. M. (1987) Disconnected: a locus required for neuronal pathway formation in the visual system of *Drosophila. Cell* **50,** 1139–1153.

109. Campos, A. R., Fischbach, K. F., and Steller, H. (1992) Survival of photoreceptor neurons in the compound eye of *Drosophila* depends on connections with the optic ganglia. *Development* **114,** 355–366.

110. Campbell, G., Goring, H., Lin, T., Spana, E., Andersson, S., Doe, C. Q., and Tomlinson, A. (1994) RK2, a glial-specific homeodomain protein required for embryonic nerve cord condensation and viability in *Drosophila. Development* **120,** 2957–2966.

111. Min, K. T. and Benzer, S. (1997) Spongecake and eggroll: two hereditary diseases in *Drosophila* resemble patterns of human brain degeneration. *Curr. Biol.* **7,** 885–888.

112. Kretzschmar, D., Hasan, G., Sharma, S., Heisenberg, M., and Benzer, S. (1997) The swiss cheese mutant causes glial hyperwrapping and brain degeneration in *Drosophila. J. Neurosci.* **17,** 7425–7432.

113. Buchanan, R. L. and Benzer, S. (1993) Defective glia in the *Drosophila* brain degeneration mutant drop-dead. *Neuron* **10,** 839–850.

114. Zuker, C. S. (1996) The biology of vision of *Drosophila. Proc. Natl. Acad. Sci. USA* **93,** 571–576.

115. Hays, R., Craig, C., and Cagan, R. (2002) Programmed death in eye development, in *Drosophila Eye Development,* vol. 37 (Moses, K., ed.). Berlin: Springer-Verlag, pp. 169–189.

116. Davidson, F. F. and Steller, H. (1998) Blocking apoptosis prevents blindness in *Drosophila* retinal degeneration mutants. *Nature* **391,** 587–591.

117. Alloway, P. G., Howard, L., and Dolph, P. J. (2000) The formation of stable rhodopsin-arrestin complexes induces apoptosis and photoreceptor cell degeneration. *Neuron* **28,** 129–138.

118. Feany, M. B. and Bender, W. W. (2000) A *Drosophila* model of Parkinson's disease. *Nature* **404,** 394–398.
119. Jackson, G. R., Salecker, I., Dong, X., Yao, X., Arnheim, N., Faber, P. W., et al. (1998) Polyglutamine-expanded human huntingtin transgenes induce degeneration of *Drosophila* photoreceptor neurons. *Neuron* **21,** 633–642.
120. Warrick, J. M., Paulson, H. L., Gray-Board, G. L., Bui, Q. T., Fischbeck, K. H., Pittman, R. N., and Bonini, N. M. (1998) Expanded polyglutamine protein forms nuclear inclusions and causes neural degeneration in *Drosophila. Cell* **93,** 939–949.
121. Fernandez-Funez, P., Nino-Rosales, M. L., de Gouyon, B., She, W. C., Luchak, J. M., Martinez, P., et al. (2000) Identification of genes that modify ataxin-1-induced neurodegeneration. *Nature* **408,** 101–106.
122. Bonini, N. M. and Fortini, M. E. (2002) Applications of the *Drosophila* retina to human disease modeling, in *Drosophila Eye Development*, vol. 37 (Moses, K., ed.), Berlin: Springer, pp. 257–275.
123. Spradling, A. (1993) Developmental genetics of oogenesis, in *The Development of Drosophila melanogaster*, vol. 1 (Bate, M. and Martinez-Arias, A., eds.), Cold Spring Harbor, New York: Laboratory Press, pp. 1–70.

11
Cell Death in Mammalian Development

Chia-Yi Kuan and Keisuke Kuida

INTRODUCTION

Physiological mechanisms of cell death are required throughout the development and adulthood of all multicellular organisms *(1,2)*. From primitive multicellular organisms to higher vertebrates, well-orchestrated cell-death events are critical for the removal of superfluous cells as a vital part of tissue sculpting during development. Physiological cell death enables the elimination of unwanted extra cells to maintain cellular homeostasis in developed adults, as well as the elimination of organisms of harmful cells or cells with serious cellular or genomic damage. Apoptosis—physiological cell death—is characterized by a distinct set of morphological and biochemical features including chromatin condensation, internucleosomal DNA fragmentation, and perhaps most important, cell-surface alterations, which signal for the rapid recognition and engulfment of apoptotic cells by neighboring phagocytic cells, thus avoiding the induction of any pathological reactions *(3)*.

The actual term for physiological cell death, apoptosis, was introduced in the early 1970s by Kerr and colleagues to define a type of cell death distinct from necrosis, based on unique morphological characteristics *(4)*. Kerr, Wyllie, and Currie observed that liver cells underwent two different types of cell death that can be defined morphologically after ligation of the portal vein. While dying hepatocytes in areas immediately surrounding the vessel exhibited classic necrotic features including cellular swelling, mitochondria damage, and cytoplasmic-membrane rupturing, liver cells in the peripheral areas underwent slow ischemic death; instead of swelling up, those dying cells actually shrunk with blebbed, yet intact, cytoplasmic membranes *(4)*. Kerr et al. delineated the two distinct death mechanisms in a single cell type and named the latter type of cell death "apoptosis." Later, Wyllie demonstrated that apoptosis could be induced experimentally in isolated cells in vitro. By exposing thymocytes to glucocorticoid, he also discovered that cellular DNA was being fragmented to generate a ladder of DNA bands during apoptotic death, indicating the activation of an endogenous endonuclease *(5)*. Based on their study and other previous observations, Wyllie and his colleagues hypothesized that apoptosis is the common form of cell death under physiological conditions, such as naturally occurring deaths during development and homeostasis, whereas necrosis occurs only in response to pathological conditions such as injury. According to their model, the morphological and biochemical features of apoptosis, such as cell shrinkage, degradation of genetic material, and removal by phagocytosis, make it more congruous with the subtle removal of cells for remodeling of tissues during development than the far more violent necrosis *(3)*.

From: *Essentials of Apoptosis: A Guide for Basic and Clinical Research*
Edited by: X-M. Yin and Z. Dong © Humana Press Inc., Totowa, NJ

The importance of apoptosis is underscored by the finding that the process is genetically controlled and conserved in evolution. Using the genetic models of *Caenorhabditis elegans*, Horwitz and colleagues have made seminal contributions to our understanding of the evolutionarily well-conserved genetic and biochemical apoptotic pathways underlying physiological cell death *(6)*. In light of this, in this review we delineate mechanisms of apoptosis in the vertebrate revealed by various murine transgenic model systems.

APOPTOTIC PATHWAYS IN MAMMALS

The breakthrough in understanding cell death mechanisms came from genetic studies in *C. elegans* by Horwitz and his colleagues over the last decade *(6)*. The clearly defined and highly consistent lineage commitment during *C. elegans* development made it possible to trace the fate of each single cell and it was found that an appreciable number of cells are eliminated shortly after their generation. Importantly, these deaths occur in a highly reproducible manner in that the same cells die at the same exact time for every animal, indicating the existence of a precisely controlled death mechanism. Therefore, the phenomenon is often referred to as programmed cell death (PCD). Direct microscopic observations of these transparent organisms revealed that these dying cells all exhibited the diagnostic characteristics of apoptosis, providing indisputable evidence that apoptosis is indeed the preferred form of physiological cell death during development of multicellular organisms *(7)*.

Further genetic screening and molecular analysis has revealed more than a dozen genes whose mutation affects PCD in *C. elegans*. In particular, three genes, *ced-3*, *ced-4*, and *ced-9*, have been found to be absolutely required. Loss of function mutations in either *ced-3* or *ced-4* genes allow the survival of those cells that are destined to die *(8)*. In contrast, *ced-9* has been identified as the only gene for which activity is necessary and sufficient to inhibit cell death *(9)*. Inactivation of *ced-9* causes the death of cells that normally survive, and overexpression of wild-type *ced-9* results in the rescue of cells that would otherwise die. It is important to note that mutations in *ced-3* and *ced-4* can completely reverse the ectopic death caused by the lack of *ced-9* activity, indicating an epistatic pathway *(9)*.

The evolutionary conservation of the death pathway between nematodes and mammals became evident when homology was found between *ced-9* and *bcl-2*, a gene known to block apoptosis in mammalian cells *(10)*. *ced-3* was then found to encode a cysteine protease homologous to interleukin-1α (IL-1α) converting enzyme (ICE) *(11)*. With greater complexity expected in mammals, other mammalian homologs of *ced-3* and *ced-9* were subsequently discovered to form the caspase (cysteinyl aspartate-specific proteinases) family and the Bcl-2 protein family, respectively *(12)*. More recently, Apaf-1 (apoptotic protease-activating factor-1), a candidate mammalian homolog of *ced-4* was identified biochemically by its ability to activate one of the mammalian caspases, caspase-9, thus filling a gap between Bcl-2 function and caspase activation *(13)*. To date, a variety of cell death and survival signals have been shown eventually to activate or deactivate common apoptotic machinery composed of several adaptors represented by Apaf-1, caspases, and Bcl-2 family members (*see* Table 1 on pp. 171-172).

Caspases and Their Activation Cascades

Caspases are important effectors in inducing the characteristic changes seen in the course of apoptosis. Similar to many other intracellular proteases, all caspases are synthesized as dormant proenzymes containing three domains, N-terminal prodomain, large P20 subunit, and small P10 subunit, with little, if any, catalytic activity *(14)*. In response to apoptotic stimulation, these otherwise latent proteases are proteolytically activated to cleave a number of cellular proteins whose degradation leads to eventual cell death. Enzymatic activation of caspases requires proteolytic cleavages at least between large and small subunits at sites that contain caspase consensus sequences, strongly implying that caspases are activated autocatalytically and/or by other caspases in a sequential manner. Such a cascade mechanism would not only allow simultaneous regulation of the activities of

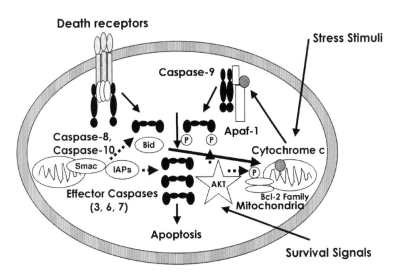

Fig. 1. Two main pathways leading to caspase activation. Analysis of mutant mice of genes implicated in caspase activation underscores the importance of caspase activation pathways during the development. Caspases are mainly activated by two pathways, one induced by death receptors and the other mediated by cytochrome *c*. Each pathway can be affected by various post-translational modifications such as phosphorylation and relocation. Those two pathways converge at the point of activation of effector caspase such as caspase-3, then leading to degradation of physiologically important molecules. Solid lines represent activation; dotted lines represent inhibition.

multiple caspases through stringent control over the initiation of upstream caspase(s), but also provide a positive feedback mechanism to amplify the activation process, ensuring the rapid elimination of the doomed cell *(15)*.

Initiation of Caspase Cascades

Although the mechanistic details as to how diverse stimuli converge into one common apoptotic machinery remains to be understood, two well-studied apoptosis-signaling pathways (Fig. 1). They respond to apoptotic signals with a limited number of adaptors, and support the overall scheme that mediate the recruitment and subsequent activation of their respective initiator caspases *(12)*.

Perhaps the best-characterized apoptotic pathway, signaling of the death receptor, Fas, relies upon the stepwise formation of the death inducing signaling complex (DISC) following the trimerization of Fas receptor *(16)*. One central component of this complex is the apoptotic adaptor molecule FADD, which is rapidly recruited to the Fas molecule through protein-protein interactions mediated by the death domain (DD) motif present both in the cytoplasmic tail of Fas and in the C-terminus of FADD *(17,18)*. Following its recruitment to the DISC complex, FADD further engages initiator caspases such as procaspases-8 and/or -10 through death effector domain (DED)–DED interactions through its N-terminal DED. This induced proximity of procaspases-8 and/or -10 proteins and, presumably due to auto-cleavage by weak intrinsic protease activity, results in their self-processing, activation, and ultimately initiation of the whole caspase cascade by cleaving and activating effector caspase such as caspases-3, -6, and -7 *(14)*. This activation mechanism is likely used by the other members of the tumor necrosis factor (TNF) receptor family containing DDs.

A second pathway capable of triggering the activation of caspase cascade requires cytochrome *c*, a component of the mitochondrial electron-transport chain *(19)*. During an early stage of apoptosis, cytochrome *c* is released from the intermembrane space of the mitochondria into the cytosol in a process that can be regulated both positively and negatively by members of the Bcl-2 family *(20–24)*.

Once in the cytosol, cytochrome *c* binds to the apoptotic adaptor Apaf-1 and, by hydrolyzing ATP, recruits procaspase-9 through interaction of caspase recruitment domains (CARDs) that are present both in Apaf-1 and in initiator caspases, to form an effector caspase-activating complex, also known as the apoptosome *(25)*. It is within this apoptosome complex where Apaf-1 potentiates initiator procaspase-9 to undergo auto- or trans-catalysis to generate active caspase-9, which is capable of activating effector caspases such as caspase-3.

Although a variety of proteins have been found to be cleaved by effector caspases, a pro-apoptotic Bcl-2 family member, Bid, requires a specific cleavage by an initiator caspase, caspase-8 *(26,27)*. Activated caspase-8 cleaves p22 Bid, and generates an active myristolyated version of Bid which is then translocated into the mitochondria membrane and induces cytochrome *c* release. As a consequence, the aforementioned caspase-9 pathway is activated to amplify the caspase cascade initiated by the death receptors.

Taken together, emerging evidence supports the model of oligomerization-induced auto- or trans-catalysis of procaspase. According to this model, procaspase are normally present in cells, but in a conformation or a complex that prevents their auto- or trans-catalysis. Upon apoptotic stimulation, recruitment by adaptors leads to dissociation of inhibitors, auto- or trans-catalysis, change of procaspase conformation, and aggregation of procaspase.

Regulation of Caspase Cascades by Bcl-2 Family Molecules

Because the caspase cascade is a self-amplifying process, it is believed that most, if not all, regulation of the caspase cascade occurs at the level of initiation. In fact, the very requirement for stepwise assembly of various components to form caspase-activating complexes such as DISC and the apoptosome implicates the existence of multiple safeguarding mechanisms that exert stringent controls at various levels over this cellular switch between life and death.

The most-studied regulatory mechanism of caspase activation is that provided by members of the Bcl-2 superfamily *(28)*. Originally *bcl-2* was identified as the gene involved in a translocation event in non-Hodgkin's follicular lymphomas to induce cell survival *(29,30)*. To date, more than a dozen Bcl-2 family members have been identified in mammalian cells. Interestingly, although many possess anti-apoptotic activity as Bcl-2 does, the other members lacking a distinct anti-apoptotic domain can antagonize Bcl-2-like functions and are death inducers *(12)*.

Two potential mechanisms by which Bcl-2 family members regulate the activation of caspases have been proposed. First, in a model analogous to biochemical studies in the *C. elegans* system where Ced-9 can directly bind to Ced-4, anti-apoptotic Bcl-2 proteins such as Bcl-x_L have been shown to bind directly to Apaf-1 *(31–33)*. Such binding may perturb the ability of Apaf-1 to bind to the CARD domain of procaspase-9 and thus inhibits the activation of caspase-9. Furthermore, according to this model, pro-apoptotic members of the Bcl-2 family, such as Bax and Bad, function at least in part by heterodimerizing with pro-survival members like Bcl-2 and thereby titrating out their anti-apoptotic activity *(34,35)*. Moreover, the most potent inducers of apoptosis in this group possessing only a BH3 (Bcl-2 homology region 3) domain such as Bak or Bik also inhibit the association of Apaf-1 with Bcl-x_L *(34)*. The similar observation has been reported for a homolog of the BH3-domain-only Bcl-2 family members, Egl-1 in *C. elegans (36)*. The second model argues that many Bcl-2 family members are localized predominantly in the outer membrane of the mitochondria and contends that their physiological function is likely either to facilitate or to block the release of cytochrome *c* and other factors capable of inducing apoptosis from the mitochondria. In support of this model, Bcl-2 and Bcl-x_L have been shown to inhibit the release of cytochrome *c* in vitro independent of caspase activity, reaffirming the notion that Bcl-2 and its relatives act upstream of caspase activation *(20–22)*. More recently, both Bax and Bid were found to translocate into the mitochondria upon apoptotic stimulation and subsequently mediate the release of cytochrome *c* in a Bcl-2-inhibitable manner *(23,24,26,27)*. Very little is known about the molecular basis of how Bcl-2 family members antagonize each other in mediating the release of cytochrome *c*. Nuclear magnetic resonance (NMR) structural analysis of Bcl-x$_L$ and other in vitro studies suggest that several Bcl-2 family members are able to form pores that regulate the transport of small molecules across the membrane *(37–40)*. More recently, pro-apoptotic Bcl-2 members such as Bax and Bak have been shown to accelerate opening of the mitochondrial porin channel termed VDAC. Moreover, cytochrome *c* can be released through this channel and the passage of cytochrome *c* can be antagonized by anti-apoptotic Bcl-2 family member, Bcl-x$_L$ *(41)*.

Regulation of Caspase Activation by Inhibitor-of-Apoptosis Proteins

A second family of proteins that regulates caspase activation is a group of endogenous caspase inhibitors known as inhibitor of apoptosis proteins (IAPs), a family of proteins characterized by the presence of at least one baculovirus IAP repeat (BIR) domain *(42)*. Interestingly, several IAPs including c-IAP1, c-IAP2, and XIAP (x-linked IAP) also contain a conserved ring finger domain at their C-termini that is also present in the viral IAP. Although the ring domain is required for the ability of viral IAP to suppress apoptosis in insect cells, NAIP, and survivin, both lacking ring domains, inhibit apoptosis in mammalian cells. Despite the uncertainty about the functional importance of the ring domain, deletion and structural studies have indicated that the BIR domain contributes to IAPs' anti-apoptotic activity by mediating direct interactions between IAPs and caspases *(43,44)*. Specifically, XIAP can potently and specifically inhibit active caspases-3 and -7 in a BIR-dependent manner while having no effect on active caspase-8 *(45)*. It has also been shown that both c-IAP1 and c-IAP2 can selectively suppress the activity of caspases-3 and -7, albeit with a much lower potency *(46)*. Recently a mitochondrial protein, Smac/Diablo, has been isolated to bind to IAP members and intervene their anti-apoptotic functions *(47,48)*, whereas structure analysis of co-complexes between the effector caspase and XIAP have revealed a steric hindrance is a mechanism for caspase inhibition *(49–51)*. Taken together, although the anti-apoptotic activities of IAPs are well-established in vitro, the biological advantage to have such a secondary anti-apoptotic mechanism is still unknown. XIAP null mice are viable without any obvious effect on caspase-mediated apoptosis *(52)*.

Other Regulations on Caspase Activities

Recent studies indicate that, in addition to regulation by Bcl-2 family proteins and IAPs, caspase activation can also be regulated through other mechanisms such as phosphorylation, nitrosylation, and compartmentalization of caspases. In studying the mechanism by which p21–Ras inhibits apoptosis, it has been shown that p21–Ras activated Akt kinase phosphorylates procaspase-9. More importantly, phosphorylation of caspase-9 by Akt reduces its proteolytic activity *(53)*. Similarly, it has been demonstrated that caspase activity can be negatively regulated through nitrosylation, whereas Fas signaling induces denitrosylation of caspase-3, thus increasing its activity *(54,55)*. To further complicate the matter, a number of studies have shown that upon apoptotic stimulation, several caspases including caspases-2, -3, -7, and -9 undergo intracellular translocation, although the functional significance of such translocations remains to be elucidated *(56–58)*.

APOPTOSIS IN VERTEBRATES DEVELOPMENT

Despite the early recognition of the existence of physiological cell death during vertebrate development, its importance in development was not fully realized until recently. Thanks to the various genetic models in *C. elegans, Drosophila*, and mice, it is now evident that physiological cell death in the form of apoptosis is critical for deleting structures that are no longer needed, for controlling cell numbers, and for molding tissue structures *(1)*. Studies using these model systems further suggest that the necessity of using apoptosis as a crucial mechanism to regulate the development of tissue and organs has increased during evolution, as defects in cell death results in much more severe phenotypes in *Drosophila* and mice than in the nematode *C. elegans*.

Physiological Roles of the Bcl-2 Family

Transgenic and gene-targeting approaches have also confirmed the involvement of Bcl-2 family members in mediating apoptosis during nervous-system development. Transgenic mice overexpressing Bcl-2 exhibit somewhat enlarged brain (12% increase in weight) and a 40–50% increase in the cell number in the facial nucleus *(59)*. In contrast, Bcl-2-deficient mice are small and have polycystic kidneys and hypopigmentation without any obvious cell death abnormality in the nervous system *(60,61)*. In addition, mature lymphocytes in the mutant mice are incapable of maintaining homeostasis and become vulnerable to apoptotic stimuli *(62)*.

The gene locus for bcl-x can produce two protein isoforms, $Bcl-x_L$ and $Bcl-x_S$, which are anti-apoptotic and pro-apoptotic molecules, respectively. Its null mutation abrogating both isoforms causes extensive apoptosis of postmitotic neurons in the embryonic nervous system *(63)*. Chimera mice generated by blastocyst complementation with recombination deficient mice reveal that immature T cells, not mature T (or B) cells, become extremely sensitive to apoptosis. These results are consistent with the expression profiles of Bcl-2 and Bcl-x in immature and mature lymophocytes. In addition, Bcl-x transgene introduced in Bcl-2 null mice can rescue the defect in T cells, suggesting that functions of the anti-apoptotic Bcl-2 family are redundant *(64)*.

Mice deficient in the other anti-apoptotic Bcl-2 members have also been reported. Bcl-w null mice have defects in spermatogenesis, despite its ubiquitous expression. The deletion of bcl-w gene results in depletion of germ cells at all stages, leading to eventual loss of supporting Sertoli cells *(65,66)*. Mcl-1-deficient embryos die at the peri-implantation period without evidence of increased apoptosis, suggesting that the bcl-w gene may have an additional function other than regulating apoptosis in hematopoietic cells *(67,68)*. Finally, mice deficient in A1–a/Bfl-1 have selected defects in neutrophil apoptosis *(69)*.

In contrast, naturally occurring neuronal death and apoptosis induced by the withdrawal of trophic factors are reduced in mice lacking the pro-apoptotic gene Bax *(70,71)*. Furthermore, the Bax deficiency is able to prevent postmitotic neurons from undergoing apoptosis in animals carrying an additional $Bcl-x_L$ null mutation, demonstrating an intracellular balance between proapoptotic and antiapoptotic effects within the Bcl-2 protein family *(72)*. Bax null mice also have defects in spermatogenesis represented by accumulation of atypical and premeiotic germ cells and lack of mutant haploid sperm *(73)*. It is noteworthy that Bax-deficient female mice have excess primordial follicles failing to undergo normal developmental apoptosis *(74)*. Again these data suggest that pro-apoptotic Bax may have a distinct role in spermatogenesis as well as in oocytegenesis.

Mice lacking another pro-apoptotic Bcl-2 member, Bak, develop normally and do not show any obvious abnormality. However, mice deficient both in Bak and in Bax exhibit interdigital webs, imperforate viginal and accumulated neurons in the central nervous system *(75)*. The vast majority of the double knockout mice die perinatally due to the developmental defects. The survived mutant mice are still sensitive to anti-Fas-induced liver damage. Interestingly, embryonic fibroblasts (EFs) from the double-mutant mice are resistant to many apoptotic stimuli including ones causing endoplasmic reticulum (ER) stress, which is strengthened by the fact that cytochrome *c* release by the truncated Bid is abrogated in the double-mutant EFs. Because neither Bak nor Bax is reported to be associated with ER, this data may suggests that ER stress causes mitochondorial dysfunction leading to apoptosis.

A half of mice deficient in the BH3-domain-only member, Bim, are embryonic lethal around E10 *(76,77)*. Survived mice have abnormalities in the hematopoietic components. Bim-deficient T and B cells become resistant to cytokine withdrawal, resulting in progressive lymphadenopathy and splenomegaly with a dramatic increase in plasma cell numbers and immunoglobulin levels *(76)*. These results suggest that Bim antagonize Bcl-2 activity in mature lymphocytes. Another BH3-domain-only molecule, Bid, requires caspase-8-dependent cleavage for its action, and Bid null mice indeed are resistant to Fas-induced cell death in the liver *(78)*. Interestingly the other types of cells are still sensitive to Fas or TNF-α with a subtle delay in cell-death kinetics, implying that Bid is not absolutely required for apoptosis to occur.

Apoptosis in Neural Progenitor Cells and the Caspase Cascade

In line with the hypothesis that caspase-9 activates caspase-3 in a linear fashion, both caspases-3 and -9 null mutant mice exhibited essentially the same developmental defects in the nervous system. Most homozygous caspases-3 and -9 knockout mice die perinatally with severe brain malformations including multiple indentations of the cerebrum and periventricular ectopic cell masses *(79–81)*. Close examination of the mutant embryonic tissues revealed a drastic reduction in the number of pyknotic cells in the neuroepithelial progenitor population around the proliferative zone. Moreover, Apaf-1 null mice exhibit the same neural phenotype, indicating that the mitochondria caspase activation pathway is indispensable in developmental apoptosis in the early neural progenitor cells *(82,83)*. These results are in contrast to those of Bcl-x_L-deficient mice of which only postmitotic neurons show excess apoptosis. In *C. elegans*, the mutation of Ced-3 suppresses the ectopic cell deaths caused by the mutation of Ced-9 *(9)*. Similarly in mice deficient both in caspase-3 and in Bcl-x_L, the aberrant neuronal apoptosis due to the Bcl-x_L deficiency is indeed abolished by the additional caspase-3 deficiency *(84)*. In fact, the neuronal phenotype of the caspase-3 and Bcl-x_L double deficiency is literally indistinguishable from the caspase-3 single null mutation. Taken together, caspase-3 and -9 knockout models not only confirmed the importance of cell death during neuronal development, but also revealed the existence of early cell death in the progenitor population that is not target-driven. Based on the various phenotypes of the Bax, Bcl-x_L, caspase-3, and caspase-9 null mutations, it appears that Bcl-x deficiency causes increased apoptosis of postmitotic immature neurons, which is prevented by the additional absence of caspase-3 *(84)*. Therefore, the null-mutation of Bax reduces the normally occurring developmental death of postmitotic matured neurons without affecting the global formation of the nervous system *(70,71)*.

Caspase-3-deficient mice have been shown to survive to adulthood when backcrossed to C57BL/6 *(85)*. Those mice exhibit a significantly delayed response to Fas-induced apoptosis in vivo *(85,86)*. In addition, lymphocytes from the mice are partially resistant to activation-induced cell death *(87)*. These results indicate that caspase-3 is the main target of caspase-8 in Fas-induced apoptosis in mature lymphocytes and liver cells.

Caspase-8 and Cardiac Development

Mice lacking caspase-8, the upstream caspase primarily involved in death receptor-mediated apoptosis, die *in utero* around E11.5 and exhibit impaired formation of cardiac muscles *(88)*. Importantly, a nearly identical developmental defect was also observed in embryos carrying a null mutation in FADD *(89,90)*, the apoptotic adaptor molecule bridging death receptors and caspase-8, suggesting that apoptosis mediated through death receptor is crucial for proper cardiac development. Contrarily, mice deficient in Casper/c-FLIP, a caspase-8 homolog lacking the catalytic residues, exhibit the similar cardiac defects despite of data supporting its role as an antagonist for the death receptor-induced caspase activation in established cell lines from the mutant mice *(91)*. These results may underscore the importance of proper regulation of activation of caspases-8/-10 during the cardiac development.

Caspase-12 and ER-Stress

Caspase-12 is a mouse caspase belonging to a caspase-1 (ICE) subfamily. Members in this subfamily are implicated in cytokine maturation rather than apoptosis *(92)*. Caspase-12 is ubiquitously expressed in mouse tissues. Further biochemical analysis revealed that significant amount of caspase-12 is localized in the ER *(93)*. Interestingly, although widely used apoptotic stimuli such as serum withdrawal, TNF, and anti-Fas antibody treatment do not cause activation of caspase-12, reagents inducing ER stress such as brefeldin A, tunicamycin, and thapsigargin initiate caspase-12 cleavage and subsequent activation in vitro. In addition, it is noteworthy that Calpain, another class of cystein protease, has been shown to be responsible for caspase-12 activation, indicating the uniqueness of this pathway *(94)*. Caspase-12 null mice are viable, and histological analysis has not revealed any gross developmental abnormality *(93)*. However, caspase-12-deficient mice are resistant to pathlogical changes induced by tunicamycin in vivo. Interestingly primary neural-cell culture prepared from the null mice exhibits resistance to toxicity induced by amyloid-β protein (Aβ) *(93)*. Given that a potential intracellular target of Aβ is ER *(95)*, it is tempting to suggest that inhibition of caspase-12 would be a therapeutic target for Alzheimer's disease.

Mitochondria Proteins and Apoptosis

Cytochrome *c*, the only water-soluble component of the electron transfer chain in the mitochondoria, has been shown to have the additional role in activation of caspase when it is released from the mitochondria into the cytosol upon apoptotic stimuli compromising integrity of the mitochondoria *(19)*. Mouse embryos deficient in cytochrome *c* can live up to E8.5 with a significant developmental delay probably due to transmission of healthy mitochondrion from oocytes *(96)*. Consistent with data from cell lines deprived of the mitochondria, however, embryonic cell lines can be established to study a role of cytochrome *c* in apoptosis. The cell lines are completely resistant to ultraviolet (UV) irradiation and staurosporine, whereas they show partial resistance to serum withdrawal. In accordance with data from Apaf-1 and caspase-9 null mice, the death-receptor pathway is intact in cells deficient in cytochrome c *(96)*.

Apoptosis-inducing factor (AIF) is an enigmatic molecule homologous to flavoproteins that also plays a role in the electron transfer in the mitochondria. In addition, AIF is translocated to the nucleus upon apototic stimulation and then induces chromatin condensation and large-scale DNA fragmentation *(97)*. Embryonic bodies derived from AIF null embryonic stem cells lack cavities inside the bodies *(98)*. None of knockout mice described earlier shows this type of phenotype though the cavitation has been shown to be mediated by an apoptotic mechanism *(99)*. These data indicate a role of AIF in apoptosis in the very early development. However, involvement of AIF in other apoptotic events remains to be investigated.

CONCLUSION

Apoptosis or physiological cell death is a process that not only plays a critical role during development for the proper formation of various organs and tissues, but that is also intimately involved in the control of cellular homeostasis throughout the life span of the organism. Following the breakthrough discovery of the genetic control of apoptosis in the nematode by Horvitz and his colleague *(6–11)*, great progress has been made in understanding the underlying mechanism by which the apoptotic pathways are regulated. The molecular dissection of the apoptosis pathways indicates that a blueprint for apoptosis is the same throughout multicellular organisms and even conserved during the evolution. Along with the analysis of apoptotic mutants in *C. elegans,* gene-disrupted mice described here have made a significant contribution to our understanding of apoptotic mechanisms and their importance in vivo. Yet many questions remain to be answered because of the functional redundancy of genes and the compensatory activation seen in some gene-targeting mice, indicating that generation of mice deficient in multiple genes is necessary. In some cases, embryonic lethality hampers us from investigating the roles of apoptotic genes in mature cells, which emphasizes the need for tissue-specific or conditional gene ablation. Moreover, genetic backgrounds of different strains of mice further complicate phenotypes of mutant mice, suggesting that more apoptotic regulators remain to be discovered. Nevertheless, mice possessing a specific gene mutation in an apoptotic pathway have been used in various pharmacological models to elucidate the pathological consequences of inhibition of a specific target. It is conceivable that, within the foreseeable future, therapeutic strategies based on modulation of apoptosis will provide new opportunities for the treatment of various human diseases.

Table 1
Summary of Knockout Mice for Various Genes Implicated in Apoptosis

Genes		Phenotypes	References
Caspases			
	Caspase-1	Viable	*100,101*
		Maturation defects in IL-1ß and IL-18	
	Caspase-2	Viable	*102*
		Increased oocytes that are resistant to cytotoxic agents	
		B cells become resistant to apoptosis induced by granzyme B	
		Accelerated motor neuron apoptosis due to lack of expression of the short anti-apoptotic isoform	
	Caspase-3	Embryonic lethal	*79,87*
		Generation of supernemenary neural cells	
		Survived knockout mice are resistant to Fas-induced apoptosis in liver as well as AICD in lymphocytes	
	Caspase-6	Viable	*86*
		No obvious defects	
	Caspase-8	Embryonic lethal (E12)	*88*
		Impaired cardica development	
		MEFs are resistant to apoptosis induced by several death receptors	
	Caspase-9	Embryonic lethal	*80,81*
		Severe morphological abnormality in CNS due to excess of nerua cells	
		MEFs and lymphocytes become resistant to various apoptotic stimuli except ones induced by death receptors	
	Caspase-11	Viable	*103*
		Resistant to LPS-induced endotoxic shock	
	Caspase-12	Viable	*93*
		Resistant to ER stress	
	Casper	Embryonic lethal	*91*
		Similar phenotypes to those of caspase-8 knockout mice	
Adaptors			
	Apaf-1	Embryonic lethal	*82,83*
		Similar to caspase-9 knockout mice	
	FADD	Embryonic lethal	*89,90*
		Similar cardiac abnormality to that of caspase-8 knockout mice	
Bcl-2 family			
	Bcl-2	Succumb by renal failure	*60–62*
		Increased apoptosis in mature lymphocytes and melanocytes	
		Accelerated apoptosis in subsets of postmitotic neurons	
	Bcl-x	Embryonic lethal	*63,104*
		Pervasive neural and immature hematopoietic cell death	
		Increased apoptosis in immature lymphocytes	
	Mcl-1	Embryonic lethal around peri-implantation	*67*
	A1a	Viable	*69*
		Accelerated neutrophil apoptosis	
	Bcl-w	Viable	*65,66*
		Male infertility due to abnormality in spermatogenesis	
	Bax	Viable	*73,74*
		Increased cell numbers in subsets of neurons; increased oocytes lifespan	
	Bak	Viable	*105*
	Bid	Viable; resistant to Fas-induced liver damage	*78*

Table 1 *(continued)*
Summary of Knockout Mice for Various Genes Implicated in Apoptosis

Genes	Phenotypes	References
Bim	Partially embryonic lethal (E10)	76,77
	Abnormal thymic development with massive plasma cell proliferation and hyperglobulinemia; premature death due to vasculit	
Mitochondorial genes		
Cytochrome *c*	Embryonic lethal (E8.5)	96
	Embryonic cells become resistant to UV- and staurpsporine-induced apoptosis	
AIF	Embryonic lethal	98
	Lack of cavitation in embryonic bodies; ES cells are resistant to growth factor withdrawal	
IAP		
XIAP	No obvious difference in caspase-mediated apoptosis	52
	Upregulation of c-IAP1 and c-IAP2	
Combination		
Caspase-3x	Embryonic lethal	84
Bcl-x	Neural phenotypes are similar to those of caspase-3 knockouts	
	Embryos still have developmental defects in hematopoiesis	
Bax x Bak	A majority of mice are perinatal lethal	105
	Persistence of intradigital webs	
	Accumlation of neurons in CNS	
	Splenomegary	
	MEFs are resistant to a variety of apoptotic stimuli except TNFs	
	Survived mice are resistant to Fas-induced liver damage	

REFERENCES

1. Vaux, D. L. and Korsmeyer, S. J. (1999) Cell death in development. *Cell* **96(2)**, 245–254.
2. Jacobson, M. D., Weil, M., and Raff, M. C. (1997) Programmed cell death in animal development. *Cell* **88(3)**, 347–354.
3. Wyllie, A. H., Kerr, J. F., and Currie, A. R. (1980) Cell death: the significance of apoptosis. *Int. Rev. Cytol.* **68**, 251–306.
4. Kerr, J. F., Wyllie, A. H., and Currie, A. R. (1972) Apoptosis: a basic biological phenomenon with wide-ranging implications in tissue kinetics. *Br. J. Cancer* **26(4)**, 239–257.
5. Morris, R. G., Hargreaves, A. D., Duvall, E., and Wyllie, A. H. (1984) Hormone-induced cell death. 2. Surface changes in thymocytes undergoing apoptosis. *Am. J. Pathol.* **115(3)**, 426–436.
6. Metzstein, M. M., Stanfield, G. M., and Horvitz, H. R. (1998) Genetics of programmed cell death in *C. elegans*: past, present and future. *Trends Genet.* **14(10)**, 410–416.
7. Ellis, H. M., and Horvitz, H. R. (1986) Genetic control of programmed cell death in the nematode *C. elegans*. *Cell* **44(6)**, 817–829.
8. Yuan, J. Y., and Horvitz, H. R. (1990) The *Caenorhabditis elegans* genes ced-3 and ced-4 act cell autonomously to cause programmed cell death. *Dev. Biol.* **138(1)**, 33–41.
9. Hengartner, M. O., Ellis, R. E., and Horvitz, H. R. (1992) *Caenorhabditis elegans* gene ced-9 protects cells from programmed cell death. *Nature* **356(6369)**, 494–499.
10. Hengartner, M. O. and Horvitz, H. R. (1994) *C. elegans* cell survival gene ced-9 encodes a functional homolog of the mammalian proto-oncogene bcl-2. *Cell* **76(4)**, 665–676.
11. Yuan, J., Shaham, S., Ledoux, S., Ellis, H. M., and Horvitz, H. R. (1993) The *C. elegans* cell death gene ced-3 encodes a protein similar to mammalian interleukin-1 beta-converting enzyme. *Cell* **75(4)**, 641–652.
12. Strasser, A., O'Connor, L., and Dixit, V. M. (2000) Apoptosis signaling. *Annu. Rev. Biochem.* **69**, 217–245.
13. Zou, H., Henzel, W. J., Liu, X., Lutschg, A., and Wang, X. (1997) Apaf-1, a human protein homologous to C. elegans CED-4, participates in cytochrome c-dependent activation of caspase-3. *Cell* **90(3)**, 405–413.
14. Salvesen, G. S. and Dixit, V. M. (1999) Caspase activation: the induced-proximity model. *Proc. Natl. Acad. Sci. USA* **96(20)**, 10964–10967.
15. Salvesen, G. S. (2002) Caspases: opening the boxes and interpreting the arrows. *Cell Death Differ.* **9(1)**, 3–5.

16. Muzio, M., Chinnaiyan, A. M., Kischkel, F. C., O'Rourke, K., Shevchenko, A., Ni, J., et al. (1996) FLICE, a novel FADD-homologous ICE/CED-3-like protease, is recruited to the CD95 (Fas/APO-1) death—inducing signaling complex. *Cell* **85(6),** 817–827.

17. Chinnaiyan, A. M., O'Rourke, K., Tewari, M., and Dixit, V. M. (1995) FADD, a novel death domain-containing protein, interacts with the death domain of Fas and initiates apoptosis. *Cell* **81(4),** 505–512.

18. Boldin, M. P., Goncharov, T. M., Goltsev, Y. V., and Wallach, D. (1996) Involvement of MACH, a novel MORT1/FADD-interacting protease, in Fas/APO-1- and TNF receptor-induced cell death. *Cell* **85(6),** 803–815.

19. Liu, X., Kim, C. N., Yang, J., Jemmerson, R., and Wang, X. (1996) Induction of apoptotic program in cell-free extracts: requirement for dATP and cytochrome c. *Cell* **86(1),** 147–157.

20. Yang, J., Liu, X., Bhalla, K., Kim, C. N., Ibrado, A. M., Cai, J., et al. (1997) Prevention of apoptosis by Bcl-2: release of cytochrome c from mitochondria blocked. *Science* **275(5303),** 1129–1132.

21. Kluck, R. M., Bossy-Wetzel, E., Green, D. R., and Newmeyer, D. D. (1997) The release of cytochrome c from mitochondria: a primary site for Bcl-2 regulation of apoptosis. *Science* **275(5303),** 1132–1136.

22. Vander Heiden, M. G., Chandel, N. S., Williamson, E. K., Schumacker, P. T., and Thompson, C. B. (1997) Bcl-x_L regulates the membrane potential and volume homeostasis of mitochondria. *Cell* **91(5),** 627–637.

23. Rosse, T., Olivier, R., Monney, L., Rager, M., Conus, S., Fellay, I., et al. (1998) Bcl-2 prolongs cell survival after Bax-induced release of cytochrome c. *Nature* **391(6666),** 496–499.

24. Jurgensmeier, J. M., Xie, Z., Deveraux, Q., Ellerby, L., Bredesen, D., and Reed, J. C. (1998) Bax directly induces release of cytochrome c from isolated mitochondria. *Proc. Natl. Acad. Sci. USA* **95(9),** 4997–5002.

25. Salvesen, G. S. and Renatus, M. (2002) Apoptosome: the seven-spoked death machine. *Dev. Cell* **2(3),** 256–257.

26. Luo, X., Budihardjo, I., Zou, H., Slaughter, C., and Wang, X. (1998) Bid, a Bcl2 interacting protein, mediates cytochrome c release from mitochondria in response to activation of cell surface death receptors. *Cell* **94(4),** 481–490.

27. Li, H., Zhu, H., Xu, C. J., and Yuan, J. (1998) Cleavage of BID by caspase 8 mediates the mitochondrial damage in the Fas pathway of apoptosis. *Cell* **94(4),** 491–501.

28. Budihardjo, I., Oliver, H., Lutter, M., Luo, X., and Wang, X. (1999) Biochemical pathways of caspase activation during apoptosis. *Annu. Rev. Cell Dev. Biol.* **15,** 269–290.

29. Bakhshi, A., Jensen, J. P., Goldman, P., Wright, J. J., McBride, O. W., Epstein, A. L., and Korsmeyer, S. J. (1985) Cloning the chromosomal breakpoint of t(14;18) human lymphomas: clustering around JH on chromosome 14 and near a transcriptional unit on 18. *Cell* **41(3),** 899–906.

30. Tsujimoto, Y., Yunis, J., Onorato-Showe, L., Erikson, J., Nowell, P. C., and Croce, C. M. (1984) Molecular cloning of the chromosomal breakpoint of B-cell lymphomas and leukemias with the t(11;14) chromosome translocation. *Science* **224(4656),** 1403–1406.

31. Chinnaiyan, A. M., O'Rourke, K., Lane, B. R., and Dixit, V. M. (1997) Interaction of CED-4 with CED-3 and CED-9: a molecular framework for cell death [see comments]. *Science* **275(5303),** 1122–1126.

32. Spector, M. S., Desnoyers, S., Hoeppner, D. J., and Hengartner, M. O. (1997) Interaction between the C. elegans cell-death regulators CED-9 and CED-4. *Nature* **385(6617),** 653–656.

33. Wu, D., Wallen, H. D., and Nunez, G. (1997) Interaction and regulation of subcellular localization of CED-4 by CED-9. *Science* **275(5303),** 1126–1129.

34. Pan, G., O'Rourke, K., and Dixit, V. M. (1998) Caspase-9, Bcl-X_L, and Apaf-1 form a ternary complex. *J. Biol. Chem.* **273(10),** 5841–5845.

35. Hu, Y., Benedict, M. A., Wu, D., Inohara, N., and Nunez, G. (1998) Bcl-X_L interacts with Apaf-1 and inhibits Apaf-1-dependent caspase-9 activation. *Proc. Natl. Acad. Sci. USA* **95(8),** 4386–4391.

36. Conradt, B. and Horvitz, H. R. (1998) The C. elegans protein EGL-1 is required for programmed cell death and interacts with the Bcl-2-like protein CED-9. *Cell* **93(4),** 519–529.

37. Muchmore, S. W., Sattler, M., Liang, H., Meadows, R. P., Harlan, J. E., Yoon, H. S., et al. (1996) X-ray and NMR structure of human Bcl-x_L, an inhibitor of programmed cell death. *Nature* **381(6580),** 335–341.

38. Minn, A. J., Velez, P., Schendel, S. L., Liang, H., Muchmore, S. W., Fesik, S. W., et al. (1997) Bcl-x(L) forms an ion channel in synthetic lipid membranes. *Nature* **385(6614),** 353–357.

39. Schendel, S. L., Xie, Z., Montal, M. O., Matsuyama, S., Montal, M., and Reed, J. C. (1997) Channel formation by antiapoptotic protein Bcl-2. *Proc. Natl. Acad. Sci. USA* **94(10),** 5113–5118.

40. Schlesinger, P. H., Gross, A., Yin, X. M., Yamamoto, K., Saito, M., Waksman, G., and Korsmeyer, S. J. (1997) Comparison of the ion channel characteristics of proapoptotic BAX and antiapoptotic BCL-2. *Proc. Natl. Acad. Sci. USA* **94(21),** 11357–11362.

41. Shimizu, S., Narita, M., and Tsujimoto, Y. (1999) Bcl-2 family proteins regulate the release of apoptogenic cytochrome c by the mitochondrial channel VDAC. *Nature* **399(6735),** 483–487.

42. Deveraux, Q. L. and Reed, J. C. (1999) IAP family proteins—suppressors of apoptosis. *Genes Dev.* **13(3),** 239–252.

43. Sun, C., Cai, M., Gunasekera, A. H., Meadows, R. P., Wang, H., Chen, J., et al. (1999) NMR structure and mutagenesis of the inhibitor-of-apoptosis protein XIAP. *Nature* **401(6755),** 818–822.

44. Liston, P., Roy, N., Tamai, K., Lefebvre, C., Baird, S., Cherton-Horvat, G., et al. (1996) Suppression of apoptosis in mammalian cells by NAIP and a related family of IAP genes. *Nature* **379(6563),** 349–353.

45. Deveraux, Q. L., Leo, E., Stennicke, H. R., Welsh, K., Salvesen, G. S., and Reed, J. C. (1999) Cleavage of human inhibitor of apoptosis protein XIAP results in fragments with distinct specificities for caspases. *EMBO J.* **18(19),** 5242–5251.

46. Roy, N., Deveraux, Q. L., Takahashi, R., Salvesen, G. S., and Reed, J. C. (1997) The c-IAP-1 and c-IAP-2 proteins are direct inhibitors of specific caspases. *EMBO J.* **16(23),** 6914–6925.

47. Du, C., Fang, M., Li, Y., Li, L., and Wang, X. (2000) Smac, a mitochondrial protein that promotes cytochrome c-dependent caspase activation by eliminating IAP inhibition. *Cell* **102(1),** 33–42.

48. Verhagen, A. M., Ekert, P. G., Pakusch, M., Silke, J., Connolly, L. M., Reid, G. E., et al. (2000) Identification of DIABLO, a mammalian protein that promotes apoptosis by binding to and antagonizing IAP proteins. *Cell* **102(1),** 43–53.

49. Chai, J., Shiozaki, E., Srinivasula, S. M., Wu, Q., Datta, P., Alnemri, E. S., Shi, Y., and Dataa, P. (2001) Structural basis of caspase-7 inhibition by XIAP. *Cell* **104(5),** 769–780.

50. Huang, Y., Park, Y. C., Rich, R. L., Segal, D., Myszka, D. G., and Wu, H. (2001) Structural basis of caspase inhibition by XIAP: differential roles of the linker versus the BIR domain. *Cell* **104(5),** 781–790.

51. Riedl, S. J., Renatus, M., Schwarzenbacher, R., Zhou, Q., Sun, C., Fesik, S. W., et al. (2001) Structural basis for the inhibition of caspase-3 by XIAP. *Cell* **104(5),** 791–800.

52. Harlin, H., Reffey, S. B., Duckett, C. S., Lindsten, T., and Thompson, C. B. (2001) Characterization of XIAP-deficient mice. *Mol. Cell Biol.* **21(10),** 3604–3608.

53. Cardone, M. H., Roy, N., Stennicke, H. R., Salvesen, G. S., Franke, T. F., Stanbridge, E., et al. (1998) Regulation of cell death protease caspase-9 by phosphorylation. *Science* **282(5392),** 1318–1321.

54. Mannick, J. B., Hausladen, A., Liu, L., Hess, D. T., Zeng, M., Miao, Q. X., et al. (1999) Fas-induced caspase denitrosylation. *Science* **284(5414),** 651–654.

55. Mannick, J. B., Schonhoff, C., Papeta, N., Ghafourifar, P., Szibor, M., Fang, K., and Gaston, B. (2001) S-Nitrosylation of mitochondrial caspases. *J. Cell Biol.* **154(6),** 1111–1116.

56. Chandler, J. M., Cohen, G. M., and MacFarlane, M. (1998) Different subcellular distribution of caspase-3 and caspase-7 following Fas-induced apoptosis in mouse liver. *J. Biol. Chem.* **273(18),** 10815–10818.

57. Mancini, M., Nicholson, D. W., Roy, S., Thornberry, N. A., Peterson, E. P., Casciola-Rosen, L. A., and Rosen, A. (1998) The caspase-3 precursor has a cytosolic and mitochondrial distribution: implications for apoptotic signaling. *J. Cell Biol.* **140(6),** 1485–1495.

58. Susin, S. A., Lorenzo, H. K., Zamzami, N., Marzo, I., Brenner, C., Larochette, N., et al. (1999) Mitochondrial release of caspases-2 and -9 during the apoptotic process. *J. Exp. Med.* **189(2),** 381–394.

59. Martinou, J. C., Dubois-Dauphin, M., Staple, J. K., Rodriguez, I., Frankowski, H., Missotten, M., et al. (1994) Overexpression of BCL-2 in transgenic mice protects neurons from naturally occurring cell death and experimental ischemia. *Neuron* **13(4),** 1017–1030.

60. Veis, D. J., Sorenson, C. M., Shutter, J. R., and Korsmeyer, S. J. (1993) Bcl-2-deficient mice demonstrate fulminant lymphoid apoptosis, polycystic kidneys, and hypopigmented hair. *Cell* **75(2),** 229–240.

61. Nakayama, K., Nakayama, K.-i., Negishi, I., Kuida, K., Sawa, H., and Loh, D. Y. (1994) Targeted disruption of Bcl-2 alpha beta in mice: occurrence of gray hair, polycystic kidney disease, and lymphocytopenia. *Proc. Natl. Acad. Sci. USA* **91(9),** 3700–3704.

62. Nakayama, K.-i., Nakayama, K., Negishi, I., Kuida, K., Shinkai, Y., Louie, M. C., Fields, L. E., et al. (1993) Disappearance of the lymphoid system in Bcl-2 homozygous mutant chimeric mice. *Science* **261(5128),** 1584–1588.

63. Motoyama, N., Wang, F., Roth, K. A., Sawa, H., Nakayama, K., Negishi, I., et al. (1995) Massive cell death of immature hematopoietic cells and neurons in Bcl-x-deficient mice. *Science* **267(5203),** 1506–1510.

64. Chao, D. T., Linette, G. P., Boise, L. H., White, L. S., Thompson, C. B., and Korsmeyer, S. J. (1995) Bcl-X$_L$ and Bcl-2 repress a common pathway of cell death. *J. Exp. Med.* **182(3),** 821–828.

65. Print, C. G., Loveland, K. L., Gibson, L., Meehan, T., Stylianou, A., Wreford, N., et al. (1998) Apoptosis regulator bcl-w is essential for spermatogenesis but appears otherwise redundant. *Proc. Natl. Acad. Sci. USA* **95(21),** 12424–12431.

66. Ross, A. J., Waymire, K. G., Moss, J. E., Parlow, A. F., Skinner, M. K., Russell, L. D., and MacGregor, G. R. (1998) Testicular degeneration in Bclw-deficient mice. *Nat. Genet.* **18(3),** 251–256.

67. Rinkenberger, J. L., Horning, S., Klocke, B., Roth, K., and Korsmeyer, S. J. (2000) Mcl-1 deficiency results in peri-implantation embryonic lethality. *Genes Dev.* **14(1),** 23–27.

68. Zhou, P., Qian, L., Bieszczad, C. K., Noelle, R., Binder, M., Levy, N. B., and Craig, R. W. (1998) Mcl-1 in transgenic mice promotes survival in a spectrum of hematopoietic cell types and immortalization in the myeloid lineage. *Blood* **92(9),** 3226–3239.

69. Hamasaki, A., Sendo, F., Nakayama, K., Ishida, N., Negishi, I., and Hatakeyama, S. (1998) Accelerated neutrophil apoptosis in mice lacking A1-a, a subtype of the bcl-2-related A1 gene. *J. Exp. Med.* **188(11),** 1985–1992.

70. Deckwerth, T. L., Elliott, J. L., Knudson, C. M., Johnson, E. M., Jr., Snider, W. D., and Korsmeyer, S. J. (1996) BAX is required for neuronal death after trophic factor deprivation and during development. *Neuron* **17(3),** 401–411.

71. White, F. A., Keller-Peck, C. R., Knudson, C. M., Korsmeyer, S. J., and Snider, W. D. (1998) Widespread elimination of naturally occurring neuronal death in Bax-deficient mice. *J. Neurosci.* **18(4),** 1428–1439.

72. Shindler, K. S., Latham, C. B., and Roth, K. A. (1997) Bax deficiency prevents the increased cell death of immature neurons in bcl-x-deficient mice. *J. Neurosci.* **17(9),** 3112–3119.

73. Knudson, C. M., Tung, K. S., Tourtellotte, W. G., Brown, G. A., and Korsmeyer, S. J. (1995) Bax-deficient mice with lymphoid hyperplasia and male germ cell death. *Science* **270(5233),** 96–99.

74. Perez, G. I., Robles, R., Knudson, C. M., Flaws, J. A., Korsmeyer, S. J., and Tilly, J. L. (1999) Prolongation of ovarian lifespan into advanced chronological age by Bax-deficiency. *Nat. Genet.* **21(2),** 200–203.

75. Wei, M. C., Zong, W. X., Cheng, E. H., Lindsten, T., Panoutsakopoulou, V., Ross, A. J., et al. (2001) Proapoptotic BAX and BAK: a requisite gateway to mitochondrial dysfunction and death. *Science* **292(5517),** 727–730.

76. Bouillet, P., Metcalf, D., Huang, D. C., Tarlinton, D. M., Kay, T. W., Kontgen, F., et al. (1999) Proapoptotic Bcl-2 relative Bim required for certain apoptotic responses, leukocyte homeostasis, and to preclude autoimmunity. *Science* **286(5445),** 1735–1738.

77. Bouillet, P., Purton, J. F., Godfrey, D. I., Zhang, L. C., Coultas, L., Puthalakath, H., et al. (2002) BH3-only Bcl-2 family member Bim is required for apoptosis of autoreactive thymocytes. *Nature* **415(6874),** 922–926.

78. Yin, X. M., Wang, K., Gross, A., Zhao, Y., Zinkel, S., Klocke, B., et al. (1999) Bid-deficient mice are resistant to Fas-induced hepatocellular apoptosis. *Nature* **400(6747)**, 886–891.

79. Kuida, K., Zheng, T. S., Na, S., Kuan, C., Yang, D., Karasuyama, H., et al. (1996) Decreased apoptosis in the brain and premature lethality in CPP32-deficient mice. *Nature* **384(6607)**, 368–372

80. Kuida, K., Haydar, T. F., Kuan, C. Y., Gu, Y., Taya, C., Karasuyama, H., et al. (1998) Reduced apoptosis and cyto-chrome c-mediated caspase activation in mice lacking caspase 9. *Cell* **94(3)**, 325–337.

81. Hakem, R., Hakem, A., Duncan, G. S., Henderson, J. T., Woo, M., Soengas, M. S., et al. (1998) Differential require-ment for caspase 9 in apoptotic pathways in vivo. *Cell* **94(3)**, 339–352.

82. Cecconi, F., Alvarez-Bolado, G., Meyer, B. I., Roth, K. A., and Gruss, P. (1998) Apaf1 (CED-4 homolog) regulates programmed cell death in mammalian development. *Cell* **94(6)**, 727–737.

83. Yoshida, H., Kong, Y. Y., Yoshida, R., Elia, A. J., Hakem, A., Hakem, R., et al. (1998) Apaf1 is required for mitochon-drial pathways of apoptosis and brain development. *Cell* **94(6)**, 739–750.

84. Roth, K. A., Kuan, C., Haydar, T. F., D'Sa-Eipper, C., Shindler, K. S., Zheng, T. S., et al. (2000) Epistatic and independent functions of caspase-3 and Bcl-X(L) in developmental programmed cell death. *Proc. Natl. Acad. Sci. USA* **97(1)**, 466–471.

85. Zheng, T. S., Schlosser, S. F., Dao, T., Hingorani, R., Crispe, I. N., Boyer, J. L., and Flavell, R. A. (1998) Caspase-3 controls both cytoplasmic and nuclear events associated with Fas-mediated apoptosis in vivo. *Proc. Natl. Acad. Sci. USA* **95(23)**, 13618–13623.

86. Zheng, T. S., Hunot, S., Kuida, K., Momoi, T., Srinivasan, A., Nicholson, D. W., et al. (2000) Deficiency in caspase-9 or caspase-3 induces compensatory caspase activation. *Nat. Med.* **6(11)**, 1241–1247.

87. Woo, M., Hakem, R., Soengas, M. S., Duncan, G. S., Shahinian, A., Kagi, D., et al. (1998) Essential contribution of caspase 3/CPP32 to apoptosis and its associated nuclear changes. *Genes Dev.* **12(6)**, 806–819.

88. Varfolomeev, E. E., Schuchmann, M., Luria, V., Chiannilkulchai, N., Beckmann, J. S., Mett, I. L., et al. (1998) Tar-geted disruption of the mouse caspase 8 gene ablates cell death induction by the TNF receptors, Fas/Apo1, and DR3 and is lethal prenatally. *Immunity* **9(2)**, 267–276.

89. Yeh, W. C., Pompa, J. L., McCurrach, M. E., Shu, H. B., Elia, A. J., Shahinian, A., et al. (1998) FADD: essential for embryo development and signaling from some, but not all, inducers of apoptosis. *Science* **279(5358)**, 1954–1958.

90. Zhang, J., Cado, D., Chen, A., Kabra, N. H., and Winoto, A. (1998) Fas-mediated apoptosis and activation-induced T-cell proliferation are defective in mice lacking FADD/Mort1. *Nature* **392(6673)**, 296–300.

91. Yeh, W. C., Itie, A., Elia, A. J., Ng, M., Shu, H. B., Wakeham, A., Mirtsos, C., et al. (2000) Requirement for Casper (c-FLIP) in regulation of death receptor-induced apoptosis and embryonic development. *Immunity* **12(6)**, 633–642.

92. Van de Craen, M., Vandenabeele, P., Declercq, W., Van den Brande, I., Van Loo, G., Molemans, F., et al. (1997) Characterization of seven murine caspase family members. *FEBS Lett.* **403(1)**, 61–69.

93. Nakagawa, T., Zhu, H., Morishima, N., Li, E., Xu, J., Yankner, B. A., and Yuan, J. (2000) Caspase-12 mediates endo-plasmic-reticulum-specific apoptosis and cytotoxicity by amyloid-ß. *Nature* **403(6765)**, 98–103.

94. Nakagawa, T. and Yuan, J. (2000) Cross-talk between two cysteine protease families. Activation of caspase-12 by calpain in apoptosis. *J. Cell Biol.* **150(4)**, 887–894.

95. Mattson, M. P., Gary, D. S., Chan, S. L., and Duan, W. (2001) Perturbed endoplasmic reticulum function, synaptic apoptosis and the pathogenesis of Alzheimer's disease. *Biochem. Soc. Symp.* **(67)**, 151–162.

96. Li, K., Li, Y., Shelton, J. M., Richardson, J. A., Spencer, E., Chen, Z. J., et al. (2000) Cytochrome c deficiency causes embryonic lethality and attenuates stress-induced apoptosis. *Cell* **101(4)**, 389–399.

97. Susin, S. A., Lorenzo, H. K., Zamzami, N., Marzo, I., Snow, B. E., Brothers, G. M., et al. (1999) Molecular character-ization of mitochondrial apoptosis-inducing factor. *Nature* **397(6718)**, 441–446.

98. Joza, N., Susin, S. A., Daugas, E., Stanford, W. L., Cho, S. K., Li, C. Y., et al. (2001) Essential role of the mitochondrial apoptosis-inducing factor in programmed cell death. *Nature* **410(6828)**, 549–554.

99. Coucouvanis, E. and Martin, G. R. (1995) Signals for death and survival: a two-step mechanism for cavitation in the vertebrate embryo. *Cell* **83(2)**, 279–287.

100. Kuida, K., Lippke, J. A., Ku, G., Harding, M. W., Livingston, D. J., Su, M. S., and Flavell, R. A. (1995) Altered cytokine export and apoptosis in mice deficient in interleukin-1ß converting enzyme. *Science* **267(5206)**, 2000–2003.

101. Li, P., Allen, H., Banerjee, S., Franklin, S., Herzog, L., Johnston, C., et al. (1995) Mice deficient in IL-1ß-converting enzyme are defective in production of mature IL-1ß and resistant to endotoxic shock. *Cell* **80(3)**, 401–411.

102. Bergeron, L., Perez, G. I., Macdonald, G., Shi, L., Sun, Y., Jurisicova, A., et al. (1998) Defects in regulation of apoptosis in caspase-2-deficient mice. *Genes Dev.* **12(9)**, 1304–1314.

103. Wang, S., Miura, M., Jung, Y. K., Zhu, H., Li, E., and Yuan, J. (1998) Murine caspase-11, an ICE-interacting protease, is essential for the activation of ICE. *Cell* **92(4)**, 501–509.

104. Ma, A., Pena, J. C., Chang, B., Margosian, E., Davidson, L., Alt, F. W., and Thompson, C. B. (1995) Bclx regulates the survival of double-positive thymocytes. *Proc. Natl. Acad. Sci. USA* **92(11)**, 4763–4767.

105. Lindsten, T., Ross, A. J., King, A., Zong, W. X., Rathmell, J. C., Shiels, H. A., et al. (2000) The combined functions of proapoptotic Bcl-2 family members bak and bax are essential for normal development of multiple tissues. *Mol. Cell* **6(6)**, 1389–1399.

12

Apoptosis and Cancer
Pathogenic and Therapeutic Implications

Sean L. O'Connor, Fermin Briones, Nikhil S. Chari, Song H. Cho,
Rebecca L. Hamm, Yoshihiko Kadowaki, Sangjun Lee,
Kevin B. Spurgers, and Timothy J. McDonnell

INTRODUCTION

Apoptotic pathways appear to have evolved along with the earliest metazoans. Homologs of human Bcl-2 and caspase family members have been identified in eukaryotes as ancient as the nematode *Caenhorhabditis elegans*. This observation suggests these pathways did not strictly evolve as a system to prevent cancer. However, it is clear that while these pathways likely evolved to aid development, they now also play a critical role in carcinogenesis in humans. In this chapter we discuss some of the examples of how alterations in the normal apoptotic pathways can lead to tumor formation and the relevance of apoptosis to cancer therapy.

Proper control of cellular growth can be viewed as an intricate network of signaling pathways starting at the cell surface and proceeding to the nucleus. A large number of genes along these pathways have been identified as potential mutational targets in human cancer. Abnormal activation of oncogenes, genes that normally stimulate cell growth, can result in an inability to respond to normal cellular growth signals and excessive proliferation. These types of mutations can activate tumor-suppressor activity and initiate apoptotic pathways leading to cell death. Loss of homeostatic control coupled with loss of tumor suppression can therefore result in tumor formation.

It is clear that many of our well-established cancer therapies require intact apoptotic pathways to destroy tumors. Gaining knowledge of the specific mutations in a tumor is leading to treatments that are specifically tailored for the causative mutation. Several groups are even developing gene-replacement therapies designed to correct these mutations. As the number of genes involved in cancer increases, so does the number of potential therapeutic targets. Advances in our understanding of oncogenes, tumor suppressors, and apoptotic regulators should certainly increase the effectiveness of cancer therapies.

This concept has recently been validated convincingly in chronic myelogenous leukemia (CML). The cytogenetic hallmark of this cancer is the t(9:22) translocation that results in constitutive bcr-abl tyrosine kinase activity. Recently completed clinical trials selectively targeting the bcr-abl tyrosine kinase demonstrates remarkable therapeutic effect in CML patients that may be attributable in part to apoptosis induction *(1–3)*.

From: *Essentials of Apoptosis: A Guide for Basic and Clinical Research*
Edited by: X-M. Yin and Z. Dong © Humana Press Inc., Totowa, NJ

This review will focus on several of the common molecular alterations in cell death regulatory proteins observed in cancer. The significance of these alterations in the context of new therapy development will be discussed.

TRAIL

The TNF family of proteins (tumor necrosis factor-α [TNF-α], FasL, and TNF-related apoptosis-inducing ligand [TRAIL]) are cytokines with multiple physiological roles. The ability of this family to induce apoptosis has been well-characterized in inflammatory and immune responses (4,5). This has led many investigators to assess the anti-cancer therapeutic capability of these molecules (4,6). However, systemic administration of TNF-α and FasL has proven to be too toxic for therapeutic use (7,8). A more recently identified member of TNF-α family, TRAIL, may be a better candidate for cancer therapy (6). Mice that have been engineered to lack TRAIL ligand are more susceptible to spontaneous and induced tumor formation supporting the role of TRAIL as a normal defense against cancer (9).

Similar to TNF-α and FasL, TRAIL induces apoptosis by binding to specific cell surface death receptors (see Chapter 5). The TRAIL receptors DR4 and DR5 contain cytoplasmic death domains (DD) required for transmitting apoptotic signals. Binding of TRAIL to DR4 and DR5 recruits the cytoplasmic adapter protein FADD (Fas-associated death domain), via the death domain. FADD then induces caspase-8, which in turn induces an apoptotic cascade through other caspases (10). The TRAIL receptors DcR1 and DcR2 are also able to bind TRAIL but lack the cytoplasmic death domains and are unable to induce apoptosis (11). These receptors are therefore referred to as decoy receptors.

Early experiments suggested TRAIL to be an excellent cancer therapeutic because it can induce apoptosis in tumors without any effect on normal tissues. Unfortunately, certain tumors appear resistant to TRAIL and the determinants of TRAIL specificity remains unclear. It was initially believed the relative distribution of DR4 and DR5 vs DcR1 and DcR2 controlled sensitivity to TRAIL (11); however, subsequent studies suggest that this may not be the only mechanism. Recent studies have suggested TRAIL-induced apoptosis is regulated by intracellular mechanisms, including currently unidentified factors (12).

The role of the FADD/caspase-8 complex in TRAIL sensitivity is also under investigation, with some evidence supporting its importance and other results suggesting the opposite (13,14). For example, cells derived from FADD$^{-/-}$ mice are still able to undergo TRAIL-induced apoptosis, suggesting the existence of another FADD-like adapter protein (13). In addition, it was reported that FLIP (FLICE-inhibitory protein), which has similar domains as caspase-8 but does not have an active enzymatic site, has some correlation with resistance to TRAIL-induced apoptosis in melanoma cell lines (15,16).

The PI3K (phosphoinositide 3-kinase)/Akt pathway may also regulate sensitivity to TRAIL. This pathway is an important intracellular mediator of survival signals (17,18). Akt phosphorylates and subsequently inactivates inducers of apoptosis, such as Bad, caspase-9, and the forkhead transcription factors, AFK and FKHR (19–23). Akt also activates the anti-apoptotic factor NF-κB. There is some evidence that PI3K-inhibitors can enhance sensitivity to TRAIL (24). Together, these experiments suggest several regulators of apoptosis are involved in controlling sensitivity to TRAIL.

It is hoped TRAIL will be a powerful new drug capable of selectively killing cancer cells. Unlike TNF and FAS ligand, TRAIL shows no toxic effects in nonhuman primate studies (25). Recombinant TRAIL exhibited effective growth suppression of human mammary adenocarcinoma xenografts with no apparent toxicity to normal tissue (26). These tumor cells show clear signs of apoptosis within 12 h of treatment. Similarly, a 2-d treatment with TRAIL prevented tumor formation in 20% of nude mice injected with a human colon carcinoma cell line (25). However, when TRIAL was used in combination with 5-FU, tumor formation was blocked in 90% of mice, demonstrating a strong synergistic effect. Moreover, TRAIL may be able to induce apoptosis even in chemotherapy resistant

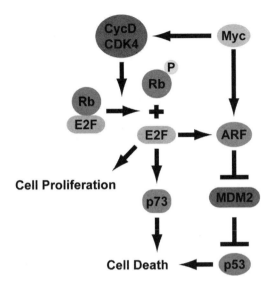

Fig. 1. Myc and E2F signaling. The proteins Myc and E2F participate in a series of signaling cascades that can result in either cell proliferation or cell death. Both Myc and E2F are largely regulated through their direct interactions with other proteins. Myc can form a heterodimer with Max, which leads to the transactivation of numerous downstream genes. E2F function is blocked by interactions with the retinoblastoma gene (RB). Phosphorylation of RB will free E2F from this repression and permit downstream signaling.

cancer cells that overexpress Bcl-2 or Bcl-X_L *(27)*. To develop TRAIL as a therapeutic drug, its side effects need to be taken into consideration. There is some evidence that human primary hepatocytes may be efficiently killed by TRAIL *(28)*. It is interesting that different recombinant versions of TRAIL appear to have varying anti-tumor abilities and normal tissue toxicity, suggesting that by manipulating the structure of TRAIL it may be possible to develop a highly effective therapeutic agent *(29)*.

E2F 1/RB

E2F1 is a critical regulator of cell cycle progression and was the first member of the E2F transcription factor family identified *(30)*. These transcription factors were first identified by their ability to bind the adenovirus E2 promoter. E2F1, like all E2F family members, functions as a heterodimer consisting of an E2F and a DP subunit. Six E2F genes and two DP genes have been identified and recent surveys of the human genome indicate many more may exist. Different forms of E2F can be found in most cell types, and these have been characterized largely by gel shift assays.

E2F1 is important for the initiation of DNA synthesis and induction of apoptosis (Fig. 1). E2F1 levels are modulated during the cell cycle and results from multiple experiments show that the level of E2F1 is rate-limiting for the G_1/S transition *(31)*. E2F1 is also regulated through its interactions with Rb proteins. E2F1/DP heterodimers can either be "free" or stably "complexed" to a member of the pRb family of proteins (pRb, P107, p130) *(31,32)*. A short, highly conserved domain near the carboxyl terminus of the E2F mediates binding to Rb family proteins *(33)*. This domain is embedded in the transactivation domain of the E2F1 subunit. The overlap of these two domains provides a model for how the association of pRb, p107, or p130 might inhibit E2F-dependent transcription.

The E2F/Rb axis is possibly the most commonly affected genetic pathway in human cancer *(34)*. Studies of E2F1 suggest it can function as both an oncogene and tumor suppressor. Overexpression of E2F1 is sufficient to drive quiescent cells into S phase *(32)*. Conversely, molecules that inhibit E2F1 dependent transcription can cause accumulation of cells in G_1. These data suggest E2F1 functions as an oncogene *(35–37)*. However, mice that lack E2F1 develop tumors and E2F1 can suppress skin carcinogenesis in a two-stage skin carcinogenesis assay, suggesting E2F1 functions as a tumor suppressor *(38–40)*.

The anticancer effects of several E2F1 pathway inhibitors are currently under investigation. The development of cyclin-dependent kinase (CDK) inhibitors like SU9516 that can deregulate E2F could serve as a plausible pharmacological strategy for cancer therapy *(41)*. DNA-binding drugs, which consist of an A + T selective DNA minor groove binding tripyrrole peptide and polyamine chains attached to a central pyrrole, can effectively inhibit E2F1 association with DNA promoter elements. Recent studies suggest E2F1 (viral delivery/overexpression) in combination with gemcitabine may be an alternative approach for treating pancreatic cancer. E2F1 gene therapy is also thought to be a promising treatment for esophageal cancer *(42)*. Further advances in the understanding of the other E2F family members could provide highly selective targets for cancer therapy in the future.

NF-κB

Nuclear factor-κB (NF-κB) is a transcription factor involved in the regulation of diverse cellular functions, including immune and inflammatory response, proliferation, apoptosis, and cell survival. In its inactive form, NF-κB exists in the cytoplasm as a dimer bound to an inhibitor protein, IκB (IκBα, β, γ, ε) *(43)*. The mammalian NF-κB family consist of five known subunits: 1) p100/p52 (NF-κB2) p100 is processed to an active p52 form, 2) p105/p50 (NF-κB1) p105 is cleaved to form p50, 3) Rel B, 4) c-Rel, and 5) p65 (Rel A). All share a conserved Rel homology domain that controls DNA binding, dimerization, and interactions with inhibitory factors *(43–45)*. The composition of the dimers in the cell is thought to control transcriptional specificity and regulation.

Activators of NF-κB include cytokines, growth factors, infectious agents, oxidative stress, pharmaceutical drugs, and γ-irradiation. These external signals stimulate kinases such as NF-κB-inducing kinase, MEKK1, and Akt, which activate the inhibitor of NF-κB protein kinase (IKK α, β, γ) *(43)*. This ser/thr kinase then phosphorylates IκB in the NH_2-terminal region, which ultimately results in the ubiquitination and degradation of IκB by a proteasome complex. This event unmasks the nuclear localization signal (NLS) sequence of NF-κB and frees the molecule for translocation to the nucleus *(43–45)*. Once in the nucleus, NF-κB dimers bind to the promoter region of genes containing κB sites, a loosely related group of DNA sequences about 10 base pairs in length, and promotes transcription of target genes. NF-κB can also be regulated by a direct phosphorylation of the subunits, which enhances the transactivating function of the transcription factor without affecting the nuclear translocation or the DNA binding activity *(46,47)*.

There are three mechanisms by which NF-κB activity becomes deregulated: 1) IκB mutation, 2) translocation/amplification of NF-κB subunits, and 3) constitutive IκB kinase activity. Each of these events can lead to the constitutive presence of NF-κB in the nucleus *(48)*. Consistent evidence from in vivo and in vitro studies exists to link these types of NF-κB mutations to tumor induction. Several members of the NF-κB and IκB are amplified or translocated in human cancers (Table 1).

NF-κB may promote oncogenesis in various ways, one of which may be to inhibit transformation-associated apoptosis. Increases in proliferation resulting from oncogene overexpression in cancer cells frequently makes the cell more susceptible to apoptotic signals. Malignant cells can overcome this adverse effect of proliferation by utilizing NF-κB activation to induce expression of genes involved in cell survival. NF-κB also inhibits cell-death induction from TNF-α, cytotoxic drugs, ionizing irradiation, and oxidative stress *(49–53)*. This may be accomplished by upregulating the expression of genes involved in cell survival, such as inhibitor of apoptosis proteins (IAPs), anti-apoptotic members of Bcl-2 family proteins, TNF-receptor associated factor 1 and 2 (TRAF1 and TRAF2), A20, and IEX-1L *(54–62)*. TRAF1 and TRAF2 interact with NF-κB-inducing kinase (NIK) to bring about phosphorylation and degradation of the I-κB family of NF-κB inhibitors *(63)*. c-IAPs block caspase-8 activation, a protease involved in execution of apoptosis *(59)*.

Table 1
Cancers Associated with NF-κB and C-myc Gene Deregulation

Cancer	Defect/Mutation	Reference
NF-κB-associated		
Hodgkin's disease	IκB-A mutations	*(207–209)*
B-cell chronic lymphocytic leukemia	Bcl-3 (t14,19) chromosomal translocation	*(210)*
B-cell neoplasmas	Bcl-3 overexpression	*(211)*
B-cell non-Hodgkin's lymphomas	NF-κB2 (t10,14) chromosomal translocation	*(212)*
Cutaneous lymphomas	NF-κB2 (t10,14) chromosomal translocation	*(213)*
Myc-associated		
Breast carcinoma	Increased expression through translation and protein stability by PI3K, and MAP kinase	*(214–217)*
Burkitt's lymphoma (occasionally large B-cell, lymphoblastic and follicular lymphomas and multiple myelomas)	Chromosomas translocations, pt. mutations	*(91,218–220)*
Prostate carcinoma	Increased copy number, RNA levels	*(221,222)*
Colon carcinoma	Increased expression (possibly transcriptional through APC)	*(223,224)*
Small cell lung cancer	Increased expression and amplification	*(225,226)*

Other oncogenic activities of NF-κB include regulation of genes involved in proliferation such as c-Myc and cyclin D1 *(64–66)*. In later stages of oncogenesis, NF-κB may promote tumorigenesis by upregulating the expression of various genes involved in cell adhesion, ICAM-1, and angiogenesis, COX-2, iNOS, GRO α *(43,67,68)*.

The effectiveness of chemotherapeutic drugs and ionizing radiation for cancer therapy is reduced by activation of NF-κB, which enhances cell survival. Inhibition of NF-κB reduces tumor growth in mouse models and sensitizes tumor cells to apoptosis induced by TNF-α, chemotherapeutic agents, and γ-irradiation *(48,69)*. These findings make NF-κB a potential target for cancer therapies.

Various strategies to prevent NF-κB activation are being developed. Nonspecific inhibitors of NF-κB including proteasome inhibitors, antioxidants, and anti-inflammatory drugs are effective in sensitizing the cancer cells to apoptosis induction from chemotherapeutic drugs and γ-irradiation *(53,70–75)*. Inhibitors with greater specificity to NF-κB are also being designed and show promising results in facilitating cell death in cancer cell lines. These include synthetic peptides that compete with NF-κB for the machinery responsible for nuclear translocation, anti-sense mRNA, and double-stranded oligonucleotides that contain promoter sequences of κB site and act as transcription decoys *(76–78)*. Even greater specificity may be obtained through the use of gene therapy to inhibit NF-κB. For example, adenovirus vector expression of IκBα "super repressor," so-called because the mutated phosphorylation site increases the half-life of the molecule, has been found to be effective in sensitizing resistant tumors to chemotherapy *(79,80)*. Although inhibition of NF-κB may not be sufficient in treating all tumors, sensitizing the cancer cells to apoptosis-inducing agents may allow reduction in dosage of chemotherapeutic drugs, thus making the treatment less toxic and more effective.

However, NF-κB is not unique to cancer cells and its activity is vital for the normal production of various cytokines and growth factors. It is also important to consider that under certain situations

NF-κB may be required for p53-mediated apoptosis by activating p53 transcription *(81,82)*. Therefore, in p53 wild-type cancers, inhibition of NF-κB may actually promote the process of tumorigenesis. In some cell systems, NF-κB is required for apoptotic response induced by Sindbis-infected cells, Fas, ischemia, and cytotoxic drugs *(83–87)*. Furthermore, NF-κB can either induce or repress apoptosis in thymocytes depending on the nature of the activation *(88)*. To make NF-κB an effective therapeutic target, further studies are needed to fully elucidate the mechanisms of NF-κB activation and its anti- and pro-apoptotic functions in cancer cells in specific contexts.

C-MYC

C-myc is a basic helix-loop-helix leucine zipper transcription factor that normally regulates cell-cycle progression and proliferation. C-myc forms a heterodimer with Max then binds to E-box sequences and transactivates effector genes. Extensive regulation of the c-myc/Max network exists through alternative binding partners to Max, including Mad, Mxi-1, Mnt, and Max itself. These alternative dimers can bind to the same E-box sequences as c-myc-Max heterodimers resulting in inhibition of transcription *(89)*.

C-myc is induced by mitogenic stimuli and is expressed throughout the entire cell cycle *(90)*. Overexpression of c-myc in quiescent fibroblasts in the absence of growth factors induces S-phase entry and c-myc-null fibroblasts proliferate at a slower rate *(91)*. Downstream targets of c-myc include: cyclin D2, ckd4, id2, and cdc25 (Fig. 1). C-myc has also been shown to repress cell-cycle inhibitors such as p21, p27, and GADD45 *(89)*. The combination of these two events enables a cell to be driven through the cell cycle even in the absence of a mitogenic stimulus.

C-myc is deregulated in many types of cancer including Burkitt's lymphoma (BL); osteosarcoma; glioblastoma; and carcinomas of the cervix, colon, breast, and prostate (Table 1) *(90)*. Deregulation occurs owing to several mechanisms including chromosomal translocation, direct mutation, and activation of upstream signaling pathways *(90)*. In BL, a highly aggressive B-cell malignancy, C-myc is overexpressed due to chromosomal translocations. The predominant translocation, (8:14)(q24:q32), results in an IgH enhancer element upstream of the c-myc coding region stimulating c-myc expression in B-cells *(91)*. Point mutations have also been identified that can dramatically increase the half-life of the C-myc protein or alter key phosphorylation sites *(92)*.

Increased expression of c-myc can also occur through alterations of upstream pathways such as the ErbB2 (HER2/neu) pathway *(92)*. ErbB2 is a receptor tyrosine kinase that is overexpressed and constitutively active in many mammary tumors. Downregulation of ErbB2 in SKBr3 breast tumor cells correlates with a decreased expression of c-myc, suggesting c-myc is a target of the ErbB2 pathway. This is associated with a downregulation of the MAP kinase and PI3K pathways. It has been demonstrated that the ras/MAP pathway stabilizes c-myc protein and the PI3K pathway is involved in the translational induction of c-myc. These results suggest that a culmination of upstream signaling pathways are important for regulation of c-myc and altered signaling through one or several pathways may lead to increased levels of c-myc and tumorigenesis. The therapeutic potential of selectively targeting these upstream pathways has recently been demonstrated in clinical trials using trastuzumab (Herceptin) in patients with metastatic breast cancer *(93)*.

C-myc can also affect tumorigenesis by controlling cellular differentiation. C-myc is usually downregulated in differentiating cells *(91)*. C-myc-fused to the IgH Eu enhancer was demonstrated to increase the number of cycling pre-B cells and decrease the number of mature B-cells *(91)*. It has been shown c-myc is able to repress transcription of C/EBP alpha, a gene involved in differentiation of many cells types. C-myc can also upregulate telomerase, which is associated with immortalization, and to downregulate LFA-1, collagen, and fibronectin, which leads to decreased cellular adhesion. C-myc also effects cellular metabolism by increasing levels of: serum lactate dehydrogenase A, enabling cells to survive under hypoxic conditions; and factors involved in protein synthesis (EIF5A, EIF4E), iron metabolism (TFRC, IRP2) and nucleotide synthesis (ODC, fibrillarin, nucleolin) *(91)*.

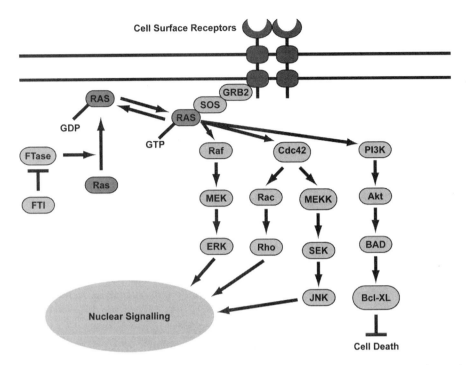

Fig. 2. Ras pathway. Ras has been demonstrated to regulate a large number of downstream pathways involved in proliferation, differentiation, and apoptosis. For Ras to be activated, it must first be transported to the cell membrane. Transport of Ras involves multiple post-translationally modifications including farnesylation, which is catalyzed by farnesyltransferase (FTase). Numerous compounds that inhibit FTase (FTI) are being tested for their chemotherapy potential. Ras is also regulated through its association with GDP and GTP. In its inactive state, Ras binds GDP, whereas in its active state, it binds GTP. Ras can hydrolyze GTP to GDP leading to the inactivation. GRB2 interacts with cell-surface receptors and forms a complex with SOS and GTP-Ras. The activated Ras complex can then stimulate multiple downstream pathways.

Although it is clear that c-myc is involved in controlling growth, it also appears to be involved in stimulating apoptosis. Initial observations demonstrated that overexpression of c-myc in an interleukin-3 (IL-3) dependent cell line in the absence of IL-3 results in the induction of apoptosis *(94)*. It was later discovered that c-myc can sensitize cells to a variety of apoptotic stimuli including heat shock, hypoxia, TNF-α, Fas, and DNA damage *(90)*. It has been suggested that c-myc-induced apoptosis can be either p53-dependent or -independent and that the induction of apoptosis may be modulated by the presence of growth/survival factors *(95)*, anti-apoptotic genes, including bcl-2 *(96)*, and p53 mutational status *(90)*.

It is apparent from the Eμ-myc transgenic mouse model of Burkitt's lymphoma that in order for tumors to develop in the context of primary c-myc gene deregulation, the affected cell must acquire compensatory mechanisms to overcome the susceptibility to undergo apoptosis. This may be achieved by one of several mechanisms. p53 is mutated in 30–40% of BL and ARF is deleted in several BL cell lines with wild-type p53 *(90)*. It has been demonstrated in animal models of lymphoma that c-myc deregulation and p53 inactivation are potently synergistic *(97)*. Caspase-8 is often methylated and silenced in aggressive neuroblastomas, which are resistant to apoptosis by Fas, TNF, and doxorubi-cin *(90)*. Many neuroblastomas also carry deletions of chromosome 1p36, which carries the gene for caspase-9 *(90)*. An alternative method through which tumorigenic cells inhibit c-myc-induced apoptosis is by upregulation of anti-apoptotic factors. Bcl-2 is able to cooperate with c-myc in tumor formation in transgenic mice *(96)*.

Current experimental therapies against c-myc-induced tumorigenesis focus on downregulation of c-myc at either the mRNA or protein level. Some initial approaches involved the use of anti-sense oligonucleotides that bind to c-myc mRNA and inhibit translation. Anti-sense oligonucleotides have been shown to downregulate c-myc levels and inhibit cellular proliferation in several cell lines including HL-60, melanoma, colon carcinoma, and breast carcinoma *(98–101)*. Recent approaches have focused on reducing c-myc levels at the transcriptional level through the use of triplex DNA formation. C-myc transcription can be blocked by using an oligonucleotide that is selective to a nuclease-hypersensitive element, upstream of the promoter P1. The mechanism involves the binding of sequence-specific oligonucleotides to a polypurine-rich strand of duplex DNA in the major groove of DNA, thus inhibiting transcription-factor access *(102)*. This strategy results in downregulation of c-myc mRNA, reduced proliferation, and induction of apoptosis *(102)*. Currently, these antisense and triplex-forming strategies are in preclinical phase studies.

The importance of c-myc in the development of the malignant phenotype is exemplified by a recent study using a tetracycline-inducible c-myc. Hematopoietic cells induced to overexpress c-myc developed leukemias and lymphomas in mice. When c-myc expression was turned off, the tumors regressed and normal differentiation and proliferation occurred *(90)*. This study suggests that therapies targeted to c-myc alone may be sufficient to treat some types of tumors, despite the multistep carcinogenesis model.

RAS

Ras proteins are some of the most important molecular switches for regulating transmission of extracellular signals such as growth factors, hormones, and cytokines (Fig. 2). Ras signaling can regulate cell proliferation, differentiation, and apoptosis. Alterations in ras signaling can lead to tumor formation and progression.

There are several ras family gene isotypes including Harvey (H), Kirsten (K), and neuroblastoma (N). These genes encode 21-kDa G proteins localized to the inner leaflet of the plasma membrane. In order for Ras to be transported to the plasma membrane, it must first go through several post-translational modifications. These modifications consist of prenylation, proteolysis, carboxylmethylation, and palmitoylation. Once Ras has been transported to the membrane, it is largely regulated by its association with either GTP (active state) or GDP (inactive state). Like other G proteins, this cycle is tightly regulated by Ras-associated proteins. GTPase-activating proteins (GAP) hydrolyze GTP to GDP and guanine nucleotide exchange factors (GEF) facilitate GDP release. Once stimulated by external signals, Ras proteins are able to activate several downstream signaling pathways (Fig. 2).

Ras has been implicated in both protection from and promotion of apoptosis depending on the type external signals and genetic background. For example, Ras-mediated activation of the PI3K-PKB/ Ark pathway in fibroblasts can inhibit c-myc-induced apoptosis by inactivating the Bad protein *(103)*. The Bad protein can also be targeted for degradation by being phosphorylated through the Ras-mediated activation of PKB *(19)*. Because Bad inhibits the anti-apoptotic protein Bcl-X_L, loss of Bad function promotes cell survival (104). Recently, raf, a Ras downstream kinase in the ERK pathway, has been shown to phosphorylate Bad *(105)*. Conversely, Ras has also been implicated in induction of apoptosis. The Ras protein was reported to function in Fas-mediated induction of apoptosis *(106)*. Furthermore, in a human T-lymphoblastoid cell line (Jurkat) and murine fibroblasts, oncogenic Ras expression induces apoptosis flowing inhibition of protein kinase C (PKC) activity *(107)*. These examples illustrate the importance of cellular context in death/survival signaling.

The large number of proteins and pathways that Ras activates and the wide range of cellular processes that it can control makes Ras a frequent site of mutation in human cancer. The two most frequent alterations of Ras in cancer are point mutations and mutations that lead to constitute Ras activity. The vast majority of Ras point mutations occur in codons 12, 13, and 61. These mutations prevent GTP hydrolysis, thus locking Ras into an active state. The second common class of Ras mutations result in high expression of the Ras protein due to deregulation at the transcriptional level. Approximately 30% of human tumors contain mutations that inappropriately activate the Ras pathway. As many as 90% of adenocarcinomas of the pancreas, 50% of colon, and 30% of lung have ras gene mutations *(108)*. Thyroid cancers (50%) and myeloid leukemias (30%) also have been shown to express mutated Ras proteins *(109)*.

Because of the high incidence of Ras mutation in human cancer, several groups are focusing on ras signaling in the development of cancer therapy. In general, three approaches are being considered: 1) inhibition of Ras protein production through antisense oligonucleotides; 2) blocking Ras activation by inhibiting required post-translational modifications; and 3) inhibition of downstream ras-signaling molecules *(110–116)*.

One approach to reduce oncogenic Ras expression is through the use of antisense oligonucleotides. These small complementary fragments are designed to hybridize with ras mRNA and block protein translation. Therapeutic efficiencies depend greatly on the delivery system of the nucleotides to the tumor sites *(110)*. Transfections using adenoviral vector or liposome carrying the nucleotides are commonly used for this purpose. It has been demonstrated that an adenoviral vector containing K-ras antisense sequence can inhibit growth of human lung tumor xenografts growing in nude mice *(111)*.

A critical step in Ras activation is its post-translational modifications required for proper trafficking of Ras to the membrane. One of the enzymes that catalyzes these modifications is farnesyl-transferase, which attaches a 15 carbon moiety to the Ras protein *(117)*. The inhibition of farnesylation prevents Ras from reaching the membrane and therefore should prevent the tumorigenic effects of activating Ras mutations. Some in vitro studies have shown this inhibition can induce apoptosis *(118–120)*. Numerous farnesyltransferase inhibitors have been identified and tested for anti-tumor effects. Several reports have shown these inhibitors have effective tumor regressions in clinical trials *(112,113)*. However, an increasing number of studies show that these compounds have anti-tumor effects in cells that do not carry activating Ras mutations *(114)*. The exact mechanism of these possible Ras-independent anti-tumor compounds remains to be determined.

Interrupting downstream ras signaling molecules is another approach in cancer therapy. Inhibitors for Raf and MEK, which are key kinases for Ras-mediated MAP kinase pathways, have also induced significant regressions in several kinds of tumors *(115)*. However, the clinical application of these compounds may be limited due to toxic effects caused by inhibiting other MAPK signaling molecules, which are critical for normal cellular functions *(116)*.

BCL-2 FAMILY

The Bcl-2 gene family consists of anti-apoptotic and pro-apoptotic proteins with opposing biological functions that either inhibit or promote cell death (*see* Chapter 2). The anti-apoptotic members such as Bcl-2, Bcl-X_L, Mcl-1, and Bcl-$_w$ suppress cell death and pro-apoptotic family members such as Bax, Bak, Bad, and Bid promote cell death *(121)*. These proteins share conserved Bcl-2 homology (BH) domains that have functional and structural significance and mediate homodimer and/or heterodimer interactions *(122)*.

The anti-apoptotic members of the Bcl-2 family have been shown to be important contributors to the development of therapeutic resistance of a wide variety of cancers. Transgenic mouse models of bcl-2 gene deregulation consistently demonstrate the oncogenic potential of bcl-2 in lymphoid, cutaneous, and prostatic malignancies *(123–125)*.

The high levels of the Bcl-2 protein in cancers may confer resistance to therapeutic cell-death induction by chemotherapeutic drugs and radiation. This effect may be mediated by inhibiting the release of cytochrome-*c* from mitochondria or by forming heterodimers with pro-apoptotic Bcl-2 family members *(122)*. Therefore, inhibiting Bcl-2 or Bcl-X_L may restore apoptotic pathways and sensitize cancer cells to chemotherapeutic agents or radiation. Several studies have demonstrated that antisense oligonucleotides targeted against Bcl-2 may be beneficial as therapeutic agents *(126,127)*. Similarly, single-chain antibodies sensitized overexpressing Bcl-2 breast cancer cells to drug-mediated cytotoxicity *(128)*. Natural products have also been found to antagonize the anti-apoptotic effect of Bcl-2, including Tetrocarcin-A *(129)*.

Bcl-X_L is homologous to Bcl-2 and overexpression of Bcl-X_L is characteristic of specific cancers, including primary colorectal carcinomas *(130)*. The ability of Bcl-X_L to inhibit apoptosis is modulated by the Akt signaling axis *(19)*. Activated Akt phosphorylates the pro-apoptotic Bcl-2 family member Bad, thereby releasing its inhibition of Bcl-X_L *(131)*. Cell lines lacking PTEN expression have elevated levels of phospho-Akt, which may account for PTEN tumor-suppressor activity in breast and prostate cancer *(132,133)*.

The pro-apoptotic members of the Bcl-2 family may be divided into two groups based on the presence of BH domains. Some members, such as Bak and Bax, possess BH1–BH3 domains. Others, including Bid and Bad, possess only the BH3 domain. Bax is a p53-regulated protein that is involved in the induction of apoptosis in response to specific stimuli *(134)*. Bax has been shown to exhibit tumor-suppressor activity in vivo *(135)* and inactivating bax mutations have been identified in selected cancers *(136)*. The absence of Bax in colorectal carcinoma cells completely abolished the apoptotic response to nonsteroidal anti-inflammatory drugs (NSAIDs) *(137)*.

Decreased expression of Bak has been observed in breast *(138)* and colorectal cancers *(130)*. The BH3 domain of pro-apoptotic Bcl-2 family members, including Bad and Bim, can interact with the hydrophobic cleft formed by the BH1–3 domains of the anti-apoptotic family members *(139)*. It is possible this interaction serves to sequester the BH3 domain-only molecules and prevent activation of bax and bak. Importantly, it has recently been demonstrated cells that are deficient in both bax and bak are resistant to cell-death induction by BH3 domain-only molecules and a variety of other agents *(140)*. Synthetic peptides that disrupt this interaction may sensitize cancer cells to apoptosis induction *(141)*. Another strategy under consideration involves adenoviral gene transfer of pro-apoptotic Bcl-2 family members *(142,143)*. The utility of tissue-specific promoters in conferring targeted transgene expression has the potential to enhance the selectivity of these vectors *(144)*.

Coincident with our understanding of the three-dimensional structure of Bcl-2 family member proteins *(145)* is the ability to manipulate the function of these proteins. This is already leading to the development of small molecular-weight molecules that have enormous potential as therapeutic agents for diseases, including cancer, characterized by disordered cell death *(146–148)*.

P53

The p53 gene is comprised of 11 exons that code for a 393 amino acid, 53kDa phosphoprotein (Fig. 3). The protein product is a transcription factor that binds to specific DNA sequences to regulate the expression of target genes. p53 possesses, in general, four functional domains. These domains include an amino terminal transactivation domain (amino acids 1–42), central sequence-specific DNA binding domain (amino acids 102–292), tetramerization domain (amino acids 320–360), and a highly basic C-terminal region (amino acids 363–393) *(149)*. p53 most efficiently binds to its consensus sequence as a tetramer *(150)*. Certain mutant p53 proteins can act in a dominant negative fashion, forming inactive tetramers with wild-type p53. The regulatory C-terminal domain can influence the ability of p53 to bind DNA in a sequence-specific manner *(149)*. Post-translational modifications or deletions of this region can activate sequence-specific DNA binding.

p53 functions as a tumor suppressor. Consistent with this function, homozygous and heterozygous p53 knockout mice are highly susceptible to early-onset tumor formation *(151)*. Similarly, Li-Fraumeni syndrome, a rare familial cancer disorder, illustrates p53's role as a tumor suppressor in humans *(152)*. Affected individuals typically develop tumors at an early age and multiple primary tumors are not uncommon. It is estimated that 50% of Li-Fraumeni syndrome family members will develop cancer by age 30, compared to 1% for the general population *(152)*. Diverse tumor types are observed including sarcomas, breast cancer, leukemia, and brain cancers. Germ-line p53 mutations

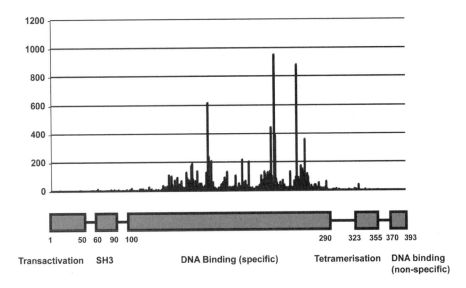

Fig. 3. p53. The p53 protein can be divided into five functional domains (transactivation, SH3, sequence-specific DNA binding, tetramerization, and sequence-nonspecific DNA binding). The location of somatic mutations of p53 isolated from tumors has been compiled by the International Agency for Research on Cancer (http://www.iarc.fr/p53/index.html). This database currently has over 15,000 entries. This histogram was generated using the data from the IARC database and uses the same N-terminus to C-terminus orientation and scale as the p53 schematic below the graph. Over 95% of tumor associated p53 somatic mutations map to the sequence-specific DNA binding domain. Several mutational hotspots have been identified in this domain. The majority of these mutations are point mutations that result in single amino acid substitutions.

were identified in affected members of Li-Fraumeni families. Importantly, tumors from these individuals exhibit loss of the remaining p53 allele *(153)*.

Under normal conditions, p53 protein is present at low levels in the cytoplasm, and is inactive for DNA binding and transactivation. These characteristics of p53 are controlled, at least in part, by the oncoprotein MDM2 *(154)*. MDM2 physically interacts with p53, acts as a ubiquitin ligase, and targets p53 for degradation by the proteosome. In addition to controlling the degradation rate and protein level of p53, MDM2 also helps maintain p53 in an inactive state. This inhibition is thought to exist because MDM2 binds directly to the transactivation domain of p53. Finally, MDM2 helps maintain the cytoplasmic localization of p53 in normal, resting cells *(155)*. MDM2 contains both a nuclear localization signal (NLS) and a nuclear export signal (NES), permitting MDM2 to constantly shuttle between the nucleus and cytoplasm. As MDM2 is exported, bound p53 shuttles to the cytoplasm, where it is degraded by the proteosome.

High levels of active p53 protein will accumulate in the nucleus in response to various forms of cell stress. The most well-characterized stimulus for nuclear p53 accumulation is DNA damage. For example, exposure of cells to either ultraviolet (UV) or ionizing radiation results in a dose-dependent increase in nuclear p53 levels *(156)*. Furthermore, many common chemotherapeutic agents, including as mitomycin C, cisplatin, doxorubicin, etoposide, and 5-FU, are known to damage DNA and induce p53 protein accumulation *(157)*. Other forms of cell stress that do not directly damage DNA, such as hypoxia, heat shock, nucleotide depletion, and oxidative stress can also lead to p53 accumulation *(158,159)*.

For active p53 to accumulate in the nucleus, the inhibition by MDM2 must be overcome. One potential mechanism by which this takes place is the phosphorylation of p53 in response to DNA damage *(160)*. Phosphorylation events at key serine residues in the N-terminus of p53 disrupt the p53/MDM2 interaction. When this occurs, p53 is temporarily free of MDM2-mediated ubiquitination, cytoplasmic shuttling, and degradation. p53 protein, active for DNA binding and transactivation, then accumulates in the nucleus.

Forced expression of exogenous p53 or activation of endogenous p53 can produce cell-cycle arrest. Furthermore, DNA damage produced in cells harboring wild-type p53 almost invariably leads to nuclear p53 accumulation and cell cycle arrest *(156)*. Importantly, cells with mutant or no p53 exhibit no cell-cycle arrest after DNA damage, suggesting the presence of p53 is required for cell-cycle arrest following DNA damage. p53 regulates cell-cycle arrest through the transactivation of the cellular gene p21$^{\text{waf1/cip1/sdi1}}$, a cyclin-dependent kinase inhibitor, and is strongly induced upon expression of exogenous p53 *(161)*.

Induced expression of exogenous p53 can also result in loss of cell viability and features of apoptosis, including chromatin condensation, nuclear disintegration, and DNA fragmentation *(162)*. Furthermore, the use of an inducible p53 vector results in regression of tumor xenografts associated with the induction of apoptosis *(162)*. Importantly, the activation of endogenous p53 is required for the induction of apoptosis in some cases, however, p53-independent apoptotic pathways do exist.

p53 can induce apoptosis by activating the mitochondrial pathway leading to caspase activation. Expression of wild-type p53 leads to cytochrome *c* release from mitochondria, processing of procaspase-9, and cleavage of caspase-3 *(163)*. Death receptor-mediated caspase activation may also be associated with p53-dependent apoptosis *(164)*. An increase in Fas protein levels is observed following adenoviral p53 gene transfer in human lung cancer cells. Importantly, northern and nuclear run-on experiments demonstrate a transcriptional regulation of Fas by p53. In contrast, p53 expression in vascular smooth-muscle cells results in increased cell-surface Fas expression due to increased Fas trafficking to the cell surface. Finally, expression of ectopic p53, or p53 activation in response to DNA damage, results in a p53-dependent upregulation of DR5, the death receptor for the ligand TRAIL.

In addition to Fas and DR5, p53 can transactivate a large number of genes that play a role in apoptosis *(165)*. For example, Bax is a direct transcriptional target of p53. During p53-dependent apoptosis, the Bax protein level increases and translocates from the cytosol to mitochondria. Noxa is another Bcl-2 protein family member directly induced by p53 *(166)*. An additional p53 target shown to have a direct role in apoptosis in p53AIP1. This gene is strongly induced after Ad-p53 infection, exposure to ionizing radiation, or treatment with adriamycin *(167)*. Like Bax and Noxa, p53AIP1 is a direct transcriptional target of p53, localizes to mitochondria and can, by itself, induce apoptosis.

p53 has been referred to as the "guardian of the genome" *(168)*. In response to DNA damage, p53 protein accumulates in the nucleus and induces a transient cell-cycle arrest, giving the cell time to make critical repairs. If the damage is too severe and/or if repair is not successful, p53 can induce apoptosis. In this way, p53 monitors the integrity of the genome and prevents the accumulation of deleterious mutations that could eventually contribute to a malignant phenotype. If p53 is mutated and cannot respond to DNA damage, then it cannot prevent the propagation of mutations in other genes. Such cells are then more easily transformed. This idea is supported by the observation that mutation of the p53 gene is the most common genetic alteration occurring in human tumors *(169)*. More specifically, it is estimated that greater than 50% of all human tumors contain a mutant p53 gene.

Mutant p53 proteins can have a greatly extended half-life, can leave altered conformation, and are capable of interfering with wild-type p53 function *(170)*. The mutations occurring in p53 are not random. The vast majority are missense mutations and over 90% of these occur in one of the four highly conserved regions within the central DNA binding domain (Fig. 3) *(169)*. These conserved regions are responsible for contacting DNA during sequence specific binding. Mutational hotspots exist at codons 175, 248, and 273. Co-transfection assays using p53 responsive reporter constructs demonstrate that codon 175, 248, and 273 mutants lose the ability to bind DNA and activate transcription *(171)*. These dominant-negative p53 proteins interfere with the transcription factor function of wild-type p53. Mutant p53 proteins can also lose the ability to induce apoptosis.

The presence of p53 mutations can have important implications for cancer therapy. Many anticancer therapies, including ionizing radiation (IR), cisplatin, mitomycin C, etoposide, doxorubicin, and 5-FU, directly or indirectly cause DNA damage, induce nuclear p53 accumulation, and kill cells by inducing apoptosis *(172)*. The presence of wild-type p53 is required for the cytotoxic action of some of these therapies. The influence of p53 status was evaluated in a National Cancer Institute drug screen using 123 standard anti-cancer agents and 60 human cancer cell lines *(173)*. In general, compared with wild-type p53 cells, cell lines harboring mutant p53 genes are less sensitive to most agents tested.

p53 status can also affect treatment outcome in vivo. Transformed mouse embryonic fibroblasts (MEFs) will form tumors in nude mice regardless of p53 status. However, only tumors formed from wild-type p53 MEFs will undergo apoptosis and regress following treatment with IR or adriamycin *(174)*. This study demonstrates a direct relationship between p53 status, induction of apoptosis, and response to therapy. Investigations of several human tumor types also find a link between p53 status, resistance to treatment, and prognosis *(175)*. Several investigators have observed that breast tumors expressing mutant p53 are more resistant than wild-type p53 tumors to treatments such as 5-FU, doxorubicin, methotrexate, and mitomycin-C. Furthermore, p53 mutation in breast cancer can be a strong predictor of relapse and shortened survival. Cancers of the colon, esophagus, ovary, and lymphoid cancers show a similar link between p53 mutation, therapy resistance, and prognosis. However, it is important to point out that some studies fail to find a correlation between p53 status and response to therapy. In some cases, the presence of mutant p53 correlates with increased sensitivity to treatment.

Advances in the field of gene therapy now allow for the introduction of exogenous wild-type p53 directly into tumors. This is being investigated as an anti-cancer treatment based on the preclinical observations that forced p53 expression can directly induce apoptosis and can increase sensitivity to other treatments. p53 gene transfer and expression is now achieved primarily by using replication-deficient adenoviral vectors engineered to drive p53 expression from a CMV promoter. This approach has proven efficacy in preclinical studies using several different human cancer cell types. For example, Ad-p53 infection can directly induce apoptosis in HeLa, colon, bladder, osteosarcoma, and prostate cancer cell lines *(163,176–180)*. Importantly, in some cell types, Ad-p53 infection can induce apoptosis independent of endogenous p53 status. Prostate cancer Tsu-pr1 and PC3 cells express no endogenous p53, whereas LNCap cells are wild-type p53 and Du145 cells express high levels of mutant p53. Despite these differences in p53 genotype, each cell line can be induced to undergo apoptosis by Ad-p53 treatment *(179,180)*. Ad-p53 treatment can also suppress tumor growth in nude mice. Ex vivo infection of prostate cancer cells or in vivo, intratumoral injection of head and neck tumors prevents tumor growth or causes tumor regression in mice *(180,181)*. However, in some reported cases, Ad-p53 treatment has minimal effect on cells harboring wild-type p53 alleles *(182)*.

Synergism between Ad-p53 infection and other cancer treatments has also been observed in preclinical development *(175)*. For example, the SW620 colon cancer cell line was treated with Ad-p53 in combination with ionizing radiation. TUNEL assay detects more apoptotic cells after combination treatment compared to either treatment alone. Additionally, Ad-p53 sensitizes lung cancer and glioma cells to radiation-induced apoptosis *(182)*. Encouraging results from preclinical experiments prompted Ad-p53 Phase I trials to be conducted. Early trials established the feasibility and safety of this approach in human patients. One of the first Phase I trials involving p53 gene transfer utilized a retroviral vector to treat lung cancer patients *(183)*. Nine patients, all with tumors harboring a mutated p53 gene, were given intratumoral injections of p53 on five consecutive days. Six patients displayed tumor regression or stabilized disease. p53 transgene expression was detected along with the induction of apoptosis. A second Phase I trial evaluated the treatment of non-small cell lung cancer (NSCLC) with Ad-p53 in combination with cisplatin *(184)*. p53 transgene mRNA was detected in

43% of patients and no severe toxicity attributable to Ad-p53 treatment was documented. Seventeen of 24 patients achieved stable disease and two demonstrated partial response.

All subsequent trials have not produced results supporting the continued use of Ad-p53 in the clinic. For example, 25 patients with nonresectable NSCLC were enrolled in a Phase II clinical trial investigating Ad-p53 treatment in combination with two chemotherapy regimens *(185)*. No patients had been previously treated with chemotherapy and all had tumors with either mutated or deleted p53. All patients received chemotherapy and intratumoral Ad-p53 injections. Tumor response was evaluated in each patient by measuring the response of two lesions, each in the same organ, one treated with Ad-p53 and one untreated. There was no difference in the clinical response of lesions treated with Ad-p53 plus chemotherapy compared to lesions treated with chemotherapy alone. Results from additional, ongoing Phase II and Phase III clinical trials are needed to fully assess the benefit of Ad-p53 treatment.

CASPASES

Caspases are a family of proteases vital for proper regulation of apoptosis *(see* Chapter 1). The term caspase is derived from the fact that these proteins are cysteine proteases with aspartate sequence specificity. Caspases are regulated in part by the translation of a proenzyme followed by cleavage to an active form. In certain cases, this cleavage is autocatalytic. At least 10 different caspases have been identified in humans with homologous proteins existing in most metazoans. Caspases can be separated into two groups, effectors and executioners. The effector caspases can be activated by multiple pathways, some involving the mitochondria and others independent of this organelle. Once effector caspases become active, they cleave and activate the downstream executioner caspases. These downstream caspases then cleave and activate a series of molecules, which are involved in terminal apoptotic events such as chromatin condensation, nuclear disintegration, and DNA fragmentation. Loss of proper caspase function is implicated in tumor formation. Without properly functioning caspases, the upstream tumor-suppressor genes may be unable to activate the apoptosis pathway. Caspases are therefore becoming an increasingly important prognostic indicator and target for therapy.

The study of multiple tumor types reveals changes in caspase expression and/or activity. For example, analysis of cancer of the colon shows decreases in caspases-1, -7, -8, and -9 relative to normal colon tissue *(186,187)*. Analysis of neuroblatomas and renal carcinomas has identified similar caspase downregulation *(188,189)*. At least one colon carcinoma study has demonstrated that alterations in caspase-3 activity correlates with a higher risk of recurrence, thus indicating that caspase-3 may be an important prognostic indicator *(190)*.

A growing body of work suggests that there are multiple mechanisms behind this loss of caspase function in tumors. Caspases can be inactivated by either direct mutation, silencing by methylation, or overexpression of an endogenous caspase inhibitor. Many caspases have promoter sequences that contain large CpG islands that can become inactivated by DNA methylation *(191)*. The inhibition of caspases by methylation of their promoter sequences has been observed in melanomas, neuroblastomas, breast carcinomas, and renal carcinomas *(188,189,192,193)*. Although the mechanisms are not fully understood, the proteins FAP-1, FLIP, and survivin are potent inhibitors of caspase activity. Analysis of neuroblastomas, melanomas, colon adenocarcinomas, and T-cell leukemias has revealed increased expression of at least one of these inhibitors in many of these tumors *(189,194–196)*. These alterations presumably prevent apoptosis in tumor cells by inhibiting the caspase pathways.

As the number of genes known to be mutated in cancer grows, researchers are becoming more interested in using gene therapy to treat cancer. Several therapies have focused on gene replacement to correct tumor-linked mutations, most notably p53 *(197)*. A major limit to gene-replacement therapy is the vast number of genes that can be mutated in cancer and the difficulty of identifying all of these mutations. It may therefore be potentially useful to develop therapies that are effective regardless of the causative defect. Because caspases are downstream of most oncogenes and tumor suppressors, they are promising candidates for gene-based therapies. Experiments using adenovirus-mediated caspase transfer have demonstrated that replacement of the appropriate caspase gene can restore normal apoptosis *(198,199)*. One approach currently being pursued is to bypass much of the normal apoptotic signaling pathway and to go directly to activating caspases in tumor cells. The normal function of caspases as the final apoptosis signal transducers makes them an obvious choice for gene therapy. This approach has many similarities to systems using HSV-tk, nitroreductase, or purine nucleoside phosphorylase, which catalyze pro-drugs into lethal compounds. Although both caspase and pro-drug based systems induced apoptosis, the later typically requires rapid cell division and can leach toxins to surrounding healthy tissue. Caspases have the advantage of functioning in a cell cycle-independent manner, do not affect neighboring cells, and activate a cell's normal death pathway. By placing the caspase under the control of tissue specific promoters and using a viral delivery system, researchers have been able to induce apoptosis in cultured tumor cells *(200,201)*. One group has added a further level of control by engineering a version of caspase-9 that also requires chemical inducers *(201)*.

Proper regulation of the caspase pathway is not only achieved through inducers but also by the activity of a family of proteins known as inhibitors of apoptosis (IAP). So far, eight members of this family have been identified in humans (NAIP, cIAP1, cIAP2, XIAP, Ts-XIAP, ML-IAP, Apollon, and Survivin), as well as several homologs in other organisms *(202)*. These proteins all contain at least one BIR (Baculovirus IAP Repeat) domain, a unique zinc-finger protein-binding motif *(202)*. The analysis of several IAP family members suggest that the BIR domain is able to inhibit caspase activity by binding directly to the caspase active site *(203)*. The mechanism of IAPs is not fully understood and certain anti-apoptotic actives of these proteins may not involve direct caspase inhibition and require other protein interactions. IAPs are normally absent from adult tissue but are present in fetal tissue and many cancers *(204)*. Presumably, the upregulation of these inhibitors gives a survival advantage to the tumor cells. These proteins have therefore attracted interest as possible therapeutic targets.

The inhibitor that has attracted the most attention is survivin. A dominant-negative form of this protein containing a threonine to alanine substitution at amino acid 34 (T34A) prevents the normal phosphorylation of the protein *(205)*. When this dominant-negative form of survivin is introduced into tumor cells using a nonreplicating adenovirus, it is an effective inducer of apoptosis *(206)*. This treatment also does not induce apoptosis in normal tissue and shows synergistic effects with some established chemotherapy agents. The effectiveness of this treatment is also independent of p53 or RB status. As with many promising anti-cancer therapies, clinical trials will need to be performed to determine whether these treatments will work in humans.

CONCLUSION

Humans have acquired numerous defenses against tumor formation. If cells lose homeostatic control, then apoptotic pathways act to eliminate the cells before tumor formation can occur. Failure to properly initiate and execute apoptosis can clearly result in cancer progression. Many of our anti-cancer treatments are able to bypass the normal signaling pathways and induce apoptosis, even when there are mutations upstream. The failure of a cancer to respond to therapy is often associated with an altered apoptotic response. As our knowledge of the molecular regulation of cell death expands, it will be possible to identify new therapeutic targets and to develop effective treatment strategies for currently refractory malignancies.

REFERENCES

1. Fang, G., Kim, C. N., Perkins, C. L., Ramadevi, N., Winton, E., Wittmann, S., and Bhalla, K. N. (2000) CGP57148B (STI-571) induces differentiation and apoptosis and sensitizes Bcr-Abl-positive human leukemia cells to apoptosis due to antileukemic drugs. *Blood* **96**, 2246–2253.

2. Mauro, M. J., O'Dwyer, M., Heinrich, M. C., and Druker, B. J. (2002) STI571: a paradigm of new agents for cancer therapeutics. *J. Clin. Oncol.* **20**, 325–334.

3. O'Dwyer, M. E., Mauro, M. J., and Druker, B. J. (2002) Recent advancements in the treatment of chronic myelogenous leukemia. *Annu. Rev. Med.* **53**, 369–381.

4. Old, L. J. (1985) Tumor necrosis factor (TNF). *Science.* **230**, 630–632.

5. Nagata, S. (1997) Apoptosis by death factor. *Cell* **88**, 355–365.

6. Wiley, S. R., Schooley, K., Smolak, P. J., Din, W. S., Huang, C. P., Nicholl, J. K., et al. (1995) Identification and characterization of a new member of the TNF family that induces apoptosis. *Immunity* **3**, 673–682.

7. Schilling, P. J., Murray, J. L., and Markowitz, A. B. (1992) Novel tumor necrosis factor toxic effects. Pulmonary hemorrhage and severe hepatic dysfunction. *Cancer* **69**, 256–260.

8. Ogasawara, J., Watanabe-Fukunaga, R., Adachi, M., Matsuzawa, A., Kasugai, T., Kitamura, Y., et al. (1993) Lethal effect of the anti-Fas antibody in mice. *Nature* **364**, 806–809.

9. Cretney, E., Takeda, K., Yagita, H., Glaccum, M., Peschon, J. J., and Smyth, M. J. (2002) Increased susceptibility to tumor initiation and metastasis in TNF-related apoptosis-inducing ligand-deficient mice. *J. Immunol.* **168**, 1356–1361.

10. Peter, M. E. (2000) The TRAIL DIScussion: it is FADD and caspase-8! *Cell Death Diff.* **7**, 759–760.

11. Griffith, T.S. and Lynch, D.H. (1998) TRAIL: a molecule with multiple receptors and control mechanisms. *Curr. Opin. Immunol.* **10**, 559–563.

12. Griffith, T. S., Rauch, C. T., Smolak, P. J., Waugh, J. Y., Boiani, N., Lynch, D. H., et al. (1999) Functional analysis of TRAIL receptors using monoclonal antibodies. *J. Immunol.* **162**, 2597–2605.

13. Yeh, W. C., Pompa, J. L., McCurrach, M. E., Shu, H. B., Elia, A. J., Shahinian, A., et al. (1998) FADD: essential for embryo development and signaling from some, but not all, inducers of apoptosis. *Science* **279**, 1954–1958.

14. Wang, J., Zheng, L., Lobito, A., Chan, F. K., Dale, J., Sneller, M., et al. (1999) Inherited human Caspase 10 mutations underlie defective lymphocyte and dendritic cell apoptosis in autoimmune lymphoproliferative syndrome type II. *Cell* **98**, 47–58.

15. Irmler, M., Thome, M., Hahne, M., Schneider, P., Hofmann, K., Steiner, V., et al. (1997) Inhibition of death receptor signals by cellular FLIP. *Nature* **388**, 190–195.

16. Griffith, T. S., Chin, W. A., Jackson, G. C., Lynch, D. H., and Kubin, M. Z. (1998) Intracellular regulation of TRAIL-induced apoptosis in human melanoma cells. *J. Immunol.* **161**, 2833–2840.

17. Kennedy, S. G., Wagner, A. J., Conzen, S. D., Jordan, J., Bellacosa, A., Tsichlis, P. N., and Hay, N. (1997) The PI 3-kinase/Akt signaling pathway delivers an anti-apoptotic signal. *Genes Dev.* **11**, 701–713.

18. Hausler, P., Papoff, G., Eramo, A., Reif, K., Cantrell, D. A., and Ruberti, G. (1998) Protection of CD95-mediated apoptosis by activation of phosphatidylinositide 3–kinase and protein kinase B. *Eur. J. Immunol.* **28**, 57–69.

19. Datta, S. R., Dudek, H., Tao, X., Masters, S., Fu, H., Gotoh, Y., and Greenberg, M. E. (1997) Akt phosphorylation of BAD couples survival signals to the cell-intrinsic death machinery. *Cell* **91**, 231–241.

20. del Peso, L., Gonzalez-Garcia, M., Page, C., Herrera, R., and Nunez, G. (1997) Interleukin-3-induced phosphorylation of BAD through the protein kinase Akt. *Science* **278**, 687–689.

21. Cardone, M. H., Roy, N., Stennicke, H. R., Salvesen, G. S., Franke, T. F., Stanbridge, E., et al. (1998) Regulation of cell death protease caspase-9 by phosphorylation. *Science* **282**, 1318–1321.

22. Brunet, A., Bonni, A., Zigmond, M. J., Lin, M. Z., Juo, P., Hu, L. S., et al. (1999) Akt promotes cell survival by phosphorylating and inhibiting a Forkhead transcription factor. *Cell* **96**, 857–868.

23. Kops, G. J., de Ruiter, N. D., De Vries-Smits, A. M., Powell, D. R., Bos, J. L., and Burgering, B. M. (1999) Direct control of the Forkhead transcription factor AFX by protein kinase B. *Nature* **398**, 630–634.

24. Nesterov, A., Lu, X., Johnson, M., Miller, G. J., Ivashchenko, Y., and Kraft, A. S. (2001) Elevated AKT activity protects the prostate cancer cell line LNCaP from TRAIL-induced apoptosis. *J. Biol. Chem.* **276**, 10767–10774.

25. Ashkenazi, A., Pai, R. C., Fong, S., Leung, S., Lawrence, D. A., Marsters, S. A., et al. (1999) Safety and antitumor activity of recombinant soluble Apo2 ligand. *J. Clin. Invest.* **104**, 155–162.

26. Walczak, H., Miller, R. E., Ariail, K., Gliniak, B., Griffith, T. S., Kubin, M., et al. (1999) Tumoricidal activity of tumor necrosis factor-related apoptosis-inducing ligand in vivo. *Nat. Med.* **5**, 157–163.

27. Walczak, H., Bouchon, A., Stahl, H., and Krammer, P. H. (2000) Tumor necrosis factor-related apoptosis-inducing ligand retains its apoptosis-inducing capacity on Bcl-2- or Bcl-x$_L$-overexpressing chemotherapy-resistant tumor cells. *Cancer Res.* **60**, 3051–3057.

28. Jo, M., Kim, T. H., Seol, D. W., Esplen, J. E., Dorko, K., Billiar, T. R., and Strom, S. C. (2000) Apoptosis induced in normal human hepatocytes by tumor necrosis factor-related apoptosis-inducing ligand. *Nat. Med.* **6**, 564–567.

29. Lawrence, D., Shahrokh, Z., Marsters, S., Achilles, K., Shih, D., Mounho, B., et al. (2001) Differential hepatocyte toxicity of recombinant Apo2L/TRAIL versions. *Nat. Med.* **7**, 383–385.

30. Nevins, J. R. (1992) E2F: a link between the Rb tumor suppressor protein and viral oncoproteins. *Science* **258**, 424–429.

31. Frolov, M. V., Huen, D. S., Stevaux, O., Dimova, D., Balczarek-Strang, K., Elsdon, M., and Dyson, N. J. (2001) Functional antagonism between E2F family members. *Genes Dev.* **15**, 2146–2160.

32. Dyson, N. (1998) The regulation of E2F by pRB-family proteins. *Genes Dev.* **12**, 2245–2262.

33. Helin, K., Harlow, E., and Fattaey, A. (1993) Inhibition of E2F-1 transactivation by direct binding of the retinoblastoma protein. *Mol. Cell Biol.* **13**, 6501–6508.

34. Sherr, C. J. (1996) Cancer cell cycles. *Science* **274**, 1672–1677.

35. Johnson, D. G., Cress, W. D., Jakoi, L., and Nevins, J. R. (1994) Oncogenic capacity of the E2F1 gene. *Proc. Natl. Acad. Sci. USA* **91**, 12823–12827.

36. Singh, P., Wong, S. H., and Hong, W. (1994) Overexpression of E2F-1 in rat embryo fibroblasts leads to neoplastic transformation. *EMBO J.* **13,** 3329–3338.
37. Xu, G., Livingston, D. M., and Krek, W. (1995) Multiple members of the E2F transcription factor family are the products of oncogenes. *Proc. Natl. Acad. Sci. USA* **92,** 1357–1361.
38. Field, S. J., Tsai, F.Y., Kuo, F., Zubiaga, A. M., Kaelin, W. G., Jr., Livingston, D. M., et al. (1996) E2F-1 functions in mice to promote apoptosis and suppress proliferation. *Cell* **85,** 549–561.
39. Rounbehler, R. J., Schneider-Broussard, R., Conti, C. J., and Johnson, D. G. (2001) Myc lacks E2F1's ability to suppress skin carcinogenesis. *Oncogene* **20,** 5341–5349.
40. Yamasaki, L., Jacks, T., Bronson, R., Goillot, E., Harlow, E., and Dyson, N. J. (1996) Tumor induction and tissue atrophy in mice lacking E2F-1. *Cell* **85,** 537–548.
41. Lane, M. E., Yu, B., Rice, A., Lipson, K. E., Liang, C., Sun, L., et al. (2001) A novel cdk2-selective inhibitor, SU9516, induces apoptosis in colon carcinoma cells. *Cancer Res.* **61,** 6170–6107.
42. Yang, H. L., Dong, Y. B., Elliott, M. J., Liu, T. J., and McMasters, K. M. (2000) Caspase activation and changes in Bcl-2 family member protein expression associated with E2F-1–mediated apoptosis in human esophageal cancer cells. *Clin. Cancer Res.* **6,** 1579–1589.
43. Ghosh, S., May, M. J., and Kopp, E. B. (1998) NF-kappa B and Rel proteins: evolutionarily conserved mediators of immune responses. *Annu. Rev. Immunol.* **16,** 225–260.
44. Mercurio, F. and Manning, A. M. (1999) Multiple signals converging on NF-kappaB. *Curr. Opin. Cell Biol.* **11,** 226–232.
45. Foo, S. Y. and Nolan, G. P. (1999) NF-kappaB to the rescue: RELs, apoptosis and cellular transformation. *Trends Genet.* **15,** 229–235.
46. Wang, D. and Baldwin, A. S., Jr. (1998) Activation of nuclear factor-kappaB-dependent transcription by tumor necrosis factor-alpha is mediated through phosphorylation of RelA/p65 on serine 529. *J. Biol. Chem.* **273,** 29411–29416.
47. Zhong, H., Voll, R. E., and Ghosh, S. (1998) Phosphorylation of NF-kappa B p65 by PKA stimulates transcriptional activity by promoting a novel bivalent interaction with the coactivator CBP/p300. *Mol. Cell.* **1,** 661–671.
48. Rayet, B. and Gelinas, C. (1999) Aberrant rel/nfkb genes and activity in human cancer. *Oncogene* **18,** 6938–6947.
49. Osborn, L., Kunkel, S., and Nabel, G. J. (1989) Tumor necrosis factor alpha and interleukin 1 stimulate the human immunodeficiency virus enhancer by activation of the nuclear factor kappa B. *Proc. Natl. Acad. Sci. USA* **86,** 2336–2340.
50. Brach, M. A., Hass, R., Sherman, M. L., Gunji, H., Weichselbaum, R., and Kufe, D. (1991) Ionizing radiation induces expression and binding activity of the nuclear factor kappa B. *J. Clin. Invest.* **88,** 691–695.
51. Das, K. C. and White, C. W. (1997) Activation of NF-kappaB by antineoplastic agents. Role of protein kinase C. *J. Biol. Chem.* **272,** 14914–14920.
52. Piret, B. and Piette, J. (1996) Topoisomerase poisons activate the transcription factor NF-kappaB in ACH-2 and CEM cells. *Nucleic Acids Res.* **24,** 4242–4248.
53. Schreck, R., Meier, B., Mannel, D. N., Droge, W., and Baeuerle, P. A. (1992) Dithiocarbamates as potent inhibitors of nuclear factor kappa B activation in intact cells. *J. Exp. Med.* **175,** 1181–1194.
54. Chen, F., Demers, L. M., Vallyathan, V., Lu, Y., Castranova, V., and Shi, X. (1999) Involvement of 5'-flanking kappaB-like sites within bcl-x gene in silica-induced Bcl-x expression. *J. Biol. Chem.* **274,** 35591–35595.
55. Grumont, R. J., Rourke, I. J., and Gerondakis, S. (1999) Rel-dependent induction of A1 transcription is required to protect B cells from antigen receptor ligation-induced apoptosis. *Genes Dev.* **13,** 400–411.
56. Krikos, A., Laherty, C. D., and Dixit, V. M. (1992) Transcriptional activation of the tumor necrosis factor alpha-inducible zinc finger protein, A20, is mediated by kappa B elements. *J. Biol. Chem.* **267,** 17971–17976.
57. Lee, H. H., Dadgostar, H., Cheng, Q., Shu, J., and Cheng, G. (1999) NF-kappaB-mediated up-regulation of Bcl-x and Bfl-1/A1 is required for CD40 survival signaling in B lymphocytes. *Proc. Natl. Acad. Sci. USA* **96,** 9136–9141.
58. Stehlik, C., de Martin, R., Binder, B. R., and Lipp, J. (1998) Cytokine induced expression of porcine inhibitor of apoptosis protein (iap) family member is regulated by NF-kappa B. *Biochem. Biophys. Res. Commun.* **243,** 827–832.
59. Wang, C. Y., Mayo, M. W., Korneluk, R. G., Goeddel, D. V., and Baldwin, A. S., Jr. (1998) NF-kappaB antiapoptosis: induction of TRAF1 and TRAF2 and c-IAP1 and c-IAP2 to suppress caspase-8 activation. *Science* **281,** 1680–1683.
60. Wu, M. X., Ao, Z., Prasad, K. V., Wu, R., and Schlossman, S. F. (1998) IEX-1L, an apoptosis inhibitor involved in NF-kappaB-mediated cell survival. *Science* **281,** 998–1001.
61. Zong, W. X., Edelstein, L. C., Chen, C., Bash, J., and Gelinas, C. (1999) The prosurvival Bcl-2 homolog Bfl-1/A1 is a direct transcriptional target of NF-kappaB that blocks TNFalpha-induced apoptosis. *Genes Dev.* **13,** 382–387.
62. Kurland, J. F., Kodym, R., Story, M. D., Spurgers, K. B., McDonnell, T. J., and Meyn, R. E. (2001) NF-kappaB1 (p50) homodimers contribute to transcription of the bcl-2 oncogene. *J. Biol. Chem.* **276,** 45380–45386.
63. Stancovski, I. and Baltimore, D. (1997) NF-kappaB activation: the I kappaB kinase revealed? *Cell* **91,** 299–302.
64. Guttridge, D. C., Albanese, C., Reuther, J. Y., Pestell, R. G., and Baldwin, A. S., Jr. (1999) NF-kappaB controls cell growth and differentiation through transcriptional regulation of cyclin D1. *Mol. Cell Biol.* **19,** 5785–5799.
65. Pahl, H. L. (1999) Activators and target genes of Rel/NF-kappaB transcription factors. *Oncogene* **18,** 6853–6866.
66. Romashkova, J. A. and Makarov, S. S. (1999) NF-kappaB is a target of AKT in anti-apoptotic PDGF signalling. *Nature* **401,** 86–90.
67. Baldwin, A. S., Jr. (1996) The NF-kappa B and I kappa B proteins: new discoveries and insights. *Annu. Rev. Immunol.* **14,** 649–683.
68. Mayo, M. W. and Baldwin, A. S. (2000) The transcription factor NF-kappaB: control of oncogenesis and cancer therapy resistance. *Biochim. Biophys. Acta* **1470,** M55–M62.
69. Herrmann, J. L., Beham, A. W., Sarkiss, M., Chiao, P. J., Rands, M. T., Bruckheimer, E. M., et al. (1997) Bcl-2 suppresses apoptosis resulting from disruption of the NF-kappa B survival pathway. *Exp. Cell Res.* **237,** 101–109.

70. Delic, J., Masdehors, P., Omura, S., Cosset, J. M., Dumont, J., Binet, J. L., and Magdelenat, H. (1998) The proteasome inhibitor lactacystin induces apoptosis and sensitizes chemo- and radioresistant human chronic lymphocytic leukaemia lymphocytes to TNF-alpha-initiated apoptosis. *Br. J. Cancer* **77,** 1103–1107.

71. Epinat, J. C. and Gilmore, T. D. (1999) Diverse agents act at multiple levels to inhibit the Rel/NF-kappaB signal transduction pathway. *Oncogene* **18,** 6896–6909.

72. McDade, T. P., Perugini, R. A., Vittimberga, F. J., Jr., Carrigan, R. C., and Callery, M. P. (1999) Salicylates inhibit NF-kappaB activation and enhance TNF-alpha-induced apoptosis in human pancreatic cancer cells. *J. Surg. Res.* **83,** 56–61.

73. Waddick, K. G. and Uckun, F. M. (1999) Innovative treatment programs against cancer: II. Nuclear factor-kappaB (NF-kappaB) as a molecular target. *Biochem. Pharmacol.* **57,** 9–17.

74. Yamamoto, Y., Yin, M. J., Lin, K. M., and Gaynor, R. B. (1999) Sulindac inhibits activation of the NF-kappaB pathway. *J. Biol. Chem.* **274,** 27307–27314.

75. Adams, J., Palombella, V. J., Sausville, E. A., Johnson, J., Destree, A., Lazarus, D. D., et al. (1999) Proteasome inhibitors: a novel class of potent and effective antitumor agents. *Cancer Res.* **59,** 2615–2622.

76. Kolenko, V., Bloom, T., Rayman, P., Bukowski, R., Hsi, E., and Finke, J. (1999) Inhibition of NF-kappa B activity in human T lymphocytes induces caspase-dependent apoptosis without detectable activation of caspase-1 and -3. *J. Immunol.* **163,** 590–598.

77. Sharma, H. W., Perez, J. R., Higgins-Sochaski, K., Hsiao, R., and Narayanan, R. (1996) Transcription factor decoy approach to decipher the role of NF-kappa B in oncogenesis. *Anticancer Res.* **16,** 61–69.

78. Sumitomo, M., Tachibana, M., Ozu, C., Asakura, H., Murai, M., Hayakawa, M., et al. (1999) Induction of apoptosis of cytokine-producing bladder cancer cells by adenovirus-mediated IkappaBalpha overexpression. *Hum. Gene Ther.* **10,** 37–47.

79. Bours, V., Dejardin, E., Goujon-Letawe, F., Merville, M. P., and Castronovo, V. (1994) The NF-kappa B transcription factor and cancer: high expression of NF-kappa B- and I kappa B-related proteins in tumor cell lines. *Biochem. Pharmacol.* **47,** 145–149.

80. Paillard, F. (1999) Induction of apoptosis with I-kappaB, the inhibitor of NF-kappaB. *Hum. Gene Ther.* **10,** 1–3.

81. Hellin, A. C., Calmant, P., Gielen, J., Bours, V., and Merville, M. P. (1998) Nuclear factor - kappaB-dependent regulation of p53 gene expression induced by daunomycin genotoxic drug. *Oncogene* **16,** 1187–1195.

82. Ryan, K. M., Ernst, M. K., Rice, N. R., and Vousden, K. H. (2000) Role of NF-kappaB in p53-mediated programmed cell death. *Nature* **404,** 892–897.

83. Grilli, M., Pizzi, M., Memo, M., and Spano, P. (1996) Neuroprotection by aspirin and sodium salicylate through blockade of NF-kappaB activation. *Science* **274,** 1383–1385.

84. Kasibhatla, S., Genestier, L., and Green, D. R. (1999) Regulation of fas-ligand expression during activation-induced cell death in T lymphocytes via nuclear factor kappaB. *J. Biol. Chem.* **274,** 987–992.

85. Lin, K. I., Lee, S. H., Narayanan, R., Baraban, J. M., Hardwick, J. M., and Ratan, R. R. (1995) Thiol agents and Bcl-2 identify an alphavirus-induced apoptotic pathway that requires activation of the transcription factor NF-kappa B. *J. Cell Biol.* **131,** 1149–1161.

86. Ouaaz, F., Li, M., and Beg, A. A. (1999) A critical role for the RelA subunit of nuclear factor kappaB in regulation of multiple immune-response genes and in Fas-induced cell death. *J. Exp. Med.* **189,** 999–1004.

87. Usami, I., Kubota, M., Bessho, R., Kataoka, A., Koishi, S., Watanabe, K., et al. (1998) Role of protein tyrosine phosphorylation in etoposide-induced apoptosis and NF-kappa B activation. *Biochem. Pharmacol.* **55,** 185–191.

88. Barkett, M. and Gilmore, T. D. (1999) Control of apoptosis by Rel/NF-kappaB transcription factors. *Oncogene* **18,** 6910–6924.

89. Nasi, S., Ciarapica, R., Jucker, R., Rosati, J., and Soucek, L. (2001) Making decisions through Myc. *FEBS Lett.* **490,** 153–162.

90. Henriksson, M., Selivanova, G., Lindstrom, M., and Wiman, K. G. (2001) Inactivation of Myc-induced p53-dependent apoptosis in human tumors. *Apoptosis* **6,** 133–137.

91. Hecht, J. L. and Aster, J. C. (2000) Molecular biology of Burkitt's lymphoma. *J. Clin. Oncol.* **18,** 3707–3721.

92. Hynes, N. E. and Lane, H. A. (2001) Myc and mammary cancer: Myc is a downstream effector of the ErbB2 receptor tyrosine kinase. *J. Mammary Gland Biol. Neoplasia* **6,** 141–150.

93. Hortobagyi, G. N. (2001) Overview of treatment results with trastuzumab (Herceptin) in metastatic breast cancer. *Semin. Oncol.* **28,** 43–47.

94. Askew, D. S., Ashmun, R. A., Simmons, B. C., and Cleveland, J. L. (1991) Constitutive c-myc expression in an IL-3-dependent myeloid cell line suppresses cell cycle arrest and accelerates apoptosis. *Oncogene* **6,** 1915–1922.

95. Harrington, E. A., Bennett, M. R., Fanidi, A., and Evan, G. I. (1994) c-Myc-induced apoptosis in fibroblasts is inhibited by specific cytokines. *EMBO J.* **13,** 3286–3295.

96. Marin, M. C., Hsu, B., Stephens, L. C., Brisbay, S., and McDonnell, T. J. (1995) The functional basis of c-myc and bcl-2 complementation during multistep lymphomagenesis in vivo. *Exp. Cell Res.* **217,** 240–247.

97. Hsu, B., Marin, M. C., el-Naggar, A. K., Stephens, L. C., Brisbay, S., and McDonnell, T. J. (1995) Evidence that c-myc mediated apoptosis does not require wild-type p53 during lymphomagenesis. *Oncogene* **11,** 175–179.

98. Leonetti, C., D'Agnano, I., Lozupone, F., Valentini, A., Geiser, T., Zon, G., et al. (1996) Antitumor effect of c-myc antisense phosphorothioate oligodeoxynucleotides on human melanoma cells in vitro and and in mice. *J. Natl. Cancer Inst.* **88,** 419–429.

99. Watson, P. H., Pon, R. T., and Shiu, R. P. (1991) Inhibition of c-myc expression by phosphorothioate antisense oligonucleotide identifies a critical role for c-myc in the growth of human breast cancer. *Cancer Res.* **51,** 3996–4000.

100. Wickstrom, E. L., Bacon, T. A., Gonzalez, A., Freeman, D. L., Lyman, G. H., and Wickstrom, E. (1988) Human promyelocytic leukemia HL-60 cell proliferation and c-myc protein expression are inhibited by an antisense pentadecadeoxynucleotide targeted against c-myc mRNA. *Proc. Natl. Acad. Sci. USA* **85,** 1028–1032.

101. Yu, B. W., Nguyen, D., Anderson, S., and Allegra, C. A. (1997) Phosphorothioated antisense c-myc oligonucleotide inhibits the growth of human colon carcinoma cells. *Anticancer Res.* **17,** 4407–4413.

102. McGuffie, E. M., Pacheco, D., Carbone, G. M., and Catapano, C. V. (2000) Antigene and antiproliferative effects of a c-myc-targeting phosphorothioate triple helix-forming oligonucleotide in human leukemia cells. *Cancer Res.* **60,** 3790–3709.

103. Kauffmann-Zeh, A., Rodriguez-Viciana, P., Ulrich, E., Gilbert, C., Coffer, P., Downward, J., and Evan, G. (1997) Suppression of c-Myc-induced apoptosis by Ras signalling through PI(3)K and PKB. *Nature* **385,** 544–548.

104. Page, C., Lin, H. J., Jin, Y., Castle, V. P., Nunez, G., Huang, M., and Lin, J. (2000) Overexpression of Akt/AKT can modulate chemotherapy-induced apoptosis. *Anticancer Res.* **20,** 407–416.

105. Wang, H. G., Rapp, U. R., and Reed, J. C. (1996) Bcl-2 targets the protein kinase Raf-1 to mitochondria. *Cell* **87,** 629–638.

106. Gulbins, E., Coggeshall, K. M., Brenner, B., Schlottmann, K., Linderkamp, O., and Lang, F. (1996) Fas-induced apoptosis is mediated by activation of a Ras and Rac protein-regulated signaling pathway. *J. Biol. Chem.* **271,** 26389–26394.

107. Chen, C. Y. and Faller, D. V. (1995) Direction of p21ras-generated signals towards cell growth or apoptosis is determined by protein kinase C and Bcl-2. *Oncogene* **11,** 1487–1498.

108. Field, J. K. and Spandidos, D. A. (1990) The role of ras and myc oncogenes in human solid tumours and their relevance in diagnosis and prognosis (review). *Anticancer Res.* **10,** 1–22.

109. Bos, J. L. (1989) ras oncogenes in human cancer: a review. *Cancer Res.* **49,** 4682–4689.

110. Georges, R. N., Mukhopadhyay, T., Zhang, Y., Yen, N., and Roth, J. A. (1993) Prevention of orthotopic human lung cancer growth by intratracheal instillation of a retroviral antisense K-ras construct. *Cancer Res.* **53,** 1743–1746.

111. Zhang, Y., Mukhopadhyay, T., Donehower, L. A., Georges, R. N., and Roth, J. A. (1993) Retroviral vector-mediated transduction of K-ras antisense RNA into human lung cancer cells inhibits expression of the malignant phenotype. *Hum. Gene Ther.* **4,** 451–460.

112. Scholten, J. D., Zimmerman, K. K., Oxender, M. G., Leonard, D., Sebolt-Leopold, J., Gowan, R., and Hupe, D. J. (1997) Synergy between anions and farnesyldiphosphate competitive inhibitors of farnesyl:protein transferase. *J. Biol. Chem.* **272,** 18077–18081.

113. Leitner, J. W., Kline, T., Carel, K., Goalstone, M., and Draznin, B. (1997) Hyperinsulinemia potentiates activation of p21Ras by growth factors. *Endocrinology* **138,** 2211–2214.

114. Kang, M. S., Stemerick, D. M., Zwolshen, J. H., Harry, B. S., Sunkara, P. S., and Harrison, B. L. (1995) Farnesyl-derived inhibitors of ras farnesyl transferase. *Biochem. Biophys. Res. Commun.* **217,** 245–249.

115. Monia, B. P., Johnston, J. F., Geiger, T., Muller, M., and Fabbro, D. (1996) Antitumor activity of a phosphorothioate antisense oligodeoxynucleotide targeted against C-raf kinase. *Nat. Med.* **2,** 668–675.

116. Sebolt-Leopold, J. S., Dudley, D. T., Herrera, R., Van Becelaere, K., Wiland, A., Gowan, R. C., et al. (1999) Blockade of the MAP kinase pathway suppresses growth of colon tumors in vivo. *Nat. Med.* **5,** 810–816.

117. Goldstein, J. L. and Brown, M. S. (1990) Regulation of the mevalonate pathway. *Nature* **343,** 425–430.

118. Lebowitz, P. F., Sakamuro, D., and Prendergast, G. C. (1997) Farnesyl transferase inhibitors induce apoptosis of Ras-transformed cells denied substratum attachment. *Cancer Res.* **57,** 708–713.

119. Suzuki, N., Urano, J., and Tamanoi, F. (1998) Farnesyltransferase inhibitors induce cytochrome c release and caspase 3 activation preferentially in transformed cells. *Proc. Natl. Acad. Sci. USA* **95,** 15356–15361.

120. Jiang, K., Coppola, D., Crespo, N. C., Nicosia, S. V., Hamilton, A. D., Sebti, S. M., and Cheng, J. Q. (2000) The phosphoinositide 3-OH kinase/AKT2 pathway as a critical target for farnesyltransferase inhibitor-induced apoptosis. *Mol. Cell Biol.* **20,** 139–148.

121. Reed, J. C. (1994) Bcl-2 and the regulation of programmed cell death. *J. Cell Biol.* **124,** 1–6.

122. Gross, A., McDonnell, J. M., and Korsmeyer, S. J. (1999) BCL-2 family members and the mitochondria in apoptosis. *Genes Dev.* **13,** 1899–1911.

123. McDonnell, T. J. and Korsmeyer, S. J. (1991) Progression from lymphoid hyperplasia to high-grade malignant lymphoma in mice transgenic for the t(14; 18). *Nature* **349,** 254–256.

124. Rodriguez-Villanueva, J., Greenhalgh, D., Wang, X. J., Bundman, D., Cho, S., Delehedde, M., et al. (1998) Human keratin-1:bcl-2 transgenic mice aberrantly express keratin 6, exhibit reduced sensitivity to keratinocyte cell death induction, and are susceptible to skin tumor formation. *Oncogene* **16,** 853–863.

125. Bruckheimer, E. M., Brisbay, S., Johnson, D. J., Gingrich, J. R., Greenberg, N., and McDonnell, T. J. (2000) Bcl-2 accelerates multistep prostate carcinogenesis in vivo. *Oncogene* **19,** 5251–5258.

126. Webb, A., Cunningham, D., Cotter, F., Clarke, P. A., di Stefano, F., Ross, P., et al. (1997) BCL-2 antisense therapy in patients with non-Hodgkin lymphoma. *Lancet* **349,** 1137–1141.

127. Banerjee, D. (1999) Technology evaluation: G-3139. *Curr. Opin. Mol. Ther.* **1,** 404–408.

128. Piche, A., Grim, J., Rancourt, C., Gomez-Navarro, J., Reed, J. C., and Curiel, D. (1998) Modulation of Bcl-2 protein levels by an intracellular anti-Bcl-2 single-chain antibody increases drug-induced cytotoxicity in the breast cancer cell line MCF-7. *Cancer Res.* **58,** 2134–2140.

129. Nakashima, T., Miura, M., and Hara, M. (2000) Tetrocarcin A inhibits mitochondrial functions of Bcl-2 and suppresses its anti-apoptotic activity. *Cancer Res.* **60,** 1229–1235.

130. Krajewska, M., Moss, S. F., Krajewski, S., Song, K., Holt, P. R., and Reed, J. C. (1996) Elevated expression of Bcl-X and reduced Bak in primary colorectal adenocarcinomas. *Cancer Res.* **56,** 2422–2427.

131. Zha, J., Harada, H., Yang, E., Jockel, J., and Korsmeyer, S. J. (1996) Serine phosphorylation of death agonist BAD in response to survival factor results in binding to 14–3–3 not BCL-X(L). *Cell* **87,** 619–628.

132. Cantley, L. C. and Neel, B. G. (1999) New insights into tumor suppression: PTEN suppresses tumor formation by restraining the phosphoinositide 3-kinase/AKT pathway. *Proc. Natl. Acad. Sci. USA* **96,** 4240–4245.

133. Ramaswamy, S., Nakamura, N., Vazquez, F., Batt, D. B., Perera, S., Roberts, T. M., and Sellers, W. R. (1999) Regulation of G1 progression by the PTEN tumor suppressor protein is linked to inhibition of the phosphatidylinositol 3-kinase/Akt pathway. *Proc. Natl. Acad. Sci. USA* **96,** 2110–2115.

134. Miyashita, T., Krajewski, S., Krajewska, M., Wang, H. G., Lin, H. K., Liebermann, D. A., et al. (1994) Tumor suppressor p53 is a regulator of bcl-2 and bax gene expression in vitro and in vivo. *Oncogene* **9,** 1799–1805.

135. Yin, C., Knudson, C. M., Korsmeyer, S. J., and Van Dyke, T. (1997) Bax suppresses tumorigenesis and stimulates apoptosis in vivo. *Nature* **385,** 637–640.

136. Brimmell, M., Mendiola, R., Mangion, J., and Packham, G. (1998) BAX frameshift mutations in cell lines derived from human haemopoietic malignancies are associated with resistance to apoptosis and microsatellite instability. *Oncogene* **16,** 1803–1812.

137. Zhang, L., Yu, J., Park, B. H., Kinzler, K. W., and Vogelstein, B. (2000) Role of BAX in the apoptotic response to anticancer agents. *Science* **290,** 989–992.

138. Eguchi, H., Suga, K., Saji, H., Toi, M., Nakachi, K., and Hayashi, S. I. (2000) Different expression patterns of Bcl-2 family genes in breast cancer by estrogen receptor status with special reference to pro-apoptotic Bak gene. *Cell Death Differ.* **7,** 439–446.

139. Sattler, M., Liang, H., Nettesheim, D., Meadows, R. P., Harlan, J. E., Eberstadt, M., et al. (1997) Structure of Bcl-xL-Bak peptide complex: recognition between regulators of apoptosis. *Science* **275,** 983–986.

140. Wei, M.C., Zong, W. X., Cheng, E.H., Lindsten, T., Panoutsakopoulou, V., Ross, A. J., et al. (2001) Proapoptotic BAX and BAK: a requisite gateway to mitochondrial dysfunction and death. *Science* **292,** 727–730.

141. Finnegan, N. M., Curtin, J. F., Prevost, G., Morgan, B., and Cotter, T. G. (2001) Induction of apoptosis in prostate carcinoma cells by BH3 peptides which inhibit Bak/Bcl-2 interactions. *Br. J. Cancer* **85,** 115–121.

142. Kagawa, S., Pearson, S. A., Ji, L., Xu, K., McDonnell, T. J., Swisher, S., et al. (2000) A binary adenoviral vector system for expressing high levels of the proapoptotic gene bax. *Gene Ther.* **7,** 75–79.

143. Honda, T., Kagawa, S., Spurgers, K. B., Gjertsen, B. T., Roth, J. A., Fang, B., et al. A recombinant adenovirus expressing wild-type bax induces apoptosis in prostate cancer cells independently of their bcl-2 status and androgen sensitivity. *Cancer Biol. Ther.,* in press.

144. Lowe, S. L., Rubinchik, S., Honda, T., McDonnell, T. J., Dong, J. Y., and Norris, J. S. (2001) Prostate-specific expression of Bax delivered by an adenoviral vector induces apoptosis in LNCaP prostate cancer cells. *Gene Ther.* **8,** 1363–1371.

145. Fesik, S. W. (2000) Insights into programmed cell death through structural biology. *Cell* **103,** 273–272.

146. Wang, J. L., Liu, D., Zhang, Z. J., Shan, S., Han, X., Srinivasula, S. M., et al. (2000) Structure-based discovery of an organic compound that binds Bcl-2 protein and induces apoptosis of tumor cells. *Proc. Natl. Acad. Sci. USA* **97,** 7124–7129.

147. Degterev, A., Lugovskoy, A., Cardone, M., Mulley, B., Wagner, G., Mitchison, T., and Yuan, J. (2001) Identification of small-molecule inhibitors of interaction between the BH3 domain and Bcl-xL. *Nat. Cell Biol.* **3,** 173–182.

148. Tzung, S. P., Kim, K. M., Basanez, G., Giedt, C. D., Simon, J., Zimmerberg, J., et al. (2001) Antimycin A mimics a cell-death-inducing Bcl-2 homology domain 3. *Nat. Cell Biol.* **3,** 183–191.

149. Levine, A. J. (1997) p53, the cellular gatekeeper for growth and division. *Cell.* **88,** 323–331.

150. Ko, L. J. and Prives, C. (1996) p53: puzzle and paradigm. *Genes Dev.* **10,** 1054–1072.

151. Donehower, L. A., Harvey, M., Slagle, B. L., McArthur, M. J., Montgomery, C. A., Jr., Butel, J. S., and Bradley, A. (1992) Mice deficient for p53 are developmentally normal but susceptible to spontaneous tumours. *Nature* **356,** 215–221.

152. Malkin, D. (1994) p53 and the Li-Fraumeni syndrome. *Biochim. Biophys. Acta* **1198,** 197–213.

153. Malkin, D., Li, F. P., Strong, L. C., Fraumeni, J. F., Jr., Nelson, C. E., Kim, D. H., et al. (1990) Germ line p53 mutations in a familial syndrome of breast cancer, sarcomas, and other neoplasms [see comments]. *Science* **250,** 1233–1238.

154. Momand, J., Wu, H. H., and Dasgupta, G. (2000) MDM2—master regulator of the p53 tumor suppressor protein. *Gene* **242,** 15–29.

155. Tao, W. and Levine, A. J. (1999) Nucleocytoplasmic shuttling of oncoprotein Hdm2 is required for Hdm2-mediated degradation of p53. *Proc. Natl. Acad. Sci. USA* **96,** 3077–3080.

156. Kastan, M. B., Onyekwere, O., Sidransky, D., Vogelstein, B., and Craig, R. W. (1991) Participation of p53 protein in the cellular response to DNA damage. *Cancer Res.* **51,** 6304–6311.

157. Fritsche, M., Haessler, C., and Brandner, G. (1993) Induction of nuclear accumulation of the tumor-suppressor protein p53 by DNA-damaging agents. *Oncogene* **8,** 307–318.

158. Graeber, T. G., Peterson, J. F., Tsai, M., Monica, K., Fornace, A. J., Jr., and Giaccia, A. J. (1994) Hypoxia induces accumulation of p53 protein, but activation of a G1- phase checkpoint by low-oxygen conditions is independent of p53 status. *Mol. Cell Biol.* **14,** 6264–6277.

159. Yin, Y., Terauchi, Y., Solomon, G. G., Aizawa, S., Rangarajan, P. N., Yazaki, Y., et al. (1998) Involvement of p85 in p53-dependent apoptotic response to oxidative stress. *Nature* **391,** 707–710.

160. Siliciano, J. D., Canman, C. E., Taya, Y., Sakaguchi, K., Appella, E., and Kastan, M. B. (1997) DNA damage induces phosphorylation of the amino terminus of p53. *Genes Dev.* **11,** 3471–3481.

161. el-Deiry, W. S., Tokino, T., Velculescu, V. E., Levy, D. B., Parsons, R., Trent, J. M., et al. (1993) WAF1, a potential mediator of p53 tumor suppression. *Cell* **75,** 817–825.

162. Shaw, P., Bovey, R., Tardy, S., Sahli, R., Sordat, B., and Costa, J. (1992) Induction of apoptosis by wild-type p53 in a human colon tumor-derived cell line. *Proc. Natl. Acad. Sci. USA* **89,** 4495–4499.

163. Schuler, M., Bossy-Wetzel, E., Goldstein, J. C., Fitzgerald, P., and Green, D. R. (2000) p53 induces apoptosis by caspase activation through mitochondrial cytochrome c release. *J. Biol. Chem.* **275,** 7337–7342.

164. Sheikh, M. S. and Fornace, A. J., Jr. (2000) Death and decoy receptors and p53-mediated apoptosis. *Leukemia* **14,** 1509–1513.

165. el-Deiry, W. S. (1998) Regulation of p53 downstream genes. *Semin. Cancer Biol.* **8,** 345–357.

166. Oda, E., Ohki, R., Murasawa, H., Nemoto, J., Shibue, T., Yamashita, T., et al. (2000) Noxa, a BH3-only member of the Bcl-2 family and candidate mediator of p53-induced apoptosis. *Science* **288,** 1053–1058.

167. Oda, K., Arakawa, H., Tanaka, T., Matsuda, K., Tanikawa, C., Mori, T., et al. (2000) p53AIP1, a potential mediator of p53-dependent apoptosis, and its regulation by Ser-46-phosphorylated p53 [In Process Citation]. *Cell* **102,** 849–862.

168. Lane, D. P. (1992) Cancer. p53, guardian of the genome. *Nature* **358,** 15–16.

169. Hollstein, M., Rice, K., Greenblatt, M. S., Soussi, T., Fuchs, R., Sorlie, T., et al. (1994) Database of p53 gene somatic mutations in human tumors and cell lines. *Nucleic Acids Res.* **22,** 3551–3555.

170. Levine, A. J., Momand, J., and Finlay, C. A. (1991) The p53 tumour suppressor gene. *Nature* **351,** 453–456.

171. Friedlander, P., Haupt, Y., Prives, C., and Oren, M. (1996) A mutant p53 that discriminates between p53-responsive genes cannot induce apoptosis. *Mol. Cell Biol.* **16,** 4961–4971.

172. Kerr, J. F., Winterford, C. M., and Harmon, B. V. (1994) Apoptosis. Its significance in cancer and cancer therapy. *Cancer* **73,** 2013–2026.

173. O'Connor, P. M., Jackman, J., Bae, I., Myers, T. G., Fan, S., Mutoh, M., et al. (1997) Characterization of the p53 tumor suppressor pathway in cell lines of the National Cancer Institute anticancer drug screen and correlations with the growth-inhibitory potency of 123 anticancer agents. *Cancer Res.* **57,** 4285–4300.

174. Lowe, S. W., Bodis, S., McClatchey, A., Remington, L., Ruley, H. E., Fisher, D. E., et al. (1994) p53 status and the efficacy of cancer therapy in vivo. *Science* **266,** 807–810.

175. Lowe, S. (1999) p53, Apoptosis and chemosensitivity, in *Apoptosis and Cancer Chemotherapy* (Hickman, C., ed.), Totowa, NJ: Humana Press Inc., pp. 21–36.

176. Li, P.F., Dietz, R., and von Harsdorf, R. (1999) p53 regulates mitochondrial membrane potential through reactive oxygen species and induces cytochrome *c*-independent apoptosis blocked by Bcl-2. *EMBO J.* **18,** 6027–6036.

177. Polyak, K., Xia, Y., Zweier, J. L., Kinzler, K. W., and Vogelstein, B. (1997) A model for p53-induced apoptosis. *Nature* **389,** 300–305.

178. Miyake, H., Hanada, N., Nakamura, H., Kagawa, S., Fujiwara, T., Hara, I., et al. (1998) Overexpression of Bcl-2 in bladder cancer cells inhibits apoptosis induced by cisplatin and adenoviral-mediated p53 gene transfer. *Oncogene* **16,** 933–943.

179. Srivastava, S., Katayose, D., Tong, Y. A., Craig, C. R., McLeod, D. G., Moul, J. W., Cowan, K. H., and Seth, P. (1995) Recombinant adenovirus vector expressing wild-type p53 is a potent inhibitor of prostate cancer cell proliferation. *Urology* **46,** 843–848.

180. Schumacher, G., Bruckheimer, E. M., Beham, A. W., Honda, T., Brisbay, S., Roth, J. A., et al. (2001) Molecular determinants of cell death induction following adenovirus-mediated gene transfer of wild-type p53 in prostate cancer cells. *Int. J. Cancer* **91,** 159–166.

181. Clayman, G. L., el-Naggar, A. K., Roth, J. A., Zhang, W. W., Goepfert, H., Taylor, D. L., and Liu, T. J. (1995) In vivo molecular therapy with p53 adenovirus for microscopic residual head and neck squamous carcinoma. *Cancer Res.* **55,** 1–6.

182. Roth, J. A., Swisher, S. G., and Meyn, R. E. (1999) p53 tumor suppressor gene therapy for cancer. *Oncology (Huntingt)* **13,** 148–154.

183. Roth, J. A., Nguyen, D., Lawrence, D. D., Kemp, B. L., Carrasco, C. H., Ferson, D. Z., et al. (1996) Retrovirus-mediated wild-type p53 gene transfer to tumors of patients with lung cancer. *Nat. Med.* **2,** 985–991.

184. Nemunaitis, J., Swisher, S.G., Timmons, T., Connors, D., Mack, M., Doerksen, L., et al. (2000) Adenovirus-mediated p53 gene transfer in sequence with cisplatin to tumors of patients with non-small-cell lung cancer. *J. Clin. Oncol.* **18,** 609–622.

185. Schuler, M., Herrmann, R., De Greve, J. L., Stewart, A. K., Gatzemeier, U., Stewart, D. J., et al. (2001) Adenovirus-mediated wild-type p53 gene transfer in patients receiving chemotherapy for advanced non-small-cell lung cancer: results of a multicenter phase II study. *J. Clin. Oncol.* **19,** 1750–1758.

186. Jarry, A., Vallette, G., Cassagnau, E., Moreau, A., Bou-Hanna, C., Lemarre, P., et al. (1999) Interleukin 1 and interleukin 1beta converting enzyme (caspase 1) expression in the human colonic epithelial barrier. Caspase 1 downregulation in colon cancer. *Gut* **45,** 246–251.

187. Palmerini, F., Devilard, E., Jarry, A., Birg, F., and Xerri, L. (2001) Caspase 7 downregulation as an immunohistochemical marker of colonic carcinoma. *Hum. Pathol.* **32,** 461–467.

188. Ueki, T., Takeuchi, T., Nishimatsu, H., Kajiwara, T., Moriyama, N., Narita, Y., et al. (2001) Silencing of the caspase-1 gene occurs in murine and human renal cancer cells and causes solid tumor growth in vivo. *Int. J. Cancer* **91,** 673–679.

189. Eggert, A., Grotzer, M. A., Zuzak, T. J., Wiewrodt, B. R., Ho, R., Ikegaki, N., and Brodeur, G. M. (2001) Resistance to tumor necrosis factor-related apoptosis-inducing ligand (TRAIL)-induced apoptosis in neuroblastoma cells correlates with a loss of caspase-8 expression. *Cancer Res.* **61,** 1314–1319.

190. Jonges, L. E., Nagelkerke, J. F., Ensink, N. G., van der Velde, E. A., Tollenaar, R. A., Fleuren, G. J., et al. (2001) Caspase-3 activity as a prognostic factor in colorectal carcinoma. *Lab. Invest.* **81,** 681–688.

191. Jones, P. A. (2001) Cancer. Death and methylation. *Nature* **409,** 141, 143–144.

192. Conway, K. E., McConnell, B. B., Bowring, C. E., Donald, C. D., Warren, S. T., and Vertino, P. M. (2000) TMS1, a novel proapoptotic caspase recruitment domain protein, is a target of methylation-induced gene silencing in human breast cancers. *Cancer Res.* **60**, 6236–6242.

193. Soengas, M. S., Capodieci, P., Polsky, D., Mora, J., Esteller, M., Opitz-Araya, X., et al. (2001) Inactivation of the apoptosis effector Apaf-1 in malignant melanoma. *Nature* **409**, 207–211.

194. Bullani, R. R., Huard, B., Viard-Leveugle, I., Byers, H. R., Irmler, M., Saurat, J. H., et al. (2001) Selective expression of FLIP in malignant melanocytic skin lesions. *J. Invest. Dermatol.* **117**, 360–364.

195. Kamihira, S., Yamada, Y., Hirakata, Y., Tomonaga, M., Sugahara, K., Hayashi, T., Dateki, N., Harasawa, H., and Nakayama, K. (2001) Aberrant expression of caspase cascade regulatory genes in adult T-cell leukaemia: survivin is an important determinant for prognosis. *Br. J. Haematol.* **114**, 63–69.

196. Ryu, B. K., Lee, M. G., Chi, S. G., Kim, Y. W., and Park, J. H. (2001) Increased expression of cFLIP(L) in colonic adenocarcinoma. *J. Pathol.* **194**, 15–19.

197. Roth, J. A., Grammer, S. F., Swisher, S. G., Komaki, R., Nemunaitis, J., Merritt, J., and Meyn, R. E. (2001) P53 gene replacement for cancer—interactions with DNA damaging agents. *Acta Oncol.* **40**, 739–744.

198. Shinoura, N., Koike, H., Furitu, T., Hashimoto, M., Asai, A., Kirino, T., and Hamada, H. (2000) Adenovirus-mediated transfer of caspase-8 augments cell death in gliomas: implication for gene therapy. *Hum. Gene Ther.* **11**, 1123–1137.

199. Shinoura, N., Muramatsu, Y., Yoshida, Y., Asai, A., Kirino, T., and Hamada, H. (2000) Adenovirus-mediated transfer of caspase-3 with Fas ligand induces drastic apoptosis in U-373MG glioma cells. *Exp. Cell Res.* **256**, 423–433.

200. Marcelli, M., Cunningham, G. R., Walkup, M., He, Z., Sturgis, L., Kagan, C., et al. (1999) Signaling pathway activated during apoptosis of the prostate cancer cell line LNCaP: overexpression of caspase-7 as a new gene therapy strategy for prostate cancer. *Cancer Res.* **59**, 382–390.

201. Xie, X., Zhao, X., Liu, Y., Zhang, J., Matusik, R. J., Slawin, K. M., and Spencer, D. M. (2001) Adenovirus-mediated tissue-targeted expression of a caspase-9-based artificial death switch for the treatment of prostate cancer. *Cancer Res.* **61**, 6795–6804.

202. Reed, J. C. (2001) The Survivin saga goes in vivo. *J. Clin. Invest.* **108**, 965–969.

203. Deveraux, Q. L., Takahashi, R., Salvesen, G. S., and Reed, J. C. (1997) X-linked IAP is a direct inhibitor of cell-death proteases. *Nature* **388**, 300–304.

204. Ambrosini, G., Adida, C., and Altieri, D. C. (1997) A novel anti-apoptosis gene, survivin, expressed in cancer and lymphoma. *Nat. Med.* **3**, 917–921.

205. O'Connor, D. S., Grossman, D., Plescia, J., Li, F., Zhang, H., Villa, A., et al. (2000) Regulation of apoptosis at cell division by p34cdc2 phosphorylation of survivin. *Proc. Natl. Acad. Sci. USA* **97**, 13103–13107.

206. Mesri, M., Wall, N. R., Li, J., Kim, R. W., and Altieri, D. C. (2001) Cancer gene therapy using a survivin mutant adenovirus. *J. Clin. Invest.* **108**, 981–990.

207. Wood, K. M., Roff, M., and Hay, R. T. (1998) Defective IkappaBalpha in Hodgkin cell lines with constitutively active NF-kappaB. *Oncogene* **16**, 2131–2139.

208. Cabannes, E., Khan, G., Aillet, F., Jarrett, R. F., and Hay, R.T. (1999) Mutations in the IκBa gene in Hodgkin's disease suggest a tumour suppressor role for IkappaBalpha. *Oncogene* **18**, 3063–3070.

209. Krappmann, D., Emmerich, F., Kordes, U., Scharschmidt, E., Dorken, B., and Scheidereit, C. (1999) Molecular mechanisms of constitutive NF-kappaB/Rel activation in Hodgkin/Reed-Sternberg cells. *Oncogene* **18**, 943–953.

210. Ohno, H., Doi, S., Yabumoto, K., Fukuhara, S., and McKeithan, T.W. (1993) Molecular characterization of the t(14;19)(q32;q13) translocation in chronic lymphocytic leukemia. *Leukemia* **7**, 2057–2063.

211. McKeithan, T. W., Takimoto, G. S., Ohno, H., Bjorling, V. S., Morgan, R., Hecht, B. K. et al. (1997) BCL3 rearrangements and t(14;19) in chronic lymphocytic leukemia and other B-cell malignancies: a molecular and cytogenetic study. *Genes Chromosomes Cancer* **20**, 64–72.

212. Neri, A., Chang, C. C., Lombardi, L., Salina, M., Corradini, P., Maiolo, A. T., Chaganti, R. S., and Dalla-Favera, R. (1991) B cell lymphoma-associated chromosomal translocation involves candidate oncogene lyt-10, homologous to NF-kappa B p50. *Cell* **67**, 1075–1087.

213. Fracchiolla, N. S., Lombardi, L., Salina, M., Migliazza, A., Baldini, L., Berti, E., et al. (1993) Structural alterations of the NF-kappa B transcription factor lyt-10 in lymphoid malignancies. *Oncogene* **8**, 2839–2845.

214. Escot, C., Theillet, C., Lidereau, R., Spyratos, F., Champeme, M. H., Gest, J., and Callahan, R. (1986) Genetic alteration of the c-myc protooncogene (MYC) in human primary breast carcinomas. *Proc. Natl. Acad. Sci. USA* **83**, 4834–4838.

215. Nass, S. J. and Dickson, R. B. (1997) Defining a role for c-Myc in breast tumorigenesis. *Breast Cancer Res. Treat.* **44**, 1–22.

216. Sears, R., Leone, G., DeGregori, J., and Nevins, J. R. (1999) Ras enhances Myc protein stability. *Mol. Cell* **3**, 169–279.

217. West, M. J., Stoneley, M., and Willis, A. E. (1998) Translational induction of the c-myc oncogene via activation of the FRAP/TOR signalling pathway. *Oncogene* **17**, 769–780.

218. Bhatia, K., Huppi, K., Spangler, G., Siwarski, D., Iyer, R., and Magrath, I. (1993) Point mutations in the c-Myc transactivation domain are common in Burkitt's lymphoma and mouse plasmacytomas. *Nat. Genet.* **5**, 56–61.

219. Dalla-Favera, R., Bregni, M., Erikson, J., Patterson, D., Gallo, R. C., and Croce, C. M. (1982) Human c-myc onc gene is located on the region of chromosome 8 that is translocated in Burkitt lymphoma cells. *Proc. Natl. Acad. Sci. USA* **79**, 7824–7827.

220. Taub, R., Kirsch, I., Morton, C., Lenoir, G., Swan, D., Tronick, S., Aaronson, S., and Leder, P. (1982) Translocation of the c-myc gene into the immunoglobulin heavy chain locus in human Burkitt lymphoma and murine plasmacytoma cells. *Proc. Natl. Acad. Sci. USA* **79**, 7837–7841.

221. Fleming, W. H., Hamel, A., MacDonald, R., Ramsey, E., Pettigrew, N. M., Johnston, B., et al. (1986) Expression of the c-myc protooncogene in human prostatic carcinoma and benign prostatic hyperplasia. *Cancer Res.* **46,** 1535–1538.

222. Jenkins, R. B., Qian, J., Lieber, M. M., and Bostwick, D. G. (1997) Detection of c-myc oncogene amplification and chromosomal anomalies in metastatic prostatic carcinoma by fluorescence in situ hybridization. *Cancer Res.* **57,** 524–531.

223. He, T. C., Sparks, A. B., Rago, C., Hermeking, H., Zawel, L., da Costa, L. T., et al. (1998) Identification of c-MYC as a target of the APC pathway. *Science* **281,** 1509–1512.

224. Sikora, K., Chan, S., Evan, G., Gabra, H., Markham, N., Stewart, J., and Watson, J. (1987) c-myc oncogene expression in colorectal cancer. *Cancer* **59,** 1289–1295.

225. Nau, M. M., Carney, D. N., Battey, J., Johnson, B., Little, C., Gazdar, A., and Minna, J. D. (1984) Amplification, expression and rearrangement of c-myc and N-myc oncogenes in human lung cancer. *Curr. Top. Microbiol. Immunol.* **113,** 172–177.

226. Saksela, K., Bergh, J., Lehto, V. P., Nilsson, K., and Alitalo, K. (1985) Amplification of the c-myc oncogene in a subpopulation of human small cell lung cancer. *Cancer Res.* **45,** 1823–1827.

13

Cell Death in Immune, Inflammatory, and Stress Responses

J. John Cohen, Maria D. Devore, Mile Cikara, and Elizabeth A. Dowling

INTRODUCTION

Cell death occurs in all tissues of the mammalian body, most prominently in three areas: the digestive tract, the skin, and the hematopoietic system, of which the immune system is part. Each of these is an interface between the body and the external world. As such, they are exposed to a variety of physical, chemical, and biological threats. It might be possible to toughen these contact surfaces in an attempt to make them impervious to injury, as is the case in many invertebrates, which possess nearly impenetrable exoskeletons. The inconveniences of this design are apparent, however, when an invertebrate tries to grow. It must soften and shed its exoskeleton, leaving it susceptible to predators. A better solution is to replace cells in the contact zones regularly and frequently, as is the case in man. This offers an additional advantage: if a cell sustains an injury that results in DNA damage, chances are that the cell will die before the mutation is locked in by cell division. If it were not for this mechanism, it is possible that skin and gastrointestinal tumors would be far more frequent than they are. Similarly, cells of the innate and acquired immune responses have, in general, short life expectancies; one of them, the polymorphonuclear neutrophil, has probably the shortest lifespan in the human body—usually less than a day. Most lymphocytes—the cells that carry out the work of acquired immunity—are also short-lived, and furthermore, owing to extraordinary demands that are made on their specificity and reactivity, risk death at many stages in their lives. We review the most interesting examples of programmed (and, to a lesser extent, unprogrammed) cell death in the immune system. We begin with an overview of immunity, so that apoptotic events may be placed in context.

INNATE IMMUNITY AND INFLAMMATION

Threats to the body's integrity can be thought of as appearing in two types: crude and subtle. Crude threats are structural motifs that are clearly different from those that make up the mammalian host, and thus can be recognized with just a few invariant receptors. Invertebrates have similar motif-recognition mechanisms, but lack the more sophisticated adaptive response that is the immune system. The *Drosophila* gene Toll plays a major role in development, but also codes for a receptor involved in host defenses. There is a family of Toll-like receptors (TLR) in mammals. They are expressed on the surfaces of many cell types, and have a crude sort of specificity. TLR2 and TLR4

From: *Essentials of Apoptosis: A Guide for Basic and Clinical Research*
Edited by: X-M. Yin and Z. Dong © Humana Press Inc., Totowa, NJ

bind peptidoglycans from Gram-positive bacteria and lipopolysaccharides from Gram-negative bacteria, respectively. TLR9 recognizes bacterial species-specific CpG DNA motifs, and TLR3 recognizes double-stranded RNA. Binding results in signal transduction, which can lead to release of proinflammatory cytokines, reactive oxygen intermediates (ROI), and other stress signals. It can also result in death by apoptosis *(1)*. It remains to be determined when and why a cell would choose to become activated (and thus initiate an inflammatory response to the invader) or to commit suicide. Both options seem to have survival advantages for the organism.

During times of physical and psychological stress, blood levels of adrenal glucocorticoids rise in mammals, in some cases by more than 10-fold. Steroids modulate lymphocyte function and thus have many effects on the immune response *(2)*. In addition, they affect nonlymphoid cells that take part in immune-mediated inflammatory responses. Of interest in this context are glucocorticoids, which prolong the survival of polymorphonuclear leukocytes (PMN), cells that normally only live a day or so. Thus, clinically, diseases in which PMN play an important role, such as the adult respiratory distress syndrome, may not be ameliorated by glucocorticoid treatment. On the other hand, eosinophils rapidly undergo apoptosis when exposed to glucocorticoids *(3)*, which may partially explain the great usefulness of these drugs in allergic disease, in which inflammation is dominated by the eosinophil.

STRESS RESPONSES AND ALTRUISTIC SUICIDE

When a cell is severely stressed or damaged, for example, by radiation or a toxin, there are three possible outcomes, depending on the intensity of the injury: 1) repair of the damage, 2) suicide by apoptosis, or 3) direct lethality by the damaging agent (usually by necrosis). It is clear that cells of the lymphoid type choose apoptosis in response to damage far more readily than do most other cells. The immune system's response to cellular stress is well-illustrated by lymphocytes exposed to ionizing radiation. It has long been known that lymphocytes are peculiarly sensitive to radiation, dying at doses several logs lower than those required to kill fibroblasts, for example. Furthermore, they are more susceptible when resting in G_0 than when in cell cycle. This death is by apoptosis *(4)* and is dependent on p53 *(5)*. There is good reason for lymphocytes to commit suicide: they are the only cells in the body that are poised to enter cell cycle rapidly in response to a relatively minor event, the binding of antigen to their receptor. When properly stimulated, lymphocytes in the body can divide every 6 h; 4 d later, one lymphocyte has given rise to a clone of 65,000 cells. A mutation in such a rapidly expanding clone could lead to a response to a self-derived antigen and possibly autoimmunity; if growth becomes unregulated, it could lead to lymphoma. In such essential but potentially dangerous cells, it makes sense to build in mechanisms that activate an altruistic suicide program when a problem is detected. The underlying instruction to the cell is: better dead than wrong.

The most dramatic illustration of this point is the thymocyte, or immature T cell. The literature contains literally hundreds of reports of agents that induce thymocyte apoptosis (*see* Table 1 for a sampling). Clearly there are not specific receptors for all these agents, nor do they all cause DNA damage and activate a p53-dependent apoptotic pathway. It is not yet known what common events could be triggered by so many different stimuli. If they all lead to a single intracellular crisis, it may be the activation of proteinases in an ever-intensifying positive feedback loop; these proteinases include the caspases, and in some cases, calpains and cathepsins. One possible mechanism by which unrelated stressors could induce apoptosis is via the activation of eIF2α (alpha subunit of eukaryotic initiation factor 2) by kinases such as GCN2, PERK, PKR, or HRI. Phosphorylated eIF2 blocks translation and also initiates transcription of many stress-response genes, which can, under the right circumstances, trigger the apoptotic process *(6–9)*. It is also possible that many stressors increase demand for ATP (which might be needed, for example, in an attempt to refold damaged proteins), leading to an exhaustion of glucose and the activation of the glucose-regulated protein (GRP) group of stress-response genes *(10)*.

Table 1
Some Agents that Induce Apoptosis in Thymocytes

1-(5-Isoquinolinesulfonyl)-2-methylpiperazine dihydrochloride	1-beta-D-arabinosylcytosine	2,2'-bis(2-aminoethyl)-4,4'-bithiazole	2,3,7,8-Tetrachlorodibenzo-p-dioxin
25-Hydroxycholesterol	27-Hydroxycholesterol	2-amino 3,5-dihydro-7-(3-thienylmethyl)-4H-pyrrolo[3,2-D]-pyrimidin-4-one HCl	2-Amino-2-(2-[4-octylphenyl]ethyl)-1,3-propanediol HCl
2-Chloroadenosine	2-Chloroethylethyl sulfide	3-aminobenzamide	5'-(N-ethyl)-carboxamide adenosine
5-Azacytidine	5-Fluorouracil	6-hydroxydopamine	7 β,25-dihydroxycholesterol
7,12-Dimethylbenz[a]anthracene	9-cis-Retinoic acid	A 23187	Acetaldehyde
Acrolein	Actinomycin D	Adenosine	Adriamycin
All trans-retinoic acid	Amsacrine	Anti-αβ T cell receptor	Anti-γδ T-cell receptor
Anti-alpha 4 integrin	Anti-CD2	Anti-CD3	Anti-CD50
Anti-CD98	Anti-Fas	Anti-Thy-1	Antigen
Antigenic peptides	Arabinosylcytosine	Arsenic	ATP
Beauvericin	β-estradiol	Bile salts	Bis(tri-N-butyltin)oxide
Butyric acid	CdCl2	Calcium ionophores	Camptothecin
Carbachol	Carbimazole	Carbon monoxide	Chicken anemia virus
Cisplatin	Cocaine	Coffee creamer (whitener)	Colchicine
Curcumin	Cyanide	Cycloheximide	Cyclophosphamide
Cyclosporin A	Cytomegalovirus	Cytotoxic T lymphocytes	DbCAMP
DDT	Deltamethrin	Deoxycholate	Deoxycorticosterone
Dexamethasone	Dibutyltin	Dimethyl sulfoxide	Dithiocarbamate
Dopamine	Doxorubicin	Epipodophyllotoxin	Ethanol
Ethylene dimethanesulphonate	Etoposide	Feline immunodeficiency virus	Flavopiridol
Forskolin	Fusarenon-X	Galectin-1	Galectin-9
Genistein	Gliotoxin	Glucocorticoids	GM1 ganglioside
H2O2	Hemorrhage	Herbimycin A	Human immunodeficiency virus
Hydrocortisone	Hyperthermia	ICRF-154	Indium
Infectious bursal disease virus	Ingenol 3, 20-dibenzoate	Interleukin-2	LIGHT (a member of the TNF family)
Linomide	Lipopolysaccharide	Listeria monocytogenes	Magnetic field
Marek's disease virus	Measles virus	Methamphetamine	Methimazole
Methyl-2,5-dihydroxycinnamate	Methylcholanthrene	Methylprednisolone	Mitomycin C
Moloney murine leukemia virus	Moloney murine sarcoma virus	Mouse hepatitis virus	Mouse mammary tumor virus

(continued)

Table 1 (continued)
Some Agents that Induce Apoptosis in Thymocytes

MST-16	N-(4–hydroxyphenyl)retinamide	N,N, N'N',-tetrakis (2-pyridylmethyl)ethylenediamine	Nifedipine
Nitric oxide	N-methyl-N-nitrosourea	Novobiocin	Organotin compounds
Oxotremorine-M	Peritoneal dialysis fluid	Peroxynitrite	PGE2
Phorbol esters	Prednisolone	Prednisone	Pregnancy
Rabies virus	Radiation, ionizing	Sepsis	Simian immunodeficiency virus
S-nitrosoglutathione	Sodium salicylate	Somatostatin	Staphylococcal enterotoxin-B
Staurosporine	Taxol	Teniposide	Testosterone
Thapsigargin	Thromboxane A2	TNF-α	TNF-β
Trauma	Trehalose 6,6'-dimycolate	Tributyltin	Trichothecene mycotoxin
Triphenyltin	Trypanosoma cruzi	Ultrasound	UV irradiation
Valinomycin	Vanadium compounds	Verapamil	Vomitoxin

THE ADAPTIVE IMMUNE SYSTEM

We find it useful to think of the immune system as a chemical sense organ, similar to the more familiar senses of taste and smell. They all function to identify novel materials, in a sense, which we then arrange to remove or inactivate. The immune system comes into play once foreign materials have achieved access to the interior. It is our cellular and molecular surveillance organ, and must not only distinguish between self and non-self, but also arrange for the inactivation, destruction, and removal of cells or molecules that it determines to be foreign. Foreign molecules are potential pathogens. The immune system makes no value judgments; harmful or inert, if it is foreign, it should be removed.

From an immunological point of view, the body consists of two regions: extracellular spaces and intracellular spaces. Intracellular parasites are dealt with by T cells (*see* below). Infectious organisms that prefer to live extracellularly are approachable by antibodies, the immunoglobulin proteins specialized for recognition of a nearly infinite array of foreign molecular configurations. Antibodies are produced by B lymphocytes (B cells), which use a sample of the antibody they will produce as a surface receptor for antigen. Each B cell makes only one of the millions of antibody specificities of which the organism is capable. When it divides, the clone makes the same antibody, although after stimulation by antigen, there is a very high rate of mutation in the genes that code for the variable part of the molecule. This results in daughter cells producing similar but not necessarily identical antibody molecules, and allows for the selection of cells whose mutated receptor has higher affinity for antigen than did the original B cell. This process is called affinity maturation *(11)*.

B CELL-RECEPTOR GENE REARRANGEMENT

There are five major classes of antibodies, each with a specialized function: IgG, IgM, IgA, IgD, and IgE. Each has as its basic structure two identical light chains and two identical heavy chains; depending on the class, this basic tetramer has a mass of 150–190 kDa. IgA, the major antibody in secretions, is comprised of two basic tetramers held together by a J (joining) chain, and is associated with a secretory component chain. IgM has five basic tetramers and a J chain; its molecular weight is about 900 kDa and it is sometime called macroglobulin. IgG and IgM are the main antibody classes in blood. IgD serves as a membrane-bound receptor on B cells. IgE, by virtue of its ability to bind to mast cells, is the mediator of parasite resistance and allergic disease.

The origin of the enormous diversity of antibodies is extremely interesting. The immune system must be able to recognize and react against virtually any molecular configuration in the universe, but not against molecules that constitute the "self." This is not easy to manage in a random-bred species, because the genetic composition of the organism is not determined until the ovum is fertilized. Nature has designed a system that randomly generates an enormous number of antibody specificities—the possibilities are close to infinite—and then suppresses potentially dangerous, autoreactive ones.

When the primary sequences of many antibody heavy and light chains are examined, it becomes clear that a single cell can secrete IgM today, and IgG, IgA, or IgE tomorrow. The class changes, but the specificity of the antibody does not. The light chains stay the same; what changes is the constant, nonantigen-binding portion of the heavy chain. The N-terminal variable domain is retained, similar to keeping the teeth of a house key, but changing its handle. This indicates that a single heavy chain must be coded for by at least two genes. In fact, at least four loci contribute to the final functional heavy-chain gene, and three to the light-chain gene. The variable domain of the heavy chain is coded for in three subloci called V, D, and J, each of which contains multiple slight variants of a similar gene segment. The light chain locus is similar, but has only V and J. During differentiation of a B cell, special recombinases are activated and the DNA of the D region on one chromosome is brought near the DNA of the J region. The DNA is cut and spliced to make a randomly chosen DJ unit, and the intervening DNA is discarded. There are other enzymes at work during this joining process, so that a few random (nontemplate-encoded) nucleotides may be added and/or subtracted. This creates an "N region" at the joining sites, and results in an almost infinite amount of variability among the proteins that will be expressed. A similar process joins a V region to the rearranged DJ. Then the light-chain DNA rearranges.

This marvelous system creates a huge diversity of antibodies, but at a cost. Because the addition and subtraction of nucleotides at the DNA splice sites is random, it is clear that two times out of three, the net effect will be a rearrangement that puts the downstream segment out-of-frame. A trial transcript detects the situation. The cell then can try again on the other chromosome, or it can try a process called receptor editing, which is a further rearrangement of V, D, and J, as long as the recombinases are still expressed. If both these mechanisms fail, the cell is useless and will die by apoptosis *(12)*. How often this happens is unknown, but it is likely common, because the production of B-cell precursors in the bone marrow far exceeds the release of mature B cells to the periphery *(13)*.

CLONAL ABORTION IN B CELLS

When a B cell successfully rearranges its immunoglobulin genes, it then expresses the corresponding immunoglobulin on its surface as a receptor. Because the gene rearrangement and N-region creation are random, the expressed surface immunoglobulin of some cells will be "anti-self." If such cells encounter antigen at an immature stage, they may attempt receptor editing if the recombinases are still present; if not, they undergo apoptosis. The process is called clonal abortion or deletion. Like other apoptotic processes in hematopoietic cells, B-cell clonal deletion depends on the proteinase calpain *(14)*.

T CELL-MEDIATED IMMUNITY

Intracellular parasites, such as viruses, protozoa, and some fungi and bacteria, are protected from antibodies by the plasma membrane's lipid bilayer. The body uses another system to deal with these parasites: the T cells (thymus-derived lymphocytes). Particles shed by invading organisms are taken up by dendritic cells, which tend to be found in skin and mucous membranes, the natural portals of pathogen entry. The dendritic cells then move towards the draining lymph nodes. As they do, they process the antigens into peptides, and these become associated with the antigen-binding groove of major histocompatibility complex class II (MHC II) molecules. Vesicles bearing peptide-loaded MHC II molecules move to the cell surface. At the same time, the dendritic cell upregulates various accessory binding and adhesion molecules, becoming a specialized antigen-presenting cell. T cells have clonally variable receptors for antigen plus MHC; the receptor is not antibody but is conceptually very similar. T cells with receptors for MHC II plus antigenic peptides pass by the dendritic cell. If its receptor has the right affinity for the particular peptide-MHC complex, the T cell becomes activated, which leads to division and cytokine secretion. There are two major types of T cells, helpers and killers. Helper T cells are the ones specialized for examining MHC II; their characteristic accessory molecule, CD4, also associates with the MHC II, strengthening the T cell–dendritic cell bond and transducing some of the activating signals. Helper cells do what their name implies: they help B cells get activated to produce antibody of different classes; they help killer T cells get activated; and they attract large numbers of macrophages, the professional phagocytes. The central role of helper T cells in immunity is underscored by the devastating effects of HIV infection; HIV targets CD4 on helper cells.

But what about the hidden intracellular parasites? All cells, when they synthesize proteins, load peptide samples of these proteins onto major histocompatibility complex class I (MHC I) molecules, which, like MHC II, have a groove specialized for binding and displaying peptides. The loaded MHC I is transported to the surface and displayed there for the inspection of T cells. T cells that are specialized for recognition of antigenic peptides on MHC I are the cytotoxic T lymphocytes (CTL), also called killer cells. For the most part, the peptides on MHC I will derive from normal endogenous molecules, and will be ignored by the immune system. When a T cell comes by that has a receptor complimentary to the MHC I and its presented peptide, there is sufficiently strong binding that the CTL becomes activated.

APOPTOSIS INDUCED BY CTL

Cells that display a foreign or novel peptide on surface MHC I molecules become targets for CTL cells. CTL express surface CD8, which strengthens the bond to the target cell, a role analogous to that of CD4. When a CTL engages MHC I bearing a peptide, other accessory interactions between the CTL and the target cell are recruited until an activation threshold is reached. Events then follow that result in the target cell's committing suicide by apoptosis. The total time required for CTL adherence to its target is short, in the order of a few minutes, during which the target is programmed to die.

There is more than one mechanism by which apoptosis is induced in targets. The best-understood involves the upregulation of CD95L, also known as the Fas ligand, on the CTL surface. This engages CD95 (Fas, Apo-1) on the target cell surface, which becomes oligomerized, allowing it to interact with the cytoplasmic signal-transduction adaptor molecule FADD. FADD binds caspase-8; the whole molecular grouping forms an active death-inducing signal complex or DISC *(15,16)*. Active caspase-8 cleaves and activates caspase-3, triggering the final apoptotic cascade. This is a well-designed system for the context in which it works: it involves only pre-existing molecules and is not dependent on new protein synthesis. The most important class of intracellular pathogens are the viruses, and CTL are essential for recovery from infection with most, if not all, viruses. Viruses suppress transcription and translation and would be likely to escape if *de novo* protein synthesis were required for CTL-induced apoptosis. Many viruses, however, have evolved (or stolen) mechanisms for interfering with the death mechanism. One such mechanism is the production of v-FLIP, the viral equivalent of cellular c-FLIP *(17)*. FLIP competitively interferes with the interaction of FADD and caspase-8, blocking the transduction of the apoptotic signal.

Probably because viruses evolve escape mechanisms, CTL have other ways to induce apoptosis. In some cases they can secrete molecules such as tumor necrosis factor-α (TNF-α) and TNF-related apoptosis-inducing ligands (TRAILs) which, like CD95L, are members of the nerve growth factor (NGF) family that bind surface receptors related to the NGF receptor, and signal an apoptotic cascade similar to that used by CD95 *(18)*.

More intensely studied, but less well-understood, is the killer mechanism involving the CTL's cytotoxic granules *(19)*. These contain the pore-forming protein perforin. Activation of the CTL by binding its target causes release of granules (modified lysosomes) into the space between the cells. Although perforin has been shown to polymerize into complement-like pores on a cell surface, there is no convincing evidence that such large pores develop early in the cytotoxic process. Rather, it may be that perforin stimulates uptake of the granule-associated serine proteinases called granzymes. These have been shown to cleave certain caspases, and are capable of initiating the apoptotic process *(20)*.

The purpose of inducing apoptosis, rather than allowing the virus itself to kill the cell it has infected, is this: peptides from viral proteins are expressed on MHC I on the cell surface before complete viruses have been assembled within the cell. If the cell can be killed in this "eclipse" phase, it will not release infectious virus, and the cycle of infection can be interrupted. Thus, death by CTL is a true example of altruistic suicide; the cell's death ensures the survival of its neighbors.

T-CELL REPERTOIRE SELECTION

T cells develop in the thymus from bone marrow-derived precursors. Their receptors (T-cell receptors; TCR) are made up of α and β chains (an uncommon subpopulation has γ and δ chains). Like the heavy and light chains of B cells, each of the TCR chains has hypervariable or complementarity-determining regions, and they are similarly assembled by DNA recombination.

However, T cells are required to recognize not just an antigenic peptide, but the MHC I or II molecule that presents it. Because the generation of TCR is by random recombination, most newly expressed TCR will not meet these stringent requirements. It is not at present completely understood how the T-cell repertoire is selected, but most research favors an affinity selection model. It states that when a developing T cell displays its TCR in the thymic environment, it will encounter MHC molecules on the stromal cells. There are three possibilities: binding will be essentially zero, or with low affinity, or with high affinity. High-affinity TCR represent an autoreactive cell; it is potentially dangerous, and the signal (which would activate a mature T cell) is somehow, at the immature intrathymic stage, interpreted as an instruction to engage the apoptotic pathway. It may be that this is regulated by accessory binding molecules, such as CD8. In the immature cell, CD8 is not sialylated, and its affinity for MHC I is high; in the mature T cell, after sialylation, its affinity is considerably less *(21)*. As a result, a cell in the thymus that happens to bind MHC I plus a "self" peptide would receive a supraoptimal signal. Such a signal would initiate an apoptotic cascade, in which the pro-apoptotic BH3–only Bcl-2 family member Bim has been strongly implicated *(22)*. The removal of highly self-reactive developing thymocytes is called negative selection, and probably involves a relatively small fraction of cells.

In a young mouse, about 98% of the T cells that begin development in the thymus, die in the thymus. Most die because their TCR turn out to have no affinity for the MHC molecules they encounter, and thus are not likely to be of use to the organism as mature T cells. These cells also die by apoptosis, and again, the mechanism by which this occurs is unclear. It has been known for decades that these cells are exquisitely sensitive to glucocorticoids; in fact, the levels achieved at the peak of the circadian cycle are sufficient to kill thymocytes in vitro by apoptosis. A model has been suggested in which thymocytes are killed by endogenous glucocorticoids unless they are also receiving a moderate signal through their TCR. This signal would not be strong enough to activate the cell by itself; but if the cell is allowed to mature, it would then focus its attention on MHC, and if that MHC contains the correct peptide, the net signal strength could now be activating *(23,24)*. Thus, most cells would die by glucocorticoid-induced apoptosis, whereas the small fraction with low but real TCR affinity for MHC would be positively selected for survival, maturation, and export to the periphery as mature T cells.

POST-THYMIC T-CELL DELETION

T cells that bind with high affinity to self-derived peptides presented on MHC molecules in the thymus are deleted from the repertoire. Not all self antigens are expressed in the thymus, and so mechanisms must exist for refining the repertoire of mature T cells that the thymus exports. How this is done is not fully established. It is known that antigens must be presented correctly to T cells. That means that not only must there be binding between the TCR and its cognate MHC-peptide combination, but several accessory binding events must also take place between the T cell and the antigen-presenting cell. Few normal cells bear all the necessary accessory ligands, which are found primarily on dendritic cells and cells of the macrophage family. Thus, a T cell that happens to bind to a self antigen on, for instance, a kidney cell, will find itself receiving some, but not all, of the necessary activating signals. T cells seem to interpret this incomplete signal as dangerous, and cease to respond (a process called anergy) or, more commonly, undergo apoptosis. The signal is transduced through cell-surface Fas (CD95). Mice that lack a functional Fas have massive lymphoaccumulation that mimics lymphoma, and they develop autoimmunity *(25)*. Recently a human counterpart of this syndrome has been described *(26)*.

GROWTH-FACTOR WITHDRAWAL

Growth factors are often in fact survival factors; that is, they keep cells alive so that they, or other factors, can stimulate growth. Removal of such factors results in death by apoptosis. The immune system, with its complex helper interactions, offers several examples of this phenomenon. During a response to antigen, CTL upregulate interleukin-2 (IL-2) receptors, and helper T cells provide IL-2. When antigen declines, less IL-2 is available, and most of the CTL, deprived of growth factor, die by apoptosis *(27)*. Similarly, B cells can become dependent on helper T cell-derived lymphokines like IL-3 and IL-4; it was in this system that the anti-apoptotic properties of Bcl-2 were first described *(28)*.

CONCLUSIONS

The immune system, perhaps better than any other system, reveals the extraordinary importance of apoptosis in the normal functioning of adult organisms. By far the most probable fate of a developing lymphocyte is an early death; the requirements for a successful lymphocyte are so stringent that few make the grade. It is not surprising that much of our knowledge of apoptosis in mammals comes from immune system models, nor that several abnormalities of apoptosis leading to disease have already been described. As in other systems, learning enough about apoptosis in immunity to be able to regulate the process will certainly lead to new therapies, for conditions as various and important as autoimmunity, allergy, and cancer.

REFERENCES

1. Aliprantis, A. O., Weiss, D. S., and Zychlinsky, A. (2001) Toll-like receptor-2 transduces signals for NF-kappa B activation, apoptosis and reactive oxygen species production. *J. Endotoxin Res.* **7,** 287–291.
2. Ashwell, J. D., Lu, F. W., and Vacchio, M. S. (2000) Glucocorticoids in T cell development and function. *Annu. Rev. Immunol.* **18,** 309–345.
3. Walsh, G. M. (1997) Mechanisms of human eosinophil survival and apoptosis. *Clin. Exp. Allergy* **27,** 482–487.
4. Sellins, K. S. and Cohen, J. J. (1987) Gene induction by gamma-irradiation leads to DNA fragmentation in lymphocytes. *J. Immunol.* **139,** 3199–3206.
5. Lowe, S. W., Schmitt, E. M., Smith, S. W., Osborne, B. A., and Jacks, T. (1993) p53 is required for radiation-induced apoptosis in mouse thymocytes. *Nature* **362,** 847–849.
6. Harding, H. P., Novoa, I., Zhang, Y., Zeng, H., Wek, R., Schapira, M., and Ron, D. (2000) Regulated translation initiation controls stress-induced gene expression in mammalian cells. *Mol. Cell.* **6,** 1099–1108.
7. Sood, R., Porter, A. C., Olsen, D. A., Cavener, D. R., and Wek, R. C. (2000) A mammalian homologue of GCN2 protein kinase important for translational control by phosphorylation of eukaryotic initiation factor-2alpha. *Genetics* **154,** 787–801.
8. Srivastava, S. P., Kumar, K. U., and Kaufman, R. J. (1998) Phosphorylation of eukaryotic translation initiation factor 2 mediates apoptosis in response to activation of the double-stranded RNA-dependent protein kinase. *J. Biol. Chem.* **273,** 2416–2423.
9. Han, A. P., Yu, C., Lu, L., Fujiwara, Y., Browne, C., Chin, G., et al. (2001) Heme-regulated eIF2alpha kinase (HRI) is required for translational regulation and survival of erythroid precursors in iron deficiency. *EMBO. J.* **20,** 6909–6918.
10. Brostrom, C. O. and Brostrom, M. A. (1998) Regulation of translational initiation during cellular responses to stress. *Prog. Nucleic Acid Res. Mol. Biol.* **58,** 79–125.
11. Yin, J., Mundorff, E. C., Yang, P. L., Wendt, K. U., Hanway, D., Stevens, R. C., and Schultz, P. G. (2001) A comparative analysis of the immunological evolution of antibody 28B4. *Biochemistry* **40,** 10764–10773.
12. Kouskoff, V. and Nemazee, D. (2001) Role of receptor editing and revision in shaping the B and T lymphocyte repertoire. *Life Sci.* **69,** 1105–1113.
13. Lu, L. and Osmond, D. G. (2000) Apoptosis and its modulation during B lymphopoiesis in mouse bone marrow. *Immunol. Rev.* **175,** 158–174.
14. Ruiz-Vela, A., Serrano, F., Gonzalez, M. A., Abad, J. L., Bernad, A., Maki, M., and Martinez, A. (2001) Transplanted long-term cultured pre-BI cells expressing calpastatin are resistant to B cell receptor-induced apoptosis. *J. Exp. Med.* **194,** 247–254.
15. Muzio, M., Chinnaiyan, A. M., Kischkel, F. C., O'Rourke, K., Shevchenko, A., Ni, J., et al. (1996) FLICE, a novel FADD-homologous ICE/CED-3-like protease, is recruited to the CD95 (Fas/APO-1) death-inducing signaling complex. *Cell* **85,** 817–827.
16. Krammer, P. H. (2000) CD95's deadly mission in the immune system [In Process Citation]. *Nature* **407,** 789–795.
17. Thome, M., Schneider, P., Hofmann, K., Fickenscher, H., Meinl, E., Neipel, F., et al. (1997) Viral FLICE-inhibitory proteins (FLIPs) prevent apoptosis induced by death receptors. *Nature* **386,** 517–521.
18. Barber, G. N. (2001) Host defense, viruses and apoptosis. *Cell Death Differ.* **8,** 113–126.
19. Metkar, S. S., Wang, B., Aguilar-Santelises, M., Raja, S. M., Uhlin-Hansen, L., Podack, E., et al. (2002) Cytotoxic cell granule-mediated apoptosis: perforin delivers granzyme B-serglycin complexes into target cells without plasma membrane pore formation. *Immunity* **16,** 417–428.

20. Smyth, M. J., Kelly, J. M., Sutton, V. R., Davis, J. E., Browne, K. A., Sayers, T. J., and Trapani, J. A. (2001) Unlocking the secrets of cytotoxic granule proteins. *J. Leukoc. Biol.* **70,** 18–29.
21. Daniels, M. A., Devine, L., Miller, J. D., Moser, J. M., Lukacher, A. E., Altman, J. D., et al. (2001) CD8 binding to MHC class I molecules is influenced by T cell maturation and glycosylation. *Immunity* **15,** 1051–1061.
22. Bouillet, P., Purton, J. F., Godfrey, D. I., Zhang, L. C., Coultas, L., Puthalakath, H., et al. (2002) BH3-only Bcl-2 family member Bim is required for apoptosis of autoreactive thymocytes. *Nature* **415,** 922–926.
23. Iwata, M., Hanaoka, S., and Sato, K. (1991) Rescue of thymocytes and T cell hybridomas from glucocorticoid-induced apoptosis by stimulation via the T cell receptor/CD3 complex: a possible in vitro model for positive selection of the T cell repertoire. *Eur. J. Immunol.* **21,** 643–648.
24. Ashwell, J. D., Vacchio, M. S., and Galon, J. (2000) Do glucocorticoids participate in thymocyte development? *Immunol. Today* **21,** 644–646.
25. Suda, T. and Nagata, S. (1997) Why do defects in the Fas-Fas ligand system cause autoimmunity? *J. Allergy Clin. Immunol.* **100,** S97–S101.
26. Straus, S. E., Sneller, M., Lenardo, M. J., Puck, J. M., and Strober, W. (1999) An inherited disorder of lymphocyte apoptosis: the autoimmune lymphoproliferative syndrome. *Ann. Intern. Med.* **130,** 591–601.
27. Duke, R. C. and Cohen, J. J. (1986) IL-2 addiction: withdrawal of growth factor activates a suicide program in dependent T cells. *Lymphokine Res.* **5,** 289–299.
28. Vaux, D. L., Cory, S., and Adams, J. M. (1988) Bcl-2 gene promotes haemopoietic cell survival and cooperates with c-myc to immortalize pre-B cells. *Nature* **335,** 440–442.

14

Cell-Death Mechanisms in Neurodegenerative Diseases

Diversity Among Cerebral Ischemia, Parkinson's Disease, and Huntington's Disease

R. Anne Stetler and Jun Chen

THE COMPLEXITY OF CELL DEATH IN NEURODEGENERATION

Early in the field of cell death, two distinct forms of cell death were noted: necrotic, or bursting of the cell membrane, and apoptotic, wherein the cell membrane remains intact and the nuclear material is digested in a methodical manner. However, morphological analyses of both human tissue and animal models of neuronal diseases have indicated a wide range of neuronal cell-death capabilities. Although some similarities to other model systems exist, an alarming amount of disparity appears to be uniquely neuronal. Additionally, many of the biochemical processes observed in non-neuronal systems is paralleled in neuronal systems, but the morphological outcome can be vastly different. Here, we will limit the discussion of neurodegenerative diseases to the morphological, biochemical, and molecular aspects of cerebral ischemia, Parkinson's disease, and Huntington's disease as diverse examples of the different modalities of cell death in the adult nervous system.

CEREBRAL ISCHEMIA

Cellular Morphological Aspects

Despite the recent tremendous advances in the basic biology of cell death, our knowledge is insufficient regarding the role of cell death, the cell-death execution programs, and the regulation of these programs in the pathophysiology of brain injury after stroke. There have been debates about whether and to what extent the apoptotic form of cell death contributes to ischemic brain injury. Ischemic neuronal death has classically been considered to be necrotic owing to several common features of cerebral ischemia including energy depletion, loss of calcium homeostasis, and cell and organelle swelling. Moreover, several studies at light or electron microscopy levels have shown that ischemic neurons failed to manifest the characteristic morphological changes identified in classical apoptosis occurring during neuronal development *(1)*. These results, however, are in contrast to the observations made by numerous other studies using similar or different animal models, in which almost all morphological changes akin to apoptosis have been described so far *(2)*, including cell shrinkage, membrane blebbing, chromatin condensation, internucleosomal DNA fragmentation, and the forma-

From: *Essentials of Apoptosis: A Guide for Basic and Clinical Research*
Edited by: X-M. Yin and Z. Dong © Humana Press Inc., Totowa, NJ

tion of apoptotic bodies. This discrepancy may reflect the complexity of the death stimuli received by the ischemic neurons under different experimental conditions.

A prominent feature of ischemic neuronal death that strongly supports a role of apoptosis in cerebral ischemia is the induction of internucleosomal DNA fragmentation, which is a hallmark of apoptosis. Using combined *in situ* detection (TUNEL) and DNA gel electrophoresis, many studies detected internucleosomal DNA fragmentation in ischemia models *(3–7)*, as well as in human postmortem tissue *(8)*. DNA fragmentation constitutes the final step in the execution of apoptosis in ischemic neurons and is often accompanied by morphological characteristics of apoptosis *(3,6,7)*. The molecular basis for this DNA degradation process in ischemic brain has remained elusive until recently. We have recently cloned the CAD/ICAD genes from rat brain and obtained direct evidence that the caspase-3-activated CAD (caspase-activated DNase) mediates internucleosomal DNA fragmentation and advanced chromatin condensation in the brain after transient ischemia *(9,10)*. Given that both the morphological aspects of DNA fragmentation and chromatin condensation exist, along with evidence of the activation of the responsible enzymes, an apoptotic-like biochemical cascade of events during cerebral ischemia is expected.

Biochemical Aspects

Brain Mitochondria and Calcium

Owing to the high density of neurotransmitters stored in neurons, the cytotoxic potential of these molecules needs to be considered when exploring the origins of cell death in neurodegenerative diseases. Glutamate has long been known to have high excitotoxic potential, and is classically characterized to induce an excitotoxic cascade of events due to excessive release, ineffective uptake, or nonspecific diffusion across cell membrane during ischemic insults *(11)*. Excessive stimulation of NMDA receptors results in prolonged calcium influx into the cell. The molecular actions of handling this influx are still unclear; however, mitochondria have long been known to have calcium-buffering capabilities *(11)*. Recently, the role of the mitochondria and calcium in mediating cell death following cerebral ischemia has become better understood. Transient inhibition of mitochondrial function *via* the respiratory-chain uncouplers was found to have neuroprotective effects against glutamate toxicity in primary cortical cultures, despite an increase in intracellular free calcium *(12)*. In vitro studies with isolated mitochondria have demonstrated that increases in mitochondrial calcium uptake can impair respiration and cause release of cytochrome *c* from the intermembrane in isolated mitochondria *(13,14)*. The mechanism of release of cytochrome *c* from the mitochondria has been under some debate. The opening of the mitochondrial permeability transition pore (PTP) was originally thought to be the major mechanism of cytochrome *c* release *(15)*; however, several studies have found release of cytochrome *c* without evidence of PTP opening. Schild et al. found that different concentrations of extramitochondrial calcium could cause distinct changes in isolated brain mitochondria *(13)*. In particular, concentrations greater than 1 μM caused cytochrome *c* release without evidence of PTP opening or mitochondrial swelling. However, at calcium concentrations greater than 200 μM, mitochondria lost the integrity of the cristae, indicative of water influx and loss of membrane potential. Loss of mitochondrial membrane potential can lead to severe depletion of ATP *(13)*, as would be the case in the ischemic core regions in stroke. Thus, the levels of calcium concentration in differential brain areas may be a contributing factor in determining apoptotic-like vs necrotic-like cell death after cerebral ischemia. Another mechanism by which mitochondrial calcium uptake contributes to cytotoxicity is via oxidative stress. Maciel et al. found that calcium uptake into isolated brain mitochondria resulted in a burst of reactive oxygen species (ROS), and that this ROS formation and the induction of mitochondrial permeability transition were potentiated with the addition of 10 mM sodium *(16)*. Increased cellular sodium concentration has been noted in a model of glutamate-induced toxicity (17). Thus, the physiological setting for mitochondrial-generated toxicity may be present in both the ischemic core and the penumbra during cerebral ischemia.

Caspases and Alternative Cell-Death Programs

It is now widely accepted that caspases, particularly caspase-3, play an important role in mediating ischemic neuronal death. Caspase-3 gene expression and activity are selectively increased in injured neurons after ischemia *(18–20)*. Caspase-3-mediated proteolytic cleavage of cell-death substrates including poly(ADP-ribose)polymerase, DNA protein kinase, and inhibitor of CAD has been reproducibly detected in the ischemic brain *(10,18,21)*. Furthermore, pharmacological inhibition or genetic blockage of caspase-3 activity significantly decreases infarct size *(20,22,23)* or attenuates the loss of hippocampal CA1 neurons following transient cerebral ischemia *(18)*. In line with these observations, Xu et al. demonstrated that overexpression of XIAP (X chromosome-linked inhibitor of apoptosis protein), which directly inhibited caspase-3 and caspase-7, attenuated hippocampal neuronal loss after global ischemia *(24)*. In a recent study by Benchoua et al., permanent focal ischemia was found to elicit a biphasic wave of caspase-3 activity, suggesting that the cell-death mechanisms in the core and the penumbra may both trigger caspase-3 activation *(25)*. Accordingly, it has been proposed that caspase-3 and caspase-7 could be a "final common pathway" by which cell death is executed after cerebral ischemia and thus could be the optimal therapeutic target.

The precise mechanism by which caspase-3 is activated in neurons after cerebral ischemia remains unclear at the present time. Several studies suggest that both extrinsic and intrinsic pathways could be involved in ischemic cell death and caspase activation. For instance, cytochrome *c* release has been reproducibly detected in selectively vulnerable neurons after transient focal or global cerebral ischemia *(26,27)* and is consistent with the effects of mitochondrial calcium uptake in isolated brain mitochondria *(13,14)*. Furthermore, there is enhanced formation of the Apaf-1/caspase-9 heterodimeric complex in the hippocampal protein extracts after transient global ischemia *(28)*, thus providing direct evidence of cytochrome *c*-mediated activation of Apaf-1. In addition, Bax translocation to the mitochondria appears to be a prominent and early feature of ischemic neuronal injury *(29)*, which could trigger or augment cytochrome *c* release and the Apaf-1/caspase-9 pathway in ischemic neurons. On the other hand, the Fas/Fas-L system has also been found to be activated after ischemia and may contribute to neuronal cell death *(30–32)*. Caspase-8, which may serve to transmit death signals from the cell membrane, has been found to be activated within 30 min of permanent focal ischemia *(25)*. Caspase-8 has been shown to cleave Bid in vitro *(33)*. Bid, a proapoptotic member of the bcl-2 family, has been found in non-neuronal systems to translocate to the mitochondria and induce cytochrome *c* release following activation by caspase-8 *(34)*. Recently, Plesnila et al. found that Bid-deficient mice were resistant to cell death induced by transient cerebral ischemia, and these genetically altered mice also showed attenuated cytochrome *c* release after ischemia *(33)*. Furthermore, in the culture model of oxygen/glucose deprivation, caspase-3 activity was greatly reduced in neurons derived from Bid-deficient mice, whereas caspase-8 processing was unaltered. Thus, caspase-8 may cleave Bid and subsequently act on the mitochondria to activate the intrinsic pathway, resulting in the activation of caspase-3 and other terminal caspases.

Despite of the biochemical evidence of the activation of both extrinsic and intrinsic pathways, the direct functional linkage between the deduced upstream pro-apoptotic events and the activation of terminal caspases in ischemic neurons has yet to be established. Based on the temporal correlation analysis, it was suggested that in focal ischemia the activation of caspase-3 during the early phase of injury might be mediated mainly via the caspase-8-dependent extrinsic pathway, whereas caspase-9 activation might contribute to caspase-3 activation associated with the secondary expansion of infarction *(25)*. In another study, however, caspase-8 and caspase-3 were expressed in different populations of neurons after ischemia *(35)*, opposing a direct role of the extrinsic pathway. These somewhat discrepant observations raised the possibility that the mechanism underlying terminal caspase activation after ischemia may vary depending on the neuronal phenotype and the stage of ischemic injury, and that the differences between ischemia models and between different ischemic regimens (permanent vs transient) may also contribute to the variables. Moreover, the activation of terminal caspases in ischemic neurons may also be mediated *via* other alternative mechanisms. Recent findings have suggested that caspase-3 can be cleaved in vitro directly by calpain *(36)* or by calpain cleavage of caspase-12 *(37)*. Further support for the role of calpain-caspase-3 interaction in ischemic cell death was evidenced by the observation that calpain inhibitors could suppress caspase-3 cleavage in a model of neonatal hypoxia/ischemia *(36)*. These studies suggest that though the final cell-death execution pathway may be similar, apoptotic ischemic cell death may occur by multiple and highly divergent mechanisms.

PARKINSON'S DISEASE

Cellular Morphological Aspects

Parkinson's disease (PD) is a debilitating neurodegenerative disorder characterized by selective loss of dopaminergic neurons in substantia nigra pars compacta and, consequently, by diminishment of dopamine levels in the striatum and development of severe motor deficits. A number of studies based on the examination of postmortem human brain tissue suggest that apoptosis, the common morphological form of programmed cell death (PCD), may play a major role in the degeneration of neurons in PD (reviewed in refs. *38–40*). Important evidence that supports the PCD hypothesis includes the biochemical detection of DNA fragmentation, along with findings of ultrastructural features of apoptosis such as chromatin condensation, nucleolus disappearance, shrinkage of cell bodies, and formation of apoptotic-like bodies in the substantia nigra neurons of PD patients *(41–46)*. However, so far not all human-subject studies have supported the PCD hypothesis. Several studies failed to obtain convincing morphological features of apoptosis in PD nigra *(47–49)*. It is believed, however, that the failure to demonstrate apoptotic morphology in a pathologic condition does not rule out a role for PCD for at least two reasons *(40)*. First, in vivo apoptotic cells are phagocytized and digested within a matter of hours, whereas PD involves a slow pathologic process; thus, the number of apoptotic cells present in a tissue section at any single time point could be negligible. Second, apoptosis is only one of the forms of PCD, and neurons could undergo PCD without demonstrating the characteristic apoptotic morphology. However, in spite of these controversies, the induction of neuronal PCD in PD has now become an important hypothesis that has grown in interest and empirical support.

Biochemical Aspects

Caspases

Additional biochemical evidence that strongly supports the PCD hypothesis comes from the observation that nigrostriatal neuronal degeneration in PD patients is associated with the regulation of several key apoptosis-regulatory gene products. For instance, the levels of Bcl-2 were found to be significantly higher in the nigrostriatal regions, but not in cerebral cortex, compared to those in age-matched control subjects *(45,50,51)*. Furthermore, increased Bax immunoreactivity was detected in neurons of the PD nigra in conjunction with increased numbers of apoptotic neuronal nuclei in the same brain region *(45)*, although another study by Hartmann et al. failed to find any difference in the overall expression of Bax between PD patients and controls, but found an increased Bax immunoreactivity in Lewy-body containing melanin neurons as compared to non-Lewy body containing melanin neurons *(52)*. Finally, the percentage of dopaminergic neurons containing the active form of caspase-3, -8, and -9, and caspase-1- and caspase-3-like protease activity were increased in the substantia nigra of PD patients compared to controls *(53–56)*. These observations provide histological evidence suggesting that the inappropriate activation of certain PCD executioners may contribute to the intracellular cascade that leads to the pathogenesis of PD.

A growing body of evidence suggests that caspases, particularly caspase-3, may be an important executor of dopaminergic neuronal cell death in experimental models of PD. Tetrapeptide inhibitors of caspase-3 have shown reproducible protective effects against MPP+- or 6-OHDA-induced cell death in several dopaminergic cell types, including dopamine neurons in the ventral mesencephalon cultures *(57–61)*, although in the MN9D dopaminergic cell line, these two toxins may elicit different cell death mechanisms *(62–64)*. However, results supporting the role for caspase involvement in dopaminergic cell death have also been obtained in in vivo models of PD. Intraperitoneal administration of MPTP resulted in increased caspases-3-, -8-, and -9-like activities, cytochrome *c* release and cleavage of Bid in the SN of wild-type mice, all of which were attenuated in p35 transgenic mice *(53)*. Eberhardt et al. have shown that a peptide caspase inhibitor or adenorviral gene transfer of a protein caspase inhibitor (XIAP) prevents MPTP-induced cell death of dopaminergic neurons in SNc *(65)*. Consistent with the deduced role of caspase-3 in dopaminergic neuronal apoptosis, markedly increased caspase-3-like protease activity is detected in cultured dopamine neurons exposed to MPP$^+$ or 6-OHDA *(58,66)*. Jeon et al. found that activated caspase-3 is expressed in three models of PCD affecting SNc neurons in the rat *(67)*. All of these results support the concept that caspase-3 may play a critical role in the molecular cascade leading to the pathogenesis of PD and that inhibition of specific caspase(s) may be a promising therapeutic strategy for PD.

Despite the potential importance of caspases as executors of dopaminergic neuronal death in PD, two important questions warrant investigation in the future: 1) What are the mechanisms leading to the activation of caspases in dopaminergic neurons selectively? and 2) Are there caspase-independent pathways of PCD that function in parallel with or synergistically with the caspase-dependent pathway to cause dopaminergic neuron death?

The molecular events upstream of caspase activation in dopaminergic neurons have not been thoroughly investigated. Bax activation and cytochrome *c* release are known initiation factors for caspase-3 activation, and both events have been detected in dopaminergic neurons under PD-relevant insults *(61,68)*. However, a causative relationship has not yet been established. Furthermore, two recent studies suggest that caspase-independent mechanisms may also be responsible for PCD in PD. Choi et al. reported a lack of caspase-3 activation or neuroprotection by caspase inhibitors in the murine dopaminergic MN9D neuronal cell line exposed to MPP$^+$ *(63)*. MPP$^+$-induced MN9D cell death was characterized by mitochondrial swelling and lack of DNA fragmentation *(64)*. Nevertheless, Bcl-2 overexpression or inhibitors of macromolecule synthesis such as cycloheximide blocked this cell death *(64)*, strongly suggesting that caspase-independent PCD mechanisms may be involved in the cell-death process. Hence, these results warrant more thorough investigation of the execution molecules and pathways in dopamine neuron PCD in PD.

Oxidative Stress

Oxidative stress has long been the leading hypothesis as the major mechanism for cell death in PD *(69,70)*. Membrane and lipid peroxidation, oxidative DNA and protein modifications, decreased antioxidants such as glutathione (GSH), glutathione peroxidase and catalase, and increased intracellular iron have all been found in post-mortem tissue from parkinsonian patients (reviewed in ref. *71,72*). In addition, some PD patients show impaired mitochondrial function *(73)*, which may lead to increased mitochondrial superoxide production, and increased vulnerability of dopaminergic neurons to glutamate toxicity *(74)*. The initial source(s) of the oxidative stress in PD is not clear, and may be a combination of several factors contributing to the overall environment. Depletion of GSH and changes in other antioxidant reserves can reduce the cellular ability to buffer physiologically produced reactive oxygen species, and thus might exacerbate oxidative stresses within the nigrostriatal system during PD pathogenesis *(75–78)*. In addition to glutathione depletion, increased iron bound to neuromelanin has been shown to enhance the formation of lipid peroxidation and free radicals, also contributing to oxidative stress *(79,80)*. Supporting a role for increased iron in the pathogenesis of PD is the finding that iron chelators are neuroprotective in a rat model of PD *(79)*.

A recent report by Good et al. demonstrated the presence of increased nitrotyrosine in the brains of PD patients *(81)*, suggesting a role for reactive nitrogen species as ROS in PD pathogenesis. Peroxynitrite is formed from the reaction of superoxide with nitric oxide and is the major species believed to be involved in nitrating protein tyrosine residues *(82)*. Peroxynitrite can also react to modify protein cysteine residues, and these nitration effects of peroxynitrite likely contribute to the inhibition of important enzymes such as tyrosine hydroxylase *(83)*.

Dopamine itself may be another source of oxidative stress in PD, which may partially explain the particular susceptibility of dopaminergic neurons to oxidative injury *(84)*. Sequestration of dopamine within acidic secretory vesicles is protective. However, once dopamine is released from nerve terminals in the striatum, the neutral pH in the extracellular space promotes autoxidation of its catechol ring, generating superoxide, hydrogen peroxide, and hydroxyl radicals. In addition, in oxidizing environments, dopamine is able to form a highly reactive dopamine quinone, which are then able to form covalent bonds with cysteine residues found in proteins, glutathione, or free cysteine *(85,86)*. The biological significance of the formation of protein-bound cysteinyl-dopamine has yet to be determined, but work by Hastings et al. has demonstrated that the toxic potential of intrastriatal injections of dopamine is closely correlated with the formation of cysteinyl-dopamine in rats *(86)*. Supporting a role for dopamine oxidation in parkinsonian pathology, Spencer et al. were able to detect increased dopamine-modified cysteine and GSH in the substantia nigra and putamen of post-mortem human tissue *(87)*. In addition to forming protein adducts, oxidized dopamine itself may have direct effects on mitochondria. In a study by Berman and Hastings *(88)*, incubation of isolated brain mitochondria with dopamine quinone altered respiration induced permeability transition and resulted in mitochondrial swelling. The treatment of mitochondria with dopamine has also been found to induce release of cytochrome *c (89)*, supporting the findings in a cell-culture model of dopamine neurotoxicity, which showed a caspase-dependent form of cell death *(90)*. In addition to auto-oxidation, the metabolism of dopamine by monoamine oxidase generates hydrogen peroxide, and thus can also act to induce oxidative stress *(72)*.

HUNTINGTON'S DISEASE

Cellular Morphological Aspects

Huntington's disease (HD) is an autosomal dominant neurodegenerative disease that is characterized by the morphological loss of neurons predominantly in the striatum and cortex. The progression of the disease occurs over decades, often not becoming behaviorally evident until the fourth decade in life. The severity of HD occurs in a dose-dependent manner; increased severity is associated with greater expansion of the glutamine repeat and the number of affected alleles. In addition to cell death, histological examination of HD brains reveals the presence of nuclear and cytoplasmic protein aggregates. The exact role of these insoluble aggregates is currently unknown—it is debated whether these aggregates are a byproduct of the toxicity, are a mechanism for sequestering away toxic fragments, or are the cause of cellular dysfunction *(91)*. However, in cell culture, neurodegeneration can be induced by nuclear-targeted expanded htt in the absence of protein-aggregate formation *(92–94)*.

The particular morphology of cell death has been termed "dark cell death" to describe dense neuronal staining described in both transgenic mice overexpressing exon 1 htt with a pathological glutamine expansion and postmortem HD brain *(95)*. Degenerating neurons were identifiable by increased affinity for, and hence darker staining with, toluidine blue or osmium. Morphologically, the nucleus and cytoplasm were both condensed and other distinct morphological processes occurred (e.g., late transient swelling of the Golgi and mitochondria), but no evidence of blebbing or the formation of apoptotic bodies was observed at the ultrastructural level. Several studies have found TUNEL-positive degenerating neurons and glia in HD postmortem tissue *(95–98)*; however, without the ultrastructural formation of apoptotic bodies, this "dark cell death" is not considered to be classical apoptosis *(95)*.

Huntington, Caspases, and Regulation of Cell Death

The protein product of htt itself contains five potential caspase cleavage sites within exon 1, located approx 400 amino acids downstream of the polyglutamine tract *(99)*. It was shown in vitro that caspase-1 is able to cleave htt *(100)*, whereas caspases-3 and -6 were found to cleave htt both in in vitro and in in vivo cell models *(101)*. Indeed, htt N-terminal fragments of similar size to the caspase-3-cleaved form have been found in human HD brain *(102,103)*. Transgenic mice expressing full-length huntingtin with a pathological number of glutamine repeats were found to produce N-terminal fragments that aggregated both in the cytoplasm and nucleus *(104)*. It has been postulated that the N-terminal fragments of htt that translocate to the nucleus are the byproducts of caspase cleavage *(105)* and may lead to decreased cell viability *(106)*. However, if this hypothesis is correct, the question still remains as to the catalyst for caspase activation in the first place. Expansion of the polyglutamine tract has been shown to expedite cleavage of htt in vitro *(107)*, though it is currently debated whether the N-terminal fragments arise from cleavage of the expanded htt or the wild-type htt *(102,103)*. Another suggested mechanism for caspase activation is *via* caspase-8 recruitment to the protein aggregates, resulting in autoactivation *(108,109)*. Alternatively, several studies have reported abnormalities in mitochondrial energy metabolism in HD brain and/or platelets *(110)*. Although these aberrations may result in cellular toxicity, the mitochondrial caspase cascade has yet to be directly elucidated in the pathogenesis of HD.

Selective Vulnerability

Excitotoxicity

Originally, striatal neuronal death in HD was thought to be due to impaired neuronal signaling *(111,112)*. The parallel finding that injections of quinolinic acid affected the same neurons as HD, leading to the excitotoxicity hypothesis, supported this presumption. Additionally, when taken from brain lysates or co-expressed in cell culture, the protein product of huntingtin was found to interact with the postsynaptic density-95 (PSD95) protein and NMDA and GluR6 receptors *(113)*, which in turn is known to bind to NMDA and kainate receptors *(114)*. Polyglutamine expansion in huntingtin disrupts the association with PSD95, and is thus thought to potentiate NMDA and kainate receptor hypersensitivity *(113,115)*. Further indication of aberrant neuronal signaling can be derived from the full-length transgenic HD mouse model (YAC72) *(104)*. These mice were found to have electro-physiological abnormalities consistent with NMDA receptor hyperactivity. Other indications of increased sensitivity to stress included the finding that primary striatal cultures taken from the exon1 transgenic mouse model were found to be hypersensitive to oxidative stress induced by addition of exogenous dopamine *(116)*. One potential mechanism for increased sensitivity to toxic stimuli may be due to a loss of huntingtin function caused by the repeat expansion. For example, wild-type huntingtin has been found to suppress activation of the mixed-lineage kinase 2 (MLK2), an upstream kinase that activates the c-jun N-terminal kinase (JNK) pathway *(117)*. Polyglutamine expansion of huntingtin weakens this suppression, and causes activation of JNK in cellular models, which may lead to cell death via signaling or transcriptional upregulation of pro-apoptotic genes *(117,118)*. These reports indicate that the selective vulnerability of striatal neurons to HD may result from a lower threshold to handle excitotoxicity.

Transcriptional Dysregulation

BRAIN-DERIVED NEUROTROPHIC FACTOR

(BDNF) has been found to be necessary for the promotion and survival of striatal cells in vivo *(119,120)*. Although BDNF mRNA production has been found to occur in striatal cells *(121)*, most studies have found that the majority of BDNF targeted to the striatum is produced in cortical areas, and subsequently transported to the striatum *via* cortico-striatal projections *(121,122)*. The potential involvement of BDNF in HD was postulated because: 1) striatal cells in particular rely on BDNF for survival in culture *(120)*; 2) BDNF was found to be able to prevent excitotoxin-induced neurodegeneration *(123)*; and 3) the BDNF content was found to be reduced in the cortex, caudate, and putamen of HD human brains *(124,125)*. These observations then led to the question raised by Zuccato et al.: is huntingtin involved in BDNF transcription? In cell-culture models and in the cortex and striatum of a full-length htt transgenic mouse model, and more relevantly in human HD cortical tissues, Zuccato et al. reproducibly found decreased BDNF mRNA and protein levels in the presence of expanded htt *(125)*. Furthermore, cells and mice that overexpressed wild-type htt were found to have increased BDNF mRNA and protein, indicating that htt plays a role in BDNF transcription in a dose-dependent manner *(125)*. Thus, therapies aimed at increasing striatal BDNF may prove to be of some value.

CREB-Binding Protein (CBP)

CBP is one of the many proteins that have been found to be present in the intracellular aggregates of HD *(126,127)*. CBP appears to be a major determinant in promoting neuronal cell survival *(128,129)*, presumably *via* regulating the transcription of gene products downstream of the CRE element *(130)*. Interestingly, CBP itself contains a stretch of polyglutamines, which were found to be required for its association with mutant huntingtin *(131)*. In various cellular models, using both gene-chip array and luciferase reporter assays, several independent groups found that CRE-mediated gene transcription in particular was consistently decreased following overexpression of huntingtin exon1 containing a pathological number of repeats *(131,132)*. Other studies have indicated that this suppression was not unique to mutant huntingtin, because glutamine-expanded DRPLA protein could also inhibit CREB-dependent transcription *(133)*. Furthermore, Nucifora et al. found that overexpression of CBP with a deletion in its polyglutamine stretch was effective at rescuing cortical neurons from toxicity induced by overexpression of mutant huntingtin (exon1) *(131)*. The mechanism by which CBP overexpression could confer neuroprotection is not clear. However, the transcription of BDNF has been shown to be regulated by CREB *(134,135)*, and CREB activation has been found to contribute to BDNF-dependent cell survival in cerebellar granule cells *(128)*. Because YAC18 transgenic mice exhibited increased BDNF transcription, there is likely a specific function of normal huntingtin that can modulate transcription. It remains to be seen whether this is *via* interaction with CREB signaling.

Loss-of-Function or Gain-of-Function

Cleavage or glutamine expansion of wild-type htt could result in either a toxic gain of function, or deleterious loss of function. As described earlier, "sticky" glutamine aggregates in the nucleus may sequester necessary transcription factors, such as CBP, which would argue for a pathological gain of function. On the other hand, evidence that normal htt overexpression is protective against apoptotic stimuli would suggest that normal htt itself may contribute to cellular survival. Additionally, researchers have also found that htt is important in many cellular functions, such as vesicular trafficking *(136,137)*. This observation and the potential dysregulation of signaling pathways owing to the glutamine expansion in htt described earlier argue for a loss of function. Quite likely, both mechanisms exist, which feed back on each other. However, the initial impetus giving rise to the pathology is still under investigation. For example, expanded htt may, *via* disinhibition of signaling, result in low-level caspase activation and cleavage of normal htt, and result in loss of function. Dyer et al. used size and proteolytic analysis to indicate that in human, N-terminal htt fragments arose primarily from normal htt, and that the expanded htt was actually more resistant to proteolysis *(103)*. Determination of the primary mechanism site of pathology—whether transcriptional, cleavage, or aggregation—is the next step in our understanding of HD.

CONCLUDING REMARKS

Emerging evidence now points to a key role for apoptosis in the pathogenesis of various neurodegenerative diseases. Thus, the control of inappropriate activation of the apoptosis execution machinery in neurons represents an attractive and important target for development of novel therapeutic interventions. Cerebral ischemia, Huntington's disease, and Parkinson's disease are representative of the better-defined neurodegenerative diseases, however, much work remains to be done to elucidate the specific cell-death mechanisms involved in the neuropathologies. Although the mechanisms responsible for the terminal cell-death execution associated with these diseases appear to share similarities, the initiation and transduction of cell-death signals in neurons are unique for each of the disorders, and even in different stages of a given disorder. Furthermore, it remains largely elusive what mechanisms are responsible for the selective vulnerability of defined cell populations in the brain for neurodegeneration seen in different disorders. Nevertheless, by studying neuronal cell death in different neuropathologies, much insight may be gained to further understand neuronal death and survival mechanisms.

REFERENCES

1. Colbourne, F., Sutherland, G. R., and Auer, R. N. (1999) Electron microscopic evidence against apoptosis as the mechanism of neuronal death in global ischemia. *J. Neurosci.* **19(11),** 4200–4210.
2. Lipton, P. (1999) Ischemic cell death in brain neurons. *Physiol. Rev. 79(4),* 1431–1568.
3. Chen, J., Jin, K., Chen, M., Pei, W., Kawaguchi, K., Greenberg, D. A., et al. (1997) Early detection of DNA strand breaks in the brain after transient focal ischemia: implications for the role of DNA damage in apoptosis and neuronal cell death. *J. Neurochem.* **69(1),** 232–245.
4. Linnik, M. D., Zobrist, R. H., and Hatfield, M. D. (1993) Evidence supporting a role for programmed cell death in focal cerebral ischemia in rats. *Stroke* **24(12),** 2002–2008; discussion 2008–2009.
5. MacManus, J. P., Buchan, A. M., Hill, I. E., Rasquinha, I., and Preston, E. (1993) Global ischemia can cause DNA fragmentation indicative of apoptosis in rat brain. *Neurosci. Lett.* **164(1–2),** 89–92.
6. Li, Y., Sharov, V. G., Jiang, N., Zaloga, C., Sabbah, H. N., and Chopp M. (1995) Ultrastructural and light microscopic evidence of apoptosis after middle cerebral artery occlusion in the rat. *Am. J. Pathol.* **146(5),** 1045–1051.
7. Charriaut-Marlangue, C., Margaill, I., Represa, A., Popovici, T., Plotkine, M., and Ben-Ari, Y. (1996) Apoptosis and necrosis after reversible focal ischemia: an in situ DNA fragmentation analysis. *J. Cereb. Blood Flow Metab.* **16(2),** 186–194.
8. Guglielmo, M. A., Chan, P. T., Cortez, S., Stopa, E. G., McMillan, P., Johanson, C. E., et al. (1998) The temporal profile and morphologic features of neuronal death in human stroke resemble those observed in experimental forebrain ischemia: the potential role of apoptosis. *Neurol. Res.* **20(4),** 283–296.
9. Chen, D., Stetler, R. A., Cao, G., Pei, W., O'Horo, C., Yin, X. M., et al. (2000) Characterization of the rat DNA fragmentation factor 35/Inhibitor of caspase-activated DNase (Short form). The endogenous inhibitor of caspase-dependent DNA fragmentation in neuronal apoptosis. *J. Biol. Chem.* **275(49),** 38508–38517.
10. Cao, G., Pei, W., Lan, J., Stetler, R. A., Luo, Y., Nagayama, T., et al. (2001) Caspase-activated DNase/DNA fragmentation factor 40 mediates apoptotic DNA fragmentation in transient cerebral ischemia and in neuronal cultures. *J. Neurosci.* **21(13),** 4678–4690.
11. Nicholls, D. G. and Budd, S. L. (1998) Mitochondria and neuronal glutamate excitotoxicity. *Biochim. Biophys. Acta* **1366(1–2),** 97–112.
12. Stout, A. K., Raphael, H. M., Kanterewicz, B. I., Klann, E., and Reynolds, I. J. (1998) Glutamate-induced neuron death requires mitochondrial calcium uptake. *Nat. Neurosci.* **1(5),** 366–373.
13. Schild, L., Keilhoff, G., Augustin, W., Reiser, G., and Striggow, F. (2001) Distinct Ca^{2+} thresholds determine cytochrome *c* release or permeability transition pore opening in brain mitochondria. *FASEB J.* **15(3),** 565–567.
14. Ghafourifar, P., Schenk, U., Klein, S. D., and Richter, C. (1999) Mitochondrial nitric-oxide synthase stimulation causes cytochrome *c* release from isolated mitochondria. Evidence for intramitochondrial peroxynitrite formation. *J. Biol. Chem.* **274(44),** 31185–31188.
15. Martinou, J. C. (1999) Apoptosis. Key to the mitochondrial gate. *Nature* **399(6735),** 411–412.
16. Maciel, E. N., Vercesi, A. E., and Castilho, R. F. (2001) Oxidative stress in Ca(2+)-induced membrane permeability transition in brain mitochondria. *J. Neurochem.* **79(6),** 1237–1245.
17. Choi, D. W. (1987) Ionic dependence of glutamate neurotoxicity. *J. Neurosci.* **7(2),** 369–379.
18. Chen, J., Nagayama, T., Jin, K., Stetler, R. A., Zhu, R. L., Graham, S. H., et al. (1998) Induction of caspase-3-like protease may mediate delayed neuronal death in the hippocampus after transient cerebral ischemia. *J. Neurosci.* **18(13),** 4914–4928.
19. Ni, B., Wu, X., Su, Y., Stephenson, D., Smalstig, E. B., Clemens, J., et al. (1998) Transient global forebrain ischemia induces a prolonged expression of the caspase-3 mRNA in rat hippocampal CA1 pyramidal neurons. *J. Cereb. Blood Flow Metab.* **18(3),** 248–256.
20. Namura, S., Zhu, J., Fink, K., Endres, M., Srinivasan, A., Tomaselli, K. J., et al. (1998) Activation and cleavage of caspase-3 in apoptosis induced by experimental cerebral ischemia. *J. Neurosci.* **18(10),** 3659–3668.
21. Shackelford, D. A., Tobaru, T., Zhang, S., and Zivin, J. A. (1999) Changes in expression of the DNA repair protein complex DNA-dependent protein kinase after ischemia and reperfusion. *J. Neurosci.* **19(12),** 4727–4738.
22. Hara, H., Fink, K., Endres, M., Friedlander, R. M., Gagliardini, V., Yuan, J., et al. (1997) Attenuation of transient focal cerebral ischemic injury in transgenic mice expressing a mutant ICE inhibitory protein. *J. Cereb. Blood Flow Metab.* **17(4),** 370–375.
23. Hara, H., Friedlander, R. M., Gagliardini, V., Ayata, C., Fink, K., Huang, Z., et al. (1997) Inhibition of interleukin

1beta converting enzyme family proteases reduces ischemic and excitotoxic neuronal damage. *Proc. Natl. Acad. Sci. USA* **94(5),** 2007–2012.

24. Xu, D., Bureau, Y., McIntyre, D. C., Nicholson, D. W., Liston, P., Zhu, Y., et al. (1999) Attenuation of ischemia-induced cellular and behavioral deficits by X chromosome-linked inhibitor of apoptosis protein overexpression in the rat hippocampus. *J. Neurosci.* **19(12),** 5026–5033.

25. Benchoua, A., Guegan, C., Couriaud, C., Hosseini, H., Sampaio, N., Morin, D., et al. (2001) Specific caspase pathways are activated in the two stages of cerebral infarction. *J. Neurosci.* **21(18),** 7127–7134.

26. Fujimura, M., Morita-Fujimura, Y., Murakami, K., Kawase, M., and Chan, P. H. (1998) Cytosolic redistribution of cytochrome *c* after transient focal cerebral ischemia in rats. *J. Cereb. Blood Flow Metab.* **18(11),** 1239–1247.

27. Sugawara, T., Fujimura, M., Morita-Fujimura, Y., Kawase, M., and Chan, P. H. (1999) Mitochondrial release of cytochrome *c* corresponds to the selective vulnerability of hippocampal CA1 neurons in rats after transient global cerebral ischemia. *J. Neurosci.* **19(22),** RC39.

28. Cao, G., Luo, Y., Nagayama, T., Pei, W., Stetler, R. A., Graham, S. H., et al. (2002) Cloning and characterization of rat caspase-9: implications for a role in mediating caspase-3 activation and hippocampal cell death after transient cerebral ischemia. *J. Cereb. Blood Flow Metab.* **22(5),** 534–546.

29. Cao, G., Minami, M., Pei, W., Yan, C., Chen, D., O'Horo, C., et al. (2001) Intracellular Bax translocation after transient cerebral ischemia: implications for a role of the mitochondrial apoptotic signaling pathway in ischemic neuronal death. *J. Cereb. Blood Flow Metab.* **21(4),** 321–333.

30. Rosenbaum, D. M., Gupta, G., D'Amore, J., Singh, M., Weidenheim, K., Zhang, H., et al. (2000) Fas (CD95/APO-1) plays a role in the pathophysiology of focal cerebral ischemia. *J. Neurosci. Res.* **61(6),** 686–692.

31. Martin-Villalba, A., Herr, I., Jeremias, I., Hahne, M., Brandt, R., Vogel, J., et al. (1999) CD95 ligand (Fas-L/APO-1L) and tumor necrosis factor-related apoptosis-inducing ligand mediate ischemia-induced apoptosis in neurons. *J. Neurosci.* **19(10),** 3809–3817.

32. Matsuyama, T., Hata, R., Yamamoto, Y., Tagaya, M., Akita, H., Uno, H., et al. (1995) Localization of Fas antigen mRNA induced in postischemic murine forebrain by in situ hybridization. *Brain Res. Mol. Brain Res.* **34(1),** 166–172.

33. Plesnila, N., Zinkel, S., Le, D. A., Amin-Hanjani, S., Wu, Y., Qiu, J., et al. (2001) BID mediates neuronal cell death after oxygen/glucose deprivation and focal cerebral ischemia. *Proc. Natl. Acad. Sci. USA* **98(26),** 15318–15323.

34. Yin, X. M. (2000) Signal transduction mediated by Bid, a pro-death Bcl-2 family proteins, connects the death receptor and mitochondria apoptosis pathways. *Cell Res.* **10(3),** 161–167.

35. Velier, J. J., Ellison, J. A., Kikly, K. K., Spera, P. A., Barone, F. C., and Feuerstein, G. Z. (1999) Caspase-8 and caspase-3 are expressed by different populations of cortical neurons undergoing delayed cell death after focal stroke in the rat. *J. Neurosci.* **19(14),** 5932–5941.

36. Blomgren, K., Zhu, C., Wang, X., Karlsson, J. O., Leverin, A. L., Bahr, B. A., et al. (2001) Synergistic activation of caspase-3 by m-calpain after neonatal hypoxia-ischemia: a mechanism of "pathological apoptosis"? *J. Biol. Chem.* **276(13),** 10191–10198.

37. Nakagawa, T. and Yuan, J. (2000) Cross-talk between two cysteine protease families. Activation of caspase-12 by calpain in apoptosis. *J. Cell Biol.* **150(4),** 887–894.

38. Tatton, W. G. and Chalmers-Redman, R. M. (1998) Mitochondria in neurodegenerative apoptosis: an opportunity for therapy? *Ann. Neurol.* **44(3 Suppl. 1),** S134–S141.

39. Olanow, C. W. and Tatton, W. G. (1999) Etiology and pathogenesis of Parkinson's disease. *Annu. Rev. Neurosci.* **22,** 123–144.

40. Burke, R. E. and Kholodilov, N. G. (1998) Programmed cell death: does it play a role in Parkinson's disease? *Ann. Neurol.* **44(3 Suppl. 1),** S126–S133.

41. Anglade, P., Vyas, S., Javoy-Agid, F., Herrero, M. T., Michel, P. P., Marquez, J., et al. (1997) Apoptosis and autophagy in nigral neurons of patients with Parkinson's disease. *Histol. Histopathol.* **12(1),** 25–31.

42. Kingsbury, A. E., Mardsen, C. D., and Foster, O. J. (1998) DNA fragmentation in human substantia nigra: apoptosis or perimortem effect? *Mov. Disord.* **13(6),** 877–884.

43. Mochizuki, H., Goto, K., Mori, H., and Mizuno, Y. (1996) Histochemical detection of apoptosis in Parkinson's disease. *J. Neurol. Sci.* **137(2),** 120–123.

44. Tatton, N. A., Maclean-Fraser, A., Tatton, W. G., Perl, D. P., and Olanow, C. W. (1998) A fluorescent double-labeling method to detect and confirm apoptotic nuclei in Parkinson's disease. *Ann. Neurol.* **44(3 Suppl. 1),** S142–S148.

45. Tatton, N. A. (2000) Increased caspase 3 and Bax immunoreactivity accompany nuclear GAPDH translocation and neuronal apoptosis in Parkinson's disease. *Exp. Neurol.* **166(1),** 29–43.

46. Tompkins, M. M., Basgall, E. J., Zamrini, E., and Hill, W. D. (1997) Apoptotic-like changes in Lewy-body-associated disorders and normal aging in substantia nigral neurons. *Am. J. Pathol.* **150(1),** 119–131.

47. Jellinger, K. A. (1999) Post mortem studies in Parkinson's disease: is it possible to detect brain areas for specific symptoms? *J. Neural Transm. Suppl.* **56,** 1–29.

48. Kosel, S., Egensperger, R., von Eitzen, U., Mehraein, P., and Graeber, M. B. (1997) On the question of apoptosis in the parkinsonian substantia nigra. *Acta Neuropathol. (Berl)* **93(2),** 105–108.

49. Wullner, U., Kornhuber, J., Weller, M., Schulz, J. B., Loschmann, P. A., Riederer, P., et al. (1999) Cell death and apoptosis regulating proteins in Parkinson's disease: a cautionary note. *Acta Neuropathol. (Berl)* **97(4),** 408–412.

50. Marshall, K. A., Daniel, S. E., Cairns, N., Jenner, P., and Halliwell, B. (1997) Upregulation of the anti-apoptotic protein Bcl-2 may be an early event in neurodegeneration: studies on Parkinson's and incidental Lewy body disease. *Biochem. Biophys. Res. Commun.* **240(1),** 84–87.

51. Mogi, M., Harada, M., Kondo, T., Mizuno, Y., Narabayashi, H., Riederer, P., et al. (1996) bcl-2 protein is increased in the brain from parkinsonian patients. *Neurosci. Lett.* **215(2)**, 137–139.
52. Hartmann, A., Michel, P. P., Troadec, J. D., Mouatt-Prigent, A., Faucheux, B. A., Ruberg, M., et al. (2001) Is Bax a mitochondrial mediator in apoptotic death of dopaminergic neurons in Parkinson's disease? *J. Neurochem.* **76(6)**, 1785–1793.
53. Viswanath, V., Wu, Y., Boonplueang, R., Chen, S., Stevenson, F. F., Yantiri, F., et al. (2001) Caspase-9 activation results in downstream caspase-8 activation and bid cleavage in 1-methyl-4-phenyl-1,2,3,6-tetrahydropyridine-induced Parkinson's disease. *J. Neurosci.* **21(24)**, 9519–9528.
54. Mogi, M., Togari, A., Kondo, T., Mizuno, Y., Komure, O., Kuno, S., et al. (2000) Caspase activities and tumor necrosis factor receptor R1 (p55) level are elevated in the substantia nigra from parkinsonian brain. *J. Neural. Transm.* **107(3)**, 335–341.
55. Hartmann, A., Troadec, J. D., Hunot, S., Kikly, K., Faucheux, B. A., Mouatt-Prigent, A., et al. (2001) Caspase-8 is an effector in apoptotic death of dopaminergic neurons in Parkinson's disease, but pathway inhibition results in neuronal necrosis. *J. Neurosci.* **21(7)**, 2247–2255.
56. Hartmann, A., Hunot, S., Michel, P. P., Muriel, M. P., Vyas, S., Faucheux, B. A., et al. (2000) Caspase-3: A vulnerability factor and final effector in apoptotic death of dopaminergic neurons in Parkinson's disease. *Proc. Natl. Acad. Sci. USA* **97(6)**, 2875–2880.
57. Cutillas, B., Espejo, M., Gil, J., Ferrer, I., and Ambrosio, S. (1999) Caspase inhibition protects nigral neurons against 6-OHDA-induced retrograde degeneration. *Neuroreport* **10(12)**, 2605–2608.
58. Ochu, E. E., Rothwell, N. J., and Waters, C. M. (1998) Caspases mediate 6-hydroxydopamine-induced apoptosis but not necrosis in PC12 cells. *J. Neurochem.* **70(6)**, 2637–2640.
59. Lotharius, J., Dugan, L. L., and O'Malley, K. L. (1999) Distinct mechanisms underlie neurotoxin-mediated cell death in cultured dopaminergic neurons. *J. Neurosci.* **19(4)**, 1284–1293.
60. Dodel, R. C., Du, Y., Bales, K. R., Ling, Z. D., Carvey, P. M., and Paul, S. M. (1998) Peptide inhibitors of caspase-3–like proteases attenuate 1-methyl-4-phenylpyridinum-induced toxicity of cultured fetal rat mesencephalic dopamine neurons. *Neuroscience* **86(3)**, 701–707.
61. Dodel, R. C., Du, Y., Bales, K. R., Ling, Z., Carvey, P. M., and Paul, S. M. (1999) Caspase-3-like proteases and 6–hydroxydopamine induced neuronal cell death. *Brain Res. Mol. Brain Res.* **64(1)**, 141–148.
62. Choi, W. S., Lee, E. H., Chung, C. W., Jung, Y. K., Jin, B. K., Kim, S. U., et al. (2001) Cleavage of Bax is mediated by caspase-dependent or -independent calpain activation in dopaminergic neuronal cells: protective role of Bcl-2. *J. Neurochem.* **77(6)**, 1531–1541.
63. Choi, W. S., Canzoniero, L. M., Sensi, S. L., O'Malley, K. L., Gwag, B. J., Sohn, S., et al. (1999) Characterization of MPP(+)-induced cell death in a dopaminergic neuronal cell line: role of macromolecule synthesis, cytosolic calcium, caspase, and Bcl-2-related proteins. *Exp. Neurol.* **159(1)**, 274–282.
64. Choi, W. S., Yoon, S. Y., Oh, T. H., Choi, E. J., O'Malley, K. L., and Oh, Y. J. (1999) Two distinct mechanisms are involved in 6-hydroxydopamine- and MPP+-induced dopaminergic neuronal cell death: role of caspases, ROS, and JNK. *J. Neurosci. Res.* **57(1)**, 86–94.
65. Eberhardt, O., Coelln, R. V., Kugler, S., Lindenau, J., Rathke-Hartlieb, S., Gerhardt, E., et al. (2000) Protection by synergistic effects of adenovirus-mediated X-chromosome-linked inhibitor of apoptosis and glial cell line-derived neurotrophic factor gene transfer in the 1-methyl-4-phenyl-1,2,3,6-tetrahydropyridine model of Parkinson's disease. *J. Neurosci.* **20(24)**, 9126–9134.
66. Kitamura, Y., Kosaka, T., Kakimura, J. I., Matsuoka, Y., Kohno, Y., Nomura, Y., et al. (1998) Protective effects of the antiparkinsonian drugs talipexole and pramipexole against 1-methyl-4-phenylpyridinium-induced apoptotic death in human neuroblastoma SH-SY5Y cells. *Mol. Pharmacol.* **54(6)**, 1046–1054.
67. Jeon, B. S., Kholodilov, N. G., Oo, T. F., Kim, S. Y., Tomaselli, K. J., Srinivasan, A., et al. (1999) Activation of caspase-3 in developmental models of programmed cell death in neurons of the substantia nigra. *J. Neurochem.* **73(1)**, 322–333.
68. Blum, D., Wu, Y., Nissou, M. F., Arnaud, S., Alim Louis, B., and Verna, J. M. (1997) p53 and Bax activation in 6-hydroxydopamine-induced apoptosis in PC12 cells. *Brain Res.* **751(1)**, 139–142.
69. Seaton, T. A., Cooper, J. M., and Schapira, A. H. (1997) Free radical scavengers protect dopaminergic cell lines from apoptosis induced by complex I inhibitors. *Brain Res.* **777(1–2)**, 110–118.
70. Spina, M. B. and Cohen, G. (1989) Dopamine turnover and glutathione oxidation: implications for Parkinson disease. *Proc. Natl. Acad. Sci. USA* **86(4)**, 1398–1400.
71. Foley, P. and Riederer, P. (2000) Influence of neurotoxins and oxidative stress on the onset and progression of Parkinson's disease. *J. Neurol.* **247 (Suppl. 2)**, II82–II94.
72. Blum, D., Torch, S., Lambeng, N., Nissou, M., Benabid, A. L., Sadoul, R., et al.(2001) Molecular pathways involved in the neurotoxicity of 6-OHDA, dopamine and MPTP: contribution to the apoptotic theory in Parkinson's disease. *Prog. Neurobiol.* **65(2)**, 135–172.
73. Swerdlow, R. H., Parks, J. K., Davis, J. N., 2nd, Cassarino, D. S., Trimmer, P. A., Currie, L. J., et al. (1998) Matrilineal inheritance of complex I dysfunction in a multigenerational Parkinson's disease family. *Ann. Neurol.* **44(6)**, 873–881.
74. Blandini, F. and Greenamyre, J. T. (1998) Prospects of glutamate antagonists in the therapy of Parkinson's disease. *Fundam. Clin. Pharmacol.* **12(1)**, 4–12.
75. Perry, T. L., Godin, D. V., and Hansen, S. (1982) Parkinson's disease: a disorder due to nigral glutathione deficiency? *Neurosci. Lett.* **33(3)**, 305–310.

76. Kish, S. J., Morito, C., and Hornykiewicz, O. (1985) Glutathione peroxidase activity in Parkinson's disease brain. *Neurosci. Lett.* **58(3)**, 343–346.
77. Riederer, P., Sofic, E., Rausch, W. D., Schmidt, B., Reynolds, G. P., Jellinger, K., et al. (1989) Transition metals, ferritin, glutathione, and ascorbic acid in parkinsonian brains. *J. Neurochem.* **52(2)**, 515–520.
78. Sofic, E., Lange, K. W., Jellinger, K., and Riederer, P. (1992) Reduced and oxidized glutathione in the substantia nigra of patients with Parkinson's disease. *Neurosci. Lett.* **142(2)**, 128–130.
79. Ben-Shachar, D., Eshel, G., Riederer, P., and Youdim, M. B. (1992) Role of iron and iron chelation in dopaminergic-induced neurodegeneration: implication for Parkinson's disease. *Ann. Neurol.* **32(Suppl.)**, S105–S110.
80. Double, K. L., Gerlach, M., Youdim, M. B., and Riederer, P. (2000) Impaired iron homeostasis in Parkinson's disease. *J. Neural. Transm. Suppl.* **60**, 37–58.
81. Good, P. F., Hsu, A., Werner, P., Perl, D. P., and Olanow, C. W. (1998) Protein nitration in Parkinson's disease. *J. Neuropathol. Exp. Neurol.* **57(4)**, 338–342.
82. Oury, T. D., Tatro, L., Ghio, A. J., and Piantadosi, C. A. (1995) Nitration of tyrosine by hydrogen peroxide and nitrite. *Free Radic. Res.* **23(6)**, 537–547.
83. Ara, J., Przedborski, S., Naini, A. B., Jackson-Lewis, V., Trifiletti, R. R., Horwitz, J., et al. (1998) Inactivation of tyrosine hydroxylase by nitration following exposure to peroxynitrite and 1-methyl-4-phenyl-1,2,3,6-tetrahydropyridine (MPTP). *Proc. Natl. Acad. Sci. USA* **95(13)**, 7659–7663.
84. Olanow, C. W. (1990) Oxidation reactions in Parkinson's disease. *Neurology* **40(10 Suppl. 3)**, Suppl. 32–37; discussion 37–39.
85. Graham, D. G., Tiffany, S. M., Bell, W.R., Jr., and Gutknecht, W. F. (1978) Autoxidation versus covalent binding of quinones as the mechanism of toxicity of dopamine, 6-hydroxydopamine, and related compounds toward C1300 neuroblastoma cells in vitro. *Mol. Pharmacol.* **14(4)**, 644–653.
86. Hastings, T. G., Lewis, D. A., and Zigmond, M. J. (1996) Role of oxidation in the neurotoxic effects of intrastriatal dopamine injections. *Proc. Natl. Acad. Sci. USA* **93(5)**, 1956–1961.
87. Spencer, J. P., Jenner, P., Daniel, S. E., Lees, A. J., Marsden, D. C., and Halliwell, B. (1998) Conjugates of catecholamines with cysteine and GSH in Parkinson's disease: possible mechanisms of formation involving reactive oxygen species. *J. Neurochem.* **71(5)**, 2112–2122.
88. Berman, S. B. and Hastings, T. G. (1999) Dopamine oxidation alters mitochondrial respiration and induces permeability transition in brain mitochondria: implications for Parkinson's disease. *J. Neurochem.* **73(3)**, 1127–1137.
89. Lee, C. S., Han, J. H., Jang, Y. Y., Song, J. H., and Han, E. S. (2002) Differential effect of catecholamines and MPP(+) on membrane permeability in brain mitochondria and cell viability in PC12 cells. *Neurochem. Int.* **40(4)**, 361–369.
90. Junn, E. and Mouradian, M. M. (2001) Apoptotic signaling in dopamine-induced cell death: the role of oxidative stress, p38 mitogen-activated protein kinase, cytochrome *c* and caspases. *J. Neurochem.* **78(2)**, 374–383.
91. Wanker, E. E. (2000) Protein aggregation and pathogenesis of Huntington's disease: mechanisms and correlations. *Biol. Chem.* **381(9–10)**, 937–942.
92. Saudou, F., Finkbeiner, S., Devys, D., and Greenberg, M. E. (1998) Huntingtin acts in the nucleus to induce apoptosis but death does not correlate with the formation of intranuclear inclusions. *Cell* **95(1)**, 55–66.
93. Kim, M., Lee, H. S., LaForet, G., McIntyre, C., Martin, E. J., Chang, P., et al. (1999) Mutant huntingtin expression in clonal striatal cells: dissociation of inclusion formation and neuronal survival by caspase inhibition. *J. Neurosci.* **19(3)**, 964–973.
94. Aronin, N., Kim, M., Laforet, G., and DiFiglia, M. (1999) Are there multiple pathways in the pathogenesis of Huntington's disease? *Philos. Trans. R. Soc. Lond. B. Biol. Sci.* **354(1386)**, 995–1003.
95. Turmaine, M., Raza, A., Mahal, A., Mangiarini, L., Bates, G. P., and Davies, S. W. (2000) Nonapoptotic neurodegeneration in a transgenic mouse model of Huntington's disease. *Proc. Natl. Acad. Sci. USA* **97(14)**, 8093–8097.
96. Portera-Cailliau, C., Hedreen, J. C., Price, D. L., and Koliatsos, V. E. (1995) Evidence for apoptotic cell death in Huntington disease and excitotoxic animal models. *J. Neurosci.* **15(5 Pt. 2)**, 3775–3787.
97. Dragunow, M., Faull, R. L., Lawlor, P., Beilharz, E. J., Singleton, K., Walker, E. B., et al. (1995) In situ evidence for DNA fragmentation in Huntington's disease striatum and Alzheimer's disease temporal lobes. *Neuroreport* **6(7)**, 1053–1057.
98. Thomas, L. B., Gates, D. J., Richfield, E. K., O'Brien, T. F., Schweitzer, J. B., and Steindler, D. A. (1995) DNA end labeling (TUNEL) in Huntington's disease and other neuropathological conditions. *Exp. Neurol.* **133(2)**, 265–272.
99. Group THsDCR. (1993) A novel gene containing a trinucleotide repeat that is expanded and unstable on Huntington's disease chromosomes. *Cell* **72(6)**, 971–983.
100. Wellington, C. L., Ellerby, L. M., Hackam, A. S., Margolis, R. L., Trifiro, M. A., Singaraja, R., et al. (1998) Caspase cleavage of gene products associated with triplet expansion disorders generates truncated fragments containing the polyglutamine tract. *J. Biol. Chem.* **273(15)**, 9158–9167.
101. Wellington, C. L., Singaraja, R., Ellerby, L., Savill, J., Roy, S., Leavitt, B., et al. (2000) Inhibiting caspase cleavage of huntingtin reduces toxicity and aggregate formation in neuronal and nonneuronal cells. *J. Biol. Chem.* **275(26)**, 19831–19838.
102. Kim, Y. J., Yi, Y., Sapp, E., Wang, Y., Cuiffo, B., Kegel, K. B., et al. (2001) Caspase 3-cleaved N-terminal fragments of wild-type and mutant huntingtin are present in normal and Huntington's disease brains, associate with membranes, and undergo calpain-dependent proteolysis. *Proc. Natl. Acad. Sci. USA* **98(22)**, 12784–12789.
103. Dyer, R. B. and McMurray, C. T. (2001) Mutant protein in Huntington disease is resistant to proteolysis in affected brain. *Nat. Genet.* **29(3)**, 270–278.

104. Hodgson, J. G., Agopyan, N., Gutekunst, C. A., Leavitt, B. R., LePiane, F., Singaraja, R., et al. (1999) A YAC mouse model for Huntington's disease with full-length mutant huntingtin, cytoplasmic toxicity, and selective striatal neurodegeneration. *Neuron* **23(1)**, 181–192.
105. Tao, T. and Tartakoff, A. M. (2001) Nuclear relocation of normal huntingtin. *Traffic* **2(6)**, 385–394.
106. Martindale, D., Hackam, A., Wieczorek, A., Ellerby, L., Wellington, C., McCutcheon, K., et al. (1998) Length of huntingtin and its polyglutamine tract influences localization and frequency of intracellular aggregates. *Nat. Genet.* **18(2)**, 150–154.
107. Goldberg, Y. P., Nicholson, D. W., Rasper, D. M., Kalchman, M. A., Koide, H. B., Graham, R. K., et al. (1996) Cleavage of huntingtin by apopain, a proapoptotic cysteine protease, is modulated by the polyglutamine tract. *Nat. Genet.* **13(4)**, 442–449.
108. U, M., Miyashita, T., Ohtsuka, Y., Okamura-Oho, Y., Shikama, Y., and Yamada, M. (2001) Extended polyglutamine selectively interacts with caspase-8 and -10 in nuclear aggregates. *Cell Death Differ.* **8(4)**, 377–386.
109. Kouroku, Y., Fujita, E., Jimbo, A., Mukasa, T., Tsuru, T., Momoi, M. Y., et al. (2000) Localization of active form of caspase-8 in mouse L929 cells induced by TNF treatment and polyglutamine aggregates. *Biochem. Biophys. Res. Commun.* **270(3)**, 972–977.
110. Sawa, A. (2001) Mechanisms for neuronal cell death and dysfunction in Huntington's disease: pathological cross-talk between the nucleus and the mitochondria? *J. Mol. Med.* **79(7)**, 375–381.
111. McGeer, E. G. and McGeer, P. L. (1976) Duplication of biochemical changes of Huntington's chorea by intrastriatal injections of glutamic and kainic acids. *Nature* **263(5577)**, 517–519.
112. Coyle, J. T. and Schwarcz, R. (1976) Lesion of striatal neurones with kainic acid provides a model for Huntington's chorea. *Nature* **263(5574)**, 244–246.
113. Sun, Y., Savanenin, A., Reddy, P. H., and Liu, Y. F. (2001) Polyglutamine-expanded huntingtin promotes sensitization of N-methyl-D-aspartate receptors via post-synaptic density 95. *J. Biol. Chem.* **276(27)**, 24713–24718.
114. Kim, E., Cho, K. O., Rothschild, A., and Sheng, M. (1996) Heteromultimerization and NMDA receptor-clustering activity of Chapsyn-110, a member of the PSD-95 family of proteins. *Neuron* **17(1)**, 103–113.
115. Davies, S. and Ramsden, D. B. (2001) Huntington's disease. *Mol. Pathol.* **54(6)**, 409–413.
116. Petersen, A., Larsen, K. E., Behr, G. G., Romero, N., Przedborski, S., Brundin, P., et al. (2001) Expanded CAG repeats in exon 1 of the Huntington's disease gene stimulate dopamine-mediated striatal neuron autophagy and degeneration. *Hum. Mol. Genet.* **10(12)**, 1243–1254.
117. Liu, Y. F., Dorow, D., and Marshall, J. (2000) Activation of MLK2-mediated signaling cascades by polyglutamine-expanded huntingtin. *J. Biol. Chem.* **275(25)**, 19035–19040.
118. Liu, Y. F. (1998) Expression of polyglutamine-expanded Huntingtin activates the SEK1–JNK pathway and induces apoptosis in a hippocampal neuronal cell line. *J. Biol. Chem.* **273(44)**, 28873–28877.
119. Ventimiglia, R., Mather, P. E., Jones, B. E., and Lindsay, R. M. (1995) The neurotrophins BDNF, NT-3 and NT-4/5 promote survival and morphological and biochemical differentiation of striatal neurons in vitro. *Eur. J. Neurosci.* **7(2)**, 213–222.
120. Ivkovic, S. and Ehrlich, M. E. (1999) Expression of the striatal DARPP-32/ARPP-21 phenotype in GABAergic neurons requires neurotrophins in vivo and in vitro. *J. Neurosci.* **19(13)**, 5409–5419.
121. Hofer, M., Pagliusi, S. R., Hohn, A., Leibrock, J., and Barde, Y. A. (1990) Regional distribution of brain-derived neurotrophic factor mRNA in the adult mouse brain. *EMBO J.* **9(8)**, 2459–2464.
122. Conner, J. M., Lauterborn, J. C., and Gall, C. M. (1998) Anterograde transport of neurotrophin proteins in the CNS: a reassessment of the neurotrophic hypothesis. *Rev. Neurosci.* **9(2)**, 91–103.
123. Canals, J. M., Checa, N., Marco, S., Akerud, P., Michels, A., Perez-Navarro, E., et al. (2001) Expression of brain-derived neurotrophic factor in cortical neurons is regulated by striatal target area. *J. Neurosci.* **21(1)**, 117–124.
124. Ferrer, I., Goutan, E., Marin, C., Rey, M. J., and Ribalta, T. (2000) Brain-derived neurotrophic factor in Huntington disease. *Brain Res.* **866(1–2)**, 257–261.
125. Zuccato, C., Ciammola, A., Rigamonti, D., Leavitt, B. R., Goffredo, D., Conti, L., et al. (2001) Loss of huntingtin-mediated BDNF gene transcription in Huntington's disease. *Science* **293(5529)**, 493–498.
126. Kazantsev, A., Preisinger, E., Dranovsky, A., Goldgaber, D., and Housman, D. (1999) Insoluble detergent-resistant aggregates form between pathological and nonpathological lengths of polyglutamine in mammalian cells. *Proc. Natl. Acad. Sci. USA* **96(20)**, 11404–11409.
127. Steffan, J. S., Kazantsev, A., Spasic-Boskovic, O., Greenwald, M., Zhu, Y. Z., Gohler, H., et al. (2000) The Huntington's disease protein interacts with p53 and CREB-binding protein and represses transcription. *Proc. Natl. Acad. Sci. USA* **97(12)**, 6763–6768.
128. Bonni, A., Brunet, A., West, A. E., Datta, S. R., Takasu, M. A., and Greenberg, M. E. (1999) Cell survival promoted by the Ras-MAPK signaling pathway by transcription-dependent and -independent mechanisms. *Science* **286(5443)**, 1358–1362.
129. Riccio, A., Ahn, S., Davenport, C. M., Blendy, J. A., and Ginty, D. D. (1999) Mediation by a CREB family transcription factor of NGF-dependent survival of sympathetic neurons. *Science* **286(5448)**, 2358–2361.
130. Walton, M. R. and Dragunow, I. (2000) Is CREB a key to neuronal survival? *Trends Neurosci.* **23(2)**, 48–53.
131. Nucifora, F. C., Jr., Sasaki, M., Peters, M. F., Huang, H., Cooper, J. K., Yamada, M., et al. (2001) Interference by huntingtin and atrophin-1 with cbp-mediated transcription leading to cellular toxicity. *Science* **291(5512)**, 2423–2428.
132. Wyttenbach, A., Swartz, J., Kita, H., Thykjaer, T., Carmichael, J., Bradley, J., et al. (2001) Polyglutamine expansions

cause decreased CRE-mediated transcription and early gene expression changes prior to cell death in an inducible cell model of Huntington's disease. *Hum. Mol. Genet.* **10(17)**, 1829–1845.

133. Shimohata, T., Onodera, O., and Tsuji, S. (2000) Interaction of expanded polyglutamine stretches with nuclear transcription factors leads to aberrant transcriptional regulation in polyglutamine diseases. *Neuropathology* **20(4)**, 326–333.

134. Shieh, P. B., Hu, S. C., Bobb, K., Timmusk, T., and Ghosh, A. Identification of a signaling pathway involved in calcium regulation of BDNF expression. *Neuron* **20(4)**, 727–740.

135. Tao, X., Finkbeiner, S., Arnold, D. B., Shaywitz, A. J., and Greenberg, M. E. (1998) Ca2+ influx regulates BDNF transcription by a CREB family transcription factor-dependent mechanism. *Neuron* **20(4)**, 709–726.

136. Cattaneo, E., Rigamonti, D., Goffredo, D., Zuccato, C., Squitieri, F., and Sipione, S. (2001) Loss of normal huntingtin function: new developments in Huntington's disease research. *Trends Neurosci.* **24(3)**, 182–188.

137. Velier, J., Kim, M., Schwarz, C., Kim, T. W., Sapp, E., Chase, K., et al. (1998) Wild-type and mutant huntingtins function in vesicle trafficking in the secretory and endocytic pathways. *Exp. Neurol.* **152(1)**, 34–40.

15

Apoptosis in Ischemic Disease

Zheng Dong and Manjeri A. Venkatachalam

INTRODUCTION

Ischemia defines a condition of lack of blood supply to tissues. It takes place in vivo under situations of cardiac malfunction, shock, or vascular defects such as constriction and obstruction of blood vessels. Ischemic injury is the key determinant of tissue pathology in devastating diseases such as myocardial infarction, acute renal failure, and stroke in the brain. Ischemic diseases are also the leading cause of morbidity in industrial countries *(1,2)*. Previously, cell death during ischemia has been described as a chaotic autolytic process, or "necrosis." Indeed, impressive cell death in the necrotic form is usually found in ischemic tissues. However, recent studies have revealed apoptosis during ischemia of organs including brain, heart, liver, and kidneys.

DETRIMENTAL FACTORS ACTIVATED BY ISCHEMIA

ATP Depletion

High-energy phosphate in the form of adenosine triphosphate (ATP) is required for various activities within the cell, from maintenance of ion homeostasis and biochemical regulation to synthesis and proliferation. Without ATP, the cell will become "passive" and degradation ensues. In mammalian cells, ATP is produced in two ways. A major pathway for ATP production is oxidative phosphoylation, which takes place in mitochondria and is oxygen-dependent. Alternatively, ATP can be produced in the absence of oxygen through glycolysis, using glucose as the main metabolic substrate. During ischemia, cells are deprived not only of nutrients but also of oxygen; either pathway for ATP generation is thereby blocked. Thus, an early and drastic response of the cell to ischemia is rapid decline of ATP, caused by consumption of the nucleotide and cessation of its synthesis. ATP depletion, in turn, initiates a set of destructive processes within the cell, culminating in tissue damage *(3–5)*.

Disturbance in Ion Homeostasis

Ion homeostasis is critical for cell physiology; loss of ion homeostasis is usually associated with the development of cell injury and death. Maintaining ion homeostasis requires the consumption of ATP, because the activity of ion pumps or channels is usually energy-dependent. During ischemia, cells are deprived of ATP, and thus have lost their ability to sustain the activity of ion pumps, resulting in the loss of cellular ion homeostasis *(6)*. This is well-exemplified by ischemia-induced alterations of the cation, Ca^{2+}. Under physiological situations, free Ca^{2+} in the cytosol is kept at extremely

From: *Essentials of Apoptosis: A Guide for Basic and Clinical Research*
Edited by: X-M. Yin and Z. Dong © Humana Press Inc., Totowa, NJ

low levels (<100 n*M*) compared with extracellular Ca^{2+} of 1.25 m*M*. Such a steep gradient is sustained mainly through the activity of Ca^{2+}, Mg^{2+}-ATPase, which, in the presence of ATP, pumps Ca^{2+} out of the cell and sequesters the ions in mitochondria and endoplasmic reticulum *(7)*. Without ATP, the pump stops, leading to uncontrollable increases of cytosolic Ca^{2+}. This is followed by the activation of various deleterious processes, including the activation of hydrolytic enzymes *(3–5)*. Similarly, lack of ATP during ischemia leads to cessation of Na^+ pump activity, which is accompanied by the accumulation Na^+, Cl^-, and water in the cytoplasm, resulting in cell swelling *(6)*.

Activation of Hydrolytic Enzymes

Destructive processes activated by ischemia include the activation of hydrolytic enzymes such as phospholipases, proteases, and endonucleases. Uncontrolled activation of these enzymes may result from cellular changes associated with ischemia, such as increases of cytosolic Ca^{2+} *(3–5)*. Upon activation, these hydrolytic enzymes attack their substrates within the cell. Phospholipases degrade lipids and promote membrane damage. Proteases cut structural as well as enzymatic proteins. Endonucleases break down DNA and cause pathological alterations in the genetic core. These events may progress to the development of irreversible injury and cell death.

Loss of Cytoprotective Agents

Yet another mechanism responsible for cell injury during ischemia is the loss of cytoprotective agents *(8)*. In this aspect, the tripeptide glutathione and the small amino acid glycine are good examples. Glutathione is one of the most important molecules in cellular defense against oxidant injury. In addition, degradation of glutathione results in the production of glycine, another crucial cytoprotective molecule in ischemic cells. During ischemia, cellular concentrations of glutathione and glycine decrease to levels that are not sufficient to maintain cell viability. Addition of these cytoprotective agents during in vivo ischemia or in vitro hypoxia has been shown to prevent membrane damage and enhance cellular viability *(9)*.

Free Radical Production

Free radicals are cytotoxic at concentrations above physiological levels, owing to their high reactivity with various molecules in cells *(10)*. During ischemia, lack of oxygen leads to cease of mitochondrial respiration and reduction of ubiquinone (complexes I and III), which may generate superoxide in the presence of residual oxygen. However, production of free radicals in the ischemic period is usually limited, owing to low availability, and ultimately the lack of oxygen. On the other hand, significant amounts of free radicals are generated when the blood flow to the tissue is resumed after ischemia. This may account in large part for cell injury caused by reperfusion *(11)*. Cell damage by free radicals takes place at various levels. For example, lipid peroxidation by free radicals can lead to alterations in the plasma membrane and loss of membrane integrity. Free radicals also damage DNA of the cells, resulting in strand breaks *(12)*.

Inflammation

Release of degraded cellular constituents in ischemic tissues can lead to inflammation after reperfusion. This is indicated by accumulation of leukocytes in the ischemic zone. Although the purpose of leukocytes is to clean up dead cells and debris, inflammation by itself leads to secondary damage of tissue. Leukocytes may induce cell injury by producing numerous toxic factors, including large amount of free radicals. In addition, accumulation of inflammatory cells may occlude blood vessels and block blood flow to the ischemic region, worsening the degree of ischemia *(13,14)*.

APOPTOSIS IN ISCHEMIC INJURY

Apoptotic Activity of Ischemic Factors

Solid evidence has been collected to indicate that the deleterious factors activated in ischemic cells are capable of initiating apoptosis. Using in vitro models, ATP depletion has been shown to activate caspases through the mitochondrial pathway, involving the translocation of Bax from cytosol to mitochondria followed by the release of cytochrome *c* from the organelles *(15)*. Stepwise or gradient depletion of ATP upregulates Fas-Fas ligand system, triggering apoptosis through the death-receptor pathway *(16)*. Disturbance of Ca^{2+} homeostasis has been proposed as a trigger for apoptosis for some time *(17)*. Increases of cytosolic free Ca^{2+} have been demonstrated in apoptotic cells. Moreover, amelioration of Ca^{2+} increases within cells diminishes apoptosis under certain situations. Echoing these observations, pharmacologically induced increases in cytosolic Ca^{2+} lead to the development of apoptotic morphology in diverse types of cells. Free radicals that are produced during ischemia or reperfusion are also among the most effective inducers of apoptosis *(11,18)*. Addition of exogenous oxidants leads to apoptosis of cells cultured in vitro. During myocardial ischemia, oxidative stress was shown to be responsible for apoptosis in cadiomyocytes *(19)*. Together, these findings indicate that detrimental factors activated by ischemia can indeed induce apoptosis.

Evidence for Apoptosis During Ischemic Injury

Morphological and Histological Examinations

Development of apoptotic morphology has been and is still considered the gold standard for apoptosis identification *(20)*. However, it has proven difficult to assess apoptotic morphology in vivo *(21)*. First of all, cells undergoing apoptosis in vivo may not progress through all the typical morphological changes observed in vitro in cultured cells. In addition, apoptotic cells are usually rapidly and "quietly" (i.e., no inflammation) cleaned up through phagocytosis by macrophages or neighboring cells *(22)*. Thus, it is not very surprising that cells with typical apoptotic morphology are infrequent in ischemic and postischemic tissues. Nevertheless, by light and electron microscopy, apoptotic alterations in cell nuclei, including fragmentation and chromatin condensation, have been shown in ischemic rat kidneys *(23)*. Similar observations have also been reported for heart, liver, and brain of several species after ischemic insults *(13,24,25)*.

DNA Fragmentation

Ischemia-associated apoptosis was initially recognized by the occurrence of DNA fragmentation, which has been considered to be biochemical "hallmark" of apoptosis *(26)*. The formation of DNA breaks in vivo is usually detected by histochemical techniques such as the TUNEL assay (*see* Chapter 16 for details). Internucleosomal DNA cleavage is shown by the formation of nucleosomal fragments, which are visualized as DNA "ladders" by gel electrophoresis. By these methods, DNA fragmentation was first shown in postischemic rat kidneys and rabbit heart *(27,28)*. In the same studies, internucleosomal DNA cleavage was accompanied by apoptotic morphology, including chromatin condensation and the formation of apoptotic bodies. DNA fragmentation, as well as TUNEL-positive cells, has been subsequently detected by numerous studies in ischemic and postischemic tissues *(13,24,25)*. DNA breakdown under these conditions may be mediated by the activation of endonucleases. The apoptotic endonuclease, caspase-activated DNase or CAD/DFF40, is activated during brain ischemia and is responsible for DNA fragmentation in ischemic neurons *(29)*.

Caspase Activation

Caspases are a family of cysteine proteases that play a central role in the initiation and execution of apoptosis *(30)*. Following ischemia, activation of caspases has been shown in the brain, liver, and heart *(31–33)*. Although systematic analysis of various caspases are not available, caspase-3 activation is often detected in ischemic tissues. Increases of caspase-3 activity have been documented during or following organ ischemia, using exogenous peptide substrates *(31–33)*. Processing of procaspase-3 into active forms has been shown in experimental models of brain and liver ischemia *(31,33)*. Moreover, using antibodies specifically reactive to the active form of caspase-3, recent studies have demonstrated activation of this caspase in cardiomyocytes and neurons in ischemic heart and brain *(31,34)*. By the same method, caspase activation was visualized in human brain infarcts *(35)*. Of significance, the cells with active caspase-3 show DNA fragmentation (TUNEL-positive), and to various extent, exhibit apoptotic morphology such as chromatin condensation *(34)*. In the heart, targeted expression of caspase-3 sensitizes cardiomyocytes to ischemic injury, which is accompanied by increases in infarct size and accelerated loss of cardiac function *(36)*. Together, these observations indicate that caspases are activated during or following ischemia, and may have an important role in the development of ischemic tissue damage.

Apoptotic Gene Regulation

An important feature of ischemic injury is the accompanied regulation of apoptotic genes. These genes either participate in the execution of apoptosis or have a regulatory role. Expression of caspases, including caspases-2, -3, and -8 is induced following ischemia of the brain and kidneys in the rat *(37–39)*. In failing sheep hearts, ischemia is associated with gene expression of caspases –2 and -3 *(40)*. Induction of Fas, Fas ligand, and the adapter protein FADD has also been documented following ischemia of the brain, heart, and kidneys *(41–43)*. For ischemic liver, however, the tumor necrosis factor-α (TNF-α) system, rather than Fas, is upregulated *(44)*. Another class of apoptotic genes that are induced by ischemia is the inhibitor of apoptosis protein (IAP). When cultured kidney cells are subjected to hypoxia to simulate ischemia, IAP2 is specifically upregulated *(45)*. Upregulation of IAP2 results from hypoxic gene transcription, and yet can be dissociated from hypoxia-inducible factor 1 *(45)*. During forebrain ischemia, neuronal IAP is induced. It is interesting to note that IAP expression is localized to neurons that are resistant to ischemic damage, suggesting a protective role for this inductive response *(46)*. Finally, ischemia or ischemia-reperfusion triggers notable alterations in the expression of Bcl-2 family genes. For example, following global brain ischemia in the rat, Bcl-2, Bcl-x$_L$, and Bax were shown to be induced *(47)*. In human hearts with myocardial infarction, expression levels of Bcl-2 and Bax are significantly higher than that of normal hearts *(48)*. Upregulation of these genes was also shown during end stage heart failure *(49)*. It is puzzling why pro-apoptotic and anti-apoptotic Bcl-2 genes are upregulated following the same course of ischemic hit. This question has been addressed by recent studies localizing gene expression to specific cells. Apparently, the Bcl-2 family genes are differentially regulated in different cells, with protective genes such as Bcl-2/Bcl-x$_L$ expressed in surviving cells and death genes such as Bax induced in dying cells *(50)*. These findings suggest an important role for Bcl-2 family genes in regulating cell injury following ischemic challenge.

Amelioration of Ischemic Injury by Antagonizing Apoptosis

Direct evidence to support a role for apoptosis in ischemic injury comes from inhibition experiments. First, infusion of caspase inhibitors has been shown to diminish ischemic tissue damage in vivo. During focal ischemia of mouse brain, peptide inhibitors of caspases not only reduce infarct size, but also significantly attenuate neurological and behavioral deficits *(51)*. It is notable that caspase inhibitors afford neuroprotection in the brain, even when administered during the reperfusion period following ischemic challenge *(52,53)*. Similar beneficial effects have been shown for caspase inhibitors during ischemia of the heart and liver *(24,32,54)*. In addition, several known approaches that reduce ischemic damage are associated with the reduction of apoptosis. For example, ischemic preconditioning decreases tissue injury by subsequent ischemic episodes. As shown by recent studies, this is accompanied by significant decreases in caspase activation and apoptosis *(55,56)*. Also, growth factors including fibroblast growth factor (FGF), insulin-like growth factor-1 (IGF-1), and hepatocyte growth factor (HGF) were shown to attenuate ischemic damage in the brain and heart by regulating

Bcl-2 family proteins and reducing apoptosis *(57–60)*. Finally, a role for apoptosis in ischemic injury is suggested by studies using transgenic models. In this direction, transgenic mice overexpressing Bcl-2 are more resistant to tissue damage during permanent brain ischemia. One study showed that stroke size in Bcl-2 mice was reduced to half of that of wild-type mice *(61)*. In models of transient global ischemia, less injury was also noticed in the hippocampus of Bcl-2-expressing animals *(62,63)*. Consistent with these observations, targeted disruption of the Bcl-2 gene exacerbates brain damage during focal ischemia *(63)*. Similar protective actions have been documented for Bcl-2 during heart and liver ischemia, using transgenic models *(64,65)*. Following ischemic challenge, hearts from Bcl-2 mice demonstrate functional recovery significantly better than that of wild type animals *(64)*. In addition to Bcl-2, Bcl-x_L overexpression was also shown to protect neurons from injury during brain ischemia *(66)*. In contrast to Bcl-2 and Bcl-x_L, pro-apoptotic genes such as Bax and Bid promote cell death by apoptosis. In knockout models, Bax deficiency is associated with a reduction of caspase activation, apoptosis, and tissue damage during neonatal brain ischemia *(67)*. Similarly, Bid knockout mice are also more resistant to focal brain ischemia, with reduced infarction volume *(68)*. In caspase transgenic models, deficiency of caspases-1 and -11 is associated with reduced tissue damage following brain ischemia *(69,70)*. In line with these studies, transgenic mice expressing dominant-negative caspase-1 were shown to be resistant to ischemic brain injury *(71)*. Transfection of caspase inhibitory genes, including X-linked IAP (XIAP) and p35, also mitigated ischemic brain damage in rat and mouse, respectively *(72,73)*. The role of caspase-mediated injury in ischemic tissue damage was further suggested by the observation that targeted overexpression of caspase-3 in transgenic mice led to increased infarct size and loss of cardiac function after heart ischemia *(36)*. Using lpr mice with a loss of function mutation in the Fas gene, it has been shown that tissue damage is significantly ameliorated following ischemia of the brain and heart *(42,74)*. This is accompanied by a reduction of apoptotic cells. Together, these transgenic studies based on specifically modified apoptotic genes provide strong evidence for a role of apoptosis in ischemic injury.

APOPTOTIC PATHWAYS IN ISCHEMIC CELL INJURY

Two major pathways leading to apoptosis have been delineated *(75)*. In the intrinsic pathway, mitochondria play a pivotal role *(76)*. Upon stimulation, pathological alterations take place in mitochondrial membranes, resulting in the release of apoptogenic factors, including cytochrome *c*. In the cytosol, cytochrome *c* binds the adapter protein Apaf-1, which in turn associates with caspase-9 to form a complex called apoptosome, leading to caspase activation. The extrinsic pathway is initiated by the engagement of death receptors with their ligands *(77)*. This leads to the recruitment of adapter proteins and the association and activation of caspase-8. These two apoptotic pathways, triggered differently, ultimately converge at the level of executioner caspases that disassemble the committed cells. During ischemic injury, activation of both intrinsic and extrinsic apoptotic pathways has been documented (Fig. 1).

For the intrinsic pathway, mitochondrial damage characterized as interruption of respiration and development of permeability transition pores has been long recognized as a key event in ischemic cell injury *(78)*. Cytochrome *c*, a central molecule relaying apoptotic signals from mitochondria, is released into cytosol in ischemic tissues *(33,79–81)*. For example, global brain ischemia in rats leads to accumulation of cytochrome *c* in the cytosol of neurons in the hippocampus, which is accompanied by sequential activations of caspases-9 and -3 *(81)*. Of interest, transgenic rats overexpressing superoxide dismutase (SOD) produced less free radicals during ischemia, and apoptotic progression was significantly suppressed *(81)*. The results suggest an important role for oxygen free radicals in ischemic activation of the mitochondrial apoptosis pathway. Alternatively, development of mitochondrial pathology in ischemic cells might be caused by the dysregulation of Bcl-2 family proteins. A good example is the proapoptotic molecule Bax. In a cell-culture model of kidney ischemia, Bax was shown to translocate from the cytosol to mitochondria and form oligomers in the outer membrane,

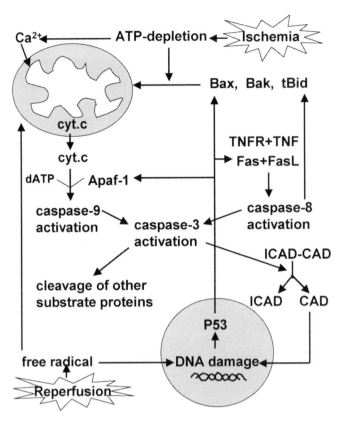

Fig. 1. Apoptotic events during tissue ischemia-reperfusion. Both intrinsic and extrinsic pathways for apoptosis can be activated. For the intrinsic pathway, ischemia leads to depletion of cellular ATP, which in turn triggers mitochondrial accumulation of Ca^{2+} and proapoptotic molecules such as Bax, resulting in the development of membrane defects in the organelle. This is followed by the release of apoptotic factors, including cytochrome c, and subsequent activation of caspase-9. For the extrinsic pathway, upregulation of Fas and TNF-α receptor and their ligands during ischemia-reperfusion leads to the formation of death-inducing signaling complex (DISC), activating caspase-8. Upon activation, the initiator caspases (-8 and -9) proteolytically process and activate executioner caspases, including caspase-3. The executioner caspases are responsible for the ultimate cleavage of enzymatic or structural proteins, culminating in the development of apoptotic morphology. One important substrate for caspase-3 is inhibitor of caspase-activated DNase (ICAD), which normally associates with CAD and keeps it in check. After being cleaved, ICAD releases CAD, leading to CAD activation, followed by cleavage of nuclear DNA. DNA damage is also induced by free radicals that are generated mainly during reperfusion. DNA damage may activate p53, which can upregulate pro-apoptotic genes such as bax, apaf-1, and fas, promoting ischemic injury.

releasing cytochrome c to activate casaspses *(15)*. It is interesting to note that Bcl-2 overexpression prevents Bax oligomerization in mitochondria, without blocking translocation of the proapoptotic molecule. This is accompanied by complete inhibition of downstream apoptotic events, including mitochondrial cytochrome c release and caspase activation *(82)*. Bax translocation was also demonstrated during transient brain ischemia. This is followed by cytochrome c release from the organelle and activation of caspase-9 *(83)*. A role for Bax in the development of mitochondrial pathology under ischemia is further supported by studies using Bax knockout mice. These animals, compared with wild-type, show significantly less apoptosis and caspase activation after hypoxic-ischemic challenge *(67)*.

With respect to the extrinsic pathway of apoptosis, upregulation of Fas and Fas ligand has been shown in ischemic regions of the brain where apoptotic cells accumulate *(42,74)*. Fas expression is also induced by ischemia-reperfusion of the kidney *(43)*. In the heart, reperfusion following ischemia leads to *de novo* synthesis of Fas ligand and its release *(42)*. Of significance, ischemic injury of the brain, heart and kidney is ameliorated in lpr mice with mutated nonfunctional Fas *(42,43)*. In the liver, however, TNF-α and not Fas mediates ischemic apoptosis, as shown by recent studies using transgenic models *(44)*. Nevertheless, these studies suggest that the extrinsic pathway is activated during ischemia, and may cooperate with the mitochondria-mediated intrinsic pathway to orchestrate cell death by apoptosis. It is possible that different populations of cells undertake different pathways to accomplish the suicide program. Such a scenario is supported by the observation that brain ischemia activates caspase-8, the apical caspase in the extrinsic pathway, in neurons that are different from those showing caspase-3 activation *(37)*.

Yet another pathway that may lead to apoptosis during ischemia is through p53 activation. p53, a well-recognized tumor-suppressor gene, is often activated in response to DNA damage *(84)*. In ischemic tissues, DNA damage in the forms of fragmentation or strand breaks prevails. Thus, it is not surprising that p53 is activated during brain and heart ischemia *(85,86)*. The role of p53 in ischemic injury is suggested by pharmacological as well as transgenic studies of brain ischemia *(87,88)*. In p53 knockout mice, infarct size after focal brain ischemia is significantly reduced, compared with wild-type animals *(88)*. However, p53 seems dispensable in ischemic damage during heart ischemia *(89)*. The mechanisms underlying p53-induced apoptosis are unclear. Pro-apoptotic genes, including bax and Apaf-1, are transcriptional targets for p53 *(90,91)*. Activation of p53 by ischemia may therefore upregulate these genes, sensitizing the cells to apoptosis. In support of this idea, p53-dependent apoptosis is attenuated in cells that are devoid of Bax or Apaf-1 *(73,91–93)*. Alternatively, p53 may promote apoptosis independent of transactivation of apoptotic genes. P53 has been implicated in the shuttling of preformed Fas to cell membrane *(94)*. Also, it may induce apoptosis by enhancing free radical production in mitochondria *(95)*. P53 mediated caspase-8 activation in a FADD-independent manner has also been documented *(96)*. Therefore, depending on experimental models, p53 may promote apoptosis in different ways. Nevertheless, little has been learned with regard to how p53 mediates apoptosis under ischemia.

A notable feature of ischemic apoptosis is the regulation of apoptotic genes. To date, such regulation has been documented for death receptors and their ligands, Bcl-2 family members, caspases, and IAP. By their respective natures, these proteins are expected to either promote apoptosis or antagonize the suicide program. However, whether and how this regulation is involved in ischemic apoptosis remain largely unknown. One important observation is that protective genes exemplified by Bcl-2 are often induced in surviving cells, whereas pro-apoptotic Bax is upregulated in dying cells *(50)*. The results point out an important role for apoptotic gene regulation in determining cellular sensitivity to apoptosis. Apparently, such regulation is involved in deciding the fate of ischemic cells, that is, whether they live or die.

CONCLUDING REMARKS

Apoptosis has been detected in ischemic tissues in numerous studies. Despite intensive investigation, the significance of apoptosis in ischemic diseases remains to be established. In particular, differential contributions by apoptotic and necrotic mechanisms to ischemic tissue damage remain controversial. For example, in experimental models of liver ischemia, as high as 80% and as low as less than 1% apoptosis has been reported *(25)*. In human heart infarcts, one study showed that 12% myocytes in the border zone and 1% in remote areas were apoptotic *(97)*; however, in another study, less than 0.5% apoptotic cells were found in the infarcts *(98)*. Discrepancy between these observations raises important questions: what is responsible for such inconsistency? How can reliable apoptosis measurements be achieved in vivo? Is apoptosis indeed a crucial contributing factor in ischemic injury?

Discrepancy between apoptosis measurements in vivo might be caused by multiple factors. First, it has proven to be difficult to monitor apoptosis in vivo. Owing to cellular heterogeneity, cells committed to apoptosis may accomplish the degenerative process at different times. In addition, end-stage apoptotic cells are rapidly and quietly absorbed through phagocytosis by neighboring cells or macrophages, without inducing inflammation *(22)*. Second, techniques for apoptosis detection are not always accurate or straightforward. For most techniques, technical difficulties are inherent, particularly when used to examine in vivo apoptosis. For example, the TUNEL assay, as useful as it is under specified conditions, may not distinguish apoptosis from necrosis (*see* Chapter 16 for discussion). Third, apoptosis and necrosis can be triggered by the same injurious process, depending on the severity of the insults. Thus, it is not surprising that mixed profiles of cell death are usually found in ischemic tissues. Finally, whether an injured cell is going to die by apoptosis or necrosis is also affected by the cellular condition. In this regard, cellular level of ATP seems to be a key; in the absence of ATP, cells committed to apoptosis may not be able to complete the suicide program and end up with necrosis *(99)*. These considerations call for caution in interpreting results obtained from specific ischemic models. On the other hand, they point out the complexity of ischemic injury and the need for systematic analysis of apoptosis in this important setting.

The association of apoptosis with ischemia reveals a new set of perspectives for treatment of ischemic diseases. Pharmacological inhibition of caspases has proven to be effective in diminishing ischemic damage in several experimental models. Promising results have also been achieved through genetic modulation of apoptotic genes, including Bcl-2. While critical evaluation and in-depth studies of these approaches are emerging, new strategies targeting other apoptotic events are expected. These, in combination with conventional methods, may lead to new therapies for ischemic diseases such as myocardial infarction, stroke, or ischemic failure of the liver and kidneys.

ACKNOWLEDGMENTS

Current work by the authors is supported by grants from the National Institutes of Health and the American Society of Nephrology. Zheng Dong is the recipient of the Lyndon Baines Johnson Research Award from the American Heart Association, the Patricia W. Robinson Young Investigator Award from the National Kidney Foundation, and the Carl W. Gottschalk Research Scholar Award from the American Society of Nephrology.

REFERENCES

1. Center for Disease Control and Prevention (1993) Cardiovascular disease surveillance: ischemic heart disease 1980–1989. Centers for Disease Control and Prevention, Washington, D.C.
2. Center for Disease Control and Prevention (1994). Cardiovascular disease surveillance: stroke 1980–1989. Centers for Disease Control and Prevention, Washington, D.C.
3. Lipton, P. (1999) Ischemic cell death in brain neurons. *Physiol. Rev.* **79**, 1431–1568.
4. Jennings, R. B. and Reimer, K. A. (1991) The cell biology of acute myocardial ischemia. *Annu. Rev. Med.* **42**, 225–246.
5. Weinberg, J. M. (1991) The cell biology of ischemic renal injury. *Kidney Int.* **39**, 476–500.
6. Pierce, G. N. and Czubryt, M. P. (1995) The contribution of ionic imbalance to ischemia/reperfusion-induced injury. *J. Mol. Cell Cardiol.* **27**, 53–63.
7. Berridge, M. J., Lipp, P., and Bootman, M. D. (2000) The versatility and universality of calcium signalling. *Nat. Rev. Mol. Cell Biol.* **1**, 11–21.
8. Venkatachalam, M. A. and Weinberg, J. M. (1993) Structural effects of intracellular amino acids during ATP depletion, in *Surviving Hypoxia* (Hochachka, P. W., Lutz, P. L., Sick, T., Rosenthal, M., and van denThillart, G., eds.), CRC Press, Boca Raton, pp. 473–494.
9. Dong, Z., Patel, Y., Saikumar, P., Weinberg, J. M., and Venkatachalam, M. A. (1998) Development of porous defects in plasma membranes of adenosine triphosphate-depleted Madin-Darby canine kidney cells and its inhibition by glycine. *Lab. Invest.* **78**, 657–668.
10. Droge, W. (2002) Free radicals in the physiological control of cell function. *Physiol. Rev.* **82**, 47–95.

11. Ferrari, R., Agnoletti, L., Comini, L., Gaia, G., Bachetti, T., Cargnoni, A., et al. (1998) Oxidative stress during myocardial ischaemia and heart failure. *Eur. Heart J.* **19**, B2–B11.
12. Yu, B. P. (1994) Cellular defenses against damage from reactive oxygen species. *Physiol. Rev.* **74**, 139–162.
13. Dirnagl, U., Iadecola, C., and Moskowitz, M. A. (1999) Pathobiology of ischaemic stroke: an integrated view. *Trends Neurosci.* **22**, 391–397.
14. Mehta, J. L. and Li, D. Y. (1999) Inflammation in ischemic heart disease: response to tissue injury or a pathogenetic villain? *Cardiovasc. Res.* **43**, 291–299.
15. Saikumar, P., Dong, Z., Patel, Y., Hall, K., Hopfer, U., Weinberg, J. M., and Venkatachalam, M. A. (1998) Role of hypoxia-induced Bax translocation and cytochrome *c* release in reoxygenation injury. *Oncogene* **17**, 3401–3415.
16. Feldenberg, L. R., Thevananther, S., del Rio, M., de Leon, M., and Devarajan, P. (1999) Partial ATP depletion induces Fas- and caspase-mediated apoptosis in MDCK cells. *Am. J. Physiol.* **276**, F837–F846.
17. McConkey, D. J. and Orrenius, S. (1997) The role of calcium in the regulation of apoptosis. *Biochem. Biophys. Res. Commun.* **239**, 357–366.
18. Love, S. (1999) Oxidative stress in brain ischemia. *Brain Pathol.* **9**, 119–131.
19. Maulik, N., Yoshida, T., and Das, D. K. (1998) Oxidative stress developed during the reperfusion of ischemic myocardium induces apoptosis. *Free Radic. Biol. Med.* **24**, 869–875.
20. Darzynkiewicz, Z., Bedner, E., and Traganos, F. (2001) Difficulties and pitfalls in analysis of apoptosis. *Methods Cell Biol.* **63**, 527–546.
21. Valavanis, C., Naber, S., and Schwartz, L. M. (2001) In situ detection of dying cells in normal and pathological tissues. *Methods Cell Biol.* **66**, 393–415.
22. Savill, J. and Fadok, V. (2000) Corpse clearance defines the meaning of cell death. *Nature* **407**, 784–788.
23. Kelly, K. J., Plotkin, Z., and Dagher, P. C. (2001) Guanosine supplementation reduces apoptosis and protects renal function in the setting of ischemic injury. *J. Clin. Invest.* **108**, 1291–1298.
24. Yaoita, H., Ogawa, K., Maehara, K., and Maruyama, Y. (2000) Apoptosis in relevant clinical situations: contribution of apoptosis in myocardial infarction. *Cardiovasc. Res.* **45**, 630–641.
25. Clavien, P. A., Rudiger, H. A., Selzner, M., Jaeschke, H., Gujral, J. S., Bucci, T. J., and Farhood, A. (2001) Mechanism of hepatocyte death after ischemia: apoptosis versus necrosis. *Hepatology* **33**, 1555–1557.
26. Arends, M. J. and Wyllie, A. H. (1991) Apoptosis: mechanisms and roles in pathology. *Int. Rev. Exp. Pathol.* **32**, 223–254.
27. Schumer, M., Colombel, M. C., Sawczuk, I. S., Gobe, G., Connor, J., O'Toole, K. M., et al. (1992) Morphologic, biochemical, and molecular evidence of apoptosis during the reperfusion phase after brief periods of renal ischemia. *Am. J. Pathol.* **140**, 831–838.
28. Gottlieb, R. A., Burleson, K. O., Kloner, R. A., Babior, B. M., and Engler, R. L. (1994) Reperfusion injury induces apoptosis in rabbit cardiomyocytes. *J. Clin. Invest.* **94**, 1621–1628.
29. Cao, G., Pei, W., Lan, J., Stetler, R. A., Luo, Y., Nagayama, T., et al. (2001) Caspase-activated DNase/DNA fragmentation factor 40 mediates apoptotic DNA fragmentation in transient cerebral ischemia and in neuronal cultures. *J. Neurosci.* **21**, 4678–4690.
30. Thornberry, N. A. and Lazebnik, Y. (1998) Caspases: enemies within. *Science* **281**, 1312–1316.
31. Namura, S., Zhu, J., Fink, K., Endres, M., Srinivasan, A., Tomaselli, K. J., et al. (1998) Activation and cleavage of caspase-3 in apoptosis induced by experimental cerebral ischemia. *J. Neurosci.* **18**, 3659–3668.
32. Holly, T. A., Drincic, A., Byun, Y., Nakamura, S., Harris, K., Klocke, F. J., and Cryns, V. L. (1999) Caspase inhibition reduces myocyte cell death induced by myocardial ischemia and reperfusion in vivo. *J. Mol. Cell Cardiol.* **31**, 1709–1715.
33. Soeda, J., Miyagawa, S., Sano, K., Masumoto, J., Taniguchi, S., and Kawasaki, S. (2001) Cytochrome *c* release into cytosol with subsequent caspase activation during warm ischemia in rat liver. *Am. J. Physiol. Gastrointest. Liver Physiol.* **281**, G1115–G1123.
34. Black, S. C., Huang, J. Q., Rezaiefar, P., Radinovic, S., Eberhart, A., Nicholson, D. W., and Rodger, I. W. (1998) Co-localization of the cysteine protease caspase-3 with apoptotic myocytes after in vivo myocardial ischemia and reperfusion in the rat. *J. Mol. Cell Cardiol.* **30**, 733–742.
35. Love, S., Barber, R., Srinivasan, A., and Wilcock, G. K. (2000) Activation of caspase-3 in permanent and transient brain ischaemia in man. *Neuroreport* **11**, 2495–2499.
36. Condorelli, G., Roncarati, R., Ross, J., Jr., Pisani, A., Stassi, G., Todaro, M., et al. (2001) Heart-targeted overexpression of caspase3 in mice increases infarct size and depresses cardiac function. *Proc. Natl. Acad. Sci. USA* **98**, 9977–9982.
37. Velier, J. J., Ellison, J. A., Kikly, K. K., Spera, P. A., Barone, F. C., and Feuerstein, G. Z. (1999) Caspase-8 and caspase-3 are expressed by different populations of cortical neurons undergoing delayed cell death after focal stroke in the rat. *J. Neurosci.* **19**, 5932–5941.
38. Asahi, M., Hoshimaru, M., Uemura, Y., Tokime, T., Kojima, M., Ohtsuka, T., et al. (1997) Expression of interleukin-1 beta converting enzyme gene family and bcl-2 gene family in the rat brain following permanent occlusion of the middle cerebral artery. *J. Cereb. Blood Flow Metab.* **17**, 11–18.
39. Kaushal, G. P., Singh, A. B., and Shah, S. V. (1998) Identification of gene family of caspases in rat kidney and altered expression in ischemia-reperfusion injury. *Am. J. Physiol.* **274**, F587–F595.
40. Jiang, L., Huang, Y., Yuasa, T., Hunyor, S., and dos Remedios, C. G. (1999) Elevated DNase activity and caspase expression in association with apoptosis in failing ischemic sheep left ventricles. *Electrophoresis* **20**, 2046–2052.
41. Jin, K., Graham, S. H., Mao, X., Nagayama, T., Simon, R. P., and Greenberg, D. A. (2001) Fas (CD95) may mediate

delayed cell death in hippocampal CA1 sector after global cerebral ischemia. *J. Cereb. Blood Flow Metab.* **21,** 1411–1421.

42. Jeremias, I., Kupatt, C., Martin-Villalba, A., Habazettl, H., Schenkel, J., Boekstegers, P., and Debatin, K. M. (2000) Involvement of CD95/Apo1/Fas in cell death after myocardial ischemia. *Circulation* **102,** 915–920.

43. Nogae, S., Miyazaki, M., Kobayashi, N., Saito, T., Abe, K., Saito, H., et al. (1998) Induction of apoptosis in ischemia-reperfusion model of mouse kidney: possible involvement of Fas. *J. Am. Soc. Nephrol.* **9,** 620–631.

44. Rudiger, H. A. and Clavien, P. A. (2002) Tumor necrosis factor alpha, but not Fas, mediates hepatocellular apoptosis in the murine ischemic liver. *Gastroenterology* **122,** 202–210.

45. Dong, Z., Venkatachalam, M. A., Wang, J., Patel, Y., Saikumar, P., Semenza, G. L., et al. (2001) Up-regulation of apoptosis inhibitory protein IAP-2 by hypoxia. Hif-1–independent mechanisms. *J. Biol. Chem.* **276,** 18702–18709.

46. Xu, D. G., Crocker, S. J., Doucet, J. P., St-Jean, M., Tamai, K., Hakim, A. M., et al. (1997) Elevation of neuronal expression of NAIP reduces ischemic damage in the rat hippocampus. *Nat. Med.* **3,** 997–1004.

47. Isenmann, S., Stoll, G., Schroeter, M., Krajewski, S., Reed, J. C., and Bahr, M. (1998) Differential regulation of Bax, Bcl-2, and Bcl-X proteins in focal cortical ischemia in the rat. *Brain Pathol.* **8,** 49–62; discussion 62–73.

48. Misao, J., Hayakawa, Y., Ohno, M., Kato, S., Fujiwara, T., and Fujiwara, H. (1996) Expression of bcl-2 protein, an inhibitor of apoptosis, and Bax, an accelerator of apoptosis, in ventricular myocytes of human hearts with myocardial infarction. *Circulation* **94,** 1506–1512.

49. Latif, N., Khan, M. A., Birks, E., O'Farrell, A., Westbrook, J., Dunn, M. J., and Yacoub, M. H. (2000) Upregulation of the Bcl-2 family of proteins in end stage heart failure. *J. Am. Coll. Cardiol.* **35,** 1769–1777.

50. Graham, S. H. and Chen, J. (2001) Programmed cell death in cerebral ischemia. *J. Cereb. Blood Flow Metab.* **21,** 99–109.

51. Hara, H., Friedlander, R. M., Gagliardini, V., Ayata, C., Fink, K., Huang, Z., Shimizet al. (1997) Inhibition of interleukin 1beta converting enzyme family proteases reduces ischemic and excitotoxic neuronal damage. *Proc. Natl. Acad. Sci. USA* **94,** 2007–2012.

52. Cheng, Y., Deshmukh, M., D'Costa, A., Demaro, J. A., Gidday, J. M., Shah, A., et al. (1998) Caspase inhibitor affords neuroprotection with delayed administration in a rat model of neonatal hypoxic-ischemic brain injury. *J. Clin. Invest.* **101,** 1992–1999.

53. Yakovlev, A. G., Knoblach, S. M., Fan, L., Fox, G. B., Goodnight, R., and Faden, A. I. (1997) Activation of CPP32-like caspases contributes to neuronal apoptosis and neurological dysfunction after traumatic brain injury. *J. Neurosci.* **17,** 7415–7424.

54. Cursio, R., Gugenheim, J., Ricci, J. E., Crenesse, D., Rostagno, P., Maulon, L., et al. (1999) A caspase inhibitor fully protects rats against lethal normothermic liver ischemia by inhibition of liver apoptosis. *FASEB J.* **13,** 253–261.

55. Yadav, S. S., Sindram, D., Perry, D. K., and Clavien, P. A. (1999) Ischemic preconditioning protects the mouse liver by inhibition of apoptosis through a caspase-dependent pathway. *Hepatology* **30,** 1223–1231.

56. Piot, C. A., Martini, J. F., Bui, S. K., and Wolfe, C. L. (1999) Ischemic preconditioning attenuates ischemia/reperfusion-induced activation of caspases and subsequent cleavage of poly(ADP-ribose) polymerase in rat hearts in vivo. *Cardiovasc. Res.* **44,** 536–542.

57. Ay, I., Sugimori, H., and Finklestein, S. P. (2001) Intravenous basic fibroblast growth factor (bFGF) decreases DNA fragmentation and prevents downregulation of Bcl-2 expression in the ischemic brain following middle cerebral artery occlusion in rats. *Brain Res. Mol. Brain Res.* **87,** 71–80.

58. Sadohara, T., Sugahara, K., Urashima, Y., Terasaki, H., and Lyama, K. (2001) Keratinocyte growth factor prevents ischemia-induced delayed neuronal death in the hippocampal CA1 field of the gerbil brain. *Neuroreport* **12,** 71–76.

59. Yamamura, T., Otani, H., Nakao, Y., Hattori, R., Osako, M., and Imamura, H. (2001) IGF-I differentially regulates Bcl-x_L and Bax and confers myocardial protection in the rat heart. *Am. J. Physiol. Heart Circ. Physiol.* **280,** H1191–H1200.

60. Nakamura, T., Mizuno, S., Matsumoto, K., Sawa, Y., and Matsuda, H. (2000) Myocardial protection from ischemia/reperfusion injury by endogenous and exogenous HGF. *J. Clin. Invest.* **106,** 1511–1519.

61. Martinou, J. C., Dubois-Dauphin, M., Staple, J. K., Rodriguez, I., Frankowski, H., Missotten, M., et al. (1994) Overexpression of BCL-2 in transgenic mice protects neurons from naturally occurring cell death and experimental ischemia. *Neuron* **13,** 1017–1030.

62. Kitagawa, K., Matsumoto, M., Tsujimoto, Y., Ohtsuki, T., Kuwabara, K., Matsushita, K., et al. (1998) Amelioration of hippocampal neuronal damage after global ischemia by neuronal overexpression of BCL-2 in transgenic mice. *Stroke* **29,** 2616–2621.

63. Hata, R., Gillardon, F., Michaelidis, T. M., and Hossmann, K. A. (1999) Targeted disruption of the bcl-2 gene in mice exacerbates focal ischemic brain injury. *Metab. Brain Dis.* **14,** 117–124.

64. Chen, Z., Chua, C. C., Ho, Y. S., Hamdy, R. C., and Chua, B. H. (2001) Overexpression of Bcl-2 attenuates apoptosis and protects against myocardial I/R injury in transgenic mice. *Am. J. Physiol. Heart Circ. Physiol.* **280,** H2313–H2320.

65. Selzner, M., Rudiger, H. A., Selzner, N., Thomas, D. W., Sindram, D., and Clavien, P. A. (2002) Transgenic mice overexpressing human Bcl-2 are resistant to hepatic ischemia and reperfusion. *J. Hepatol.* **36,** 218–225.

66. Parsadanian, A. S., Cheng, Y., Keller-Peck, C. R., Holtzman, D. M., and Snider, W. D. (1998) Bcl-xL is an antiapoptotic regulator for postnatal CNS neurons. *J. Neurosci.* **18,** 1009–1019.

67. Gibson, M. E., Han, B. H., Choi, J., Knudson, C. M., Korsmeyer, S. J., Parsadanian, M., and Holtzman, D. M. (2001) BAX contributes to apoptotic-like death following neonatal hypoxia-ischemia: evidence for distinct apoptosis pathways. *Mol. Med.* **7,** 644–655.

68. Plesnila, N., Zinkel, S., Le, D. A., Amin-Hanjani, S., Wu, Y., Qiu, J., et al. (2001) BID mediates neuronal cell death after oxygen/glucose deprivation and focal cerebral ischemia. *Proc. Natl. Acad. Sci. USA* **98,** 15318–15323.

69. Schielke, G. P., Yang, G. Y., Shivers, B. D., and Betz, A. L. (1998) Reduced ischemic brain injury in interleukin-1 beta converting enzyme-deficient mice. *J. Cereb. Blood Flow Metab.* **18,** 180–185.
70. Liu, X. H., Kwon, D., Schielke, G. P., Yang, G. Y., Silverstein, F. S., and Barks, J. D. (1999) Mice deficient in interleukin-1 converting enzyme are resistant to neonatal hypoxic-ischemic brain damage. *J. Cereb. Blood Flow Metab.* **19,** 1099–1108.
71. Friedlander, R. M., Gagliardini, V., Hara, H., Fink, K. B., Li, W., MacDonald, G., et al. (1997) Expression of a dominant negative mutant of interleukin-1 beta converting enzyme in transgenic mice prevents neuronal cell death induced by trophic factor withdrawal and ischemic brain injury. *J. Exp. Med.* **185,** 933–940.
72. Xu, D., Bureau, Y., McIntyre, D. C., Nicholson, D. W., Liston, P., Zhu, Y., et al. (1999) Attenuation of ischemia-induced cellular and behavioral deficits by X chromosome-linked inhibitor of apoptosis protein overexpression in the rat hippocampus. *J. Neurosci.* **19,** 5026–5033.
73. Shibata, M., Hisahara, S., Hara, H., Yamawaki, T., Fukuuchi, Y., Yuan, J., et al. (2000) Caspases determine the vulnerability of oligodendrocytes in the ischemic brain. *J. Clin. Invest.* **106,** 643–653.
74. Rosenbaum, D. M., Gupta, G., D'Amore, J., Singh, M., Weidenheim, K., Zhang, H., and Kessler, J. A. (2000) Fas (CD95/APO-1) plays a role in the pathophysiology of focal cerebral ischemia. *J. Neurosci. Res.* **61,** 686–692.
75. Green, D. R. (1998) Apoptotic pathways: the roads to ruin. *Cell* **94,** 695–698.
76. Green, D. R. and Reed, J. C. (1998) Mitochondria and apoptosis. *Science* **281,** 1309–1312.
77. Ashkenazi, A. and Dixit, V. M. (1998) Death receptors: signaling and modulation. *Science* **281,** 1305–1308.
78. Piper, H. M., Noll, T., and Siegmund, B. (1994) Mitochondrial function in the oxygen depleted and reoxygenated myocardial cell. *Cardiovasc. Res.* **28,** 1–15.
79. Ouyang, Y. B., Tan, Y., Comb, M., Liu, C. L., Martone, M. E., Siesjo, B. K., and Hu, B. R. (1999) Survival- and death-promoting events after transient cerebral ischemia: phosphorylation of Akt, release of cytochrome *c* and activation of caspase-like proteases. *J. Cereb. Blood Flow Metab.* **19,** 1126–1135.
80. Borutaite, V., Budriunaite, A., Morkuniene, R., and Brown, G. C. (2001) Release of mitochondrial cytochrome *c* and activation of cytosolic caspases induced by myocardial ischaemia. *Biochim. Biophys. Acta* **1537,** 101–109.
81. Sugawara, T., Noshita, N., Lewen, A., Gasche, Y., Ferrand-Drake, M., Fujimura, M., et al. (2002) Overexpression of copper/zinc superoxide dismutase in transgenic rats protects vulnerable neurons against ischemic damage by blocking the mitochondrial pathway of caspase activation. *J. Neurosci.* **22,** 209–217.
82. Mikhailov, V., Mikhailova, M., Pulkrabek, D. J., Dong, Z., Venkatachalam, M. A., and Saikumar, P. (2001) Bcl-2 prevents Bax oligomerization in the mitochondrial outer membrane. *J. Biol. Chem.* **276,** 18361–18374.
83. Cao, G., Minami, M., Pei, W., Yan, C., Chen, D., O'Horo, C., Graham, S. H., and Chen, J. (2001) Intracellular Bax translocation after transient cerebral ischemia: implications for a role of the mitochondrial apoptotic signaling pathway in ischemic neuronal death. *J. Cereb. Blood Flow Metab.* **21,** 321–333.
84. Levine, A. J. (1997) p53, the cellular gatekeeper for growth and division. *Cell* **88,** 323–331.
85. Li, Y., Chopp, M., Zhang, Z. G., Zaloga, C., Niewenhuis, L., and Gautam, S. (1994) p53-immunoreactive protein and p53 mRNA expression after transient middle cerebral artery occlusion in rats. *Stroke* **25,** 849–855; discussion 855–866.
86. Tomasevic, G., Shamloo, M., Israeli, D., and Wieloch, T. (1999) Activation of p53 and its target genes p21(WAF1/Cip1) and PAG608/Wig-1 in ischemic preconditioning. *Brain Res. Mol. Brain Res.* **70,** 304–313.
87. Culmsee, C., Zhu, X., Yu, Q. S., Chan, S. L., Camandola, S., Guo, Z., et al. (2001) A synthetic inhibitor of p53 protects neurons against death induced by ischemic and excitotoxic insults, and amyloid beta-peptide. *J. Neurochem.* **77,** 220–228.
88. Crumrine, R. C., Thomas, A. L., and Morgan, P. F. (1994) Attenuation of p53 expression protects against focal ischemic damage in transgenic mice. *J. Cereb. Blood Flow Metab.* **14,** 887–891.
89. Bialik, S., Geenen, D. L., Sasson, I. E., Cheng, R., Horner, J. W., Evans, S. M., Lord, E. M., et al. (1997) Myocyte apoptosis during acute myocardial infarction in the mouse localizes to hypoxic regions but occurs independently of p53. *J. Clin. Invest.* **100,** 1363–1372.
90. Thornborrow, E. C., Patel, S., Mastropietro, A. E., Schwartzfarb, E. M., and Manfredi, J. J. (2002) A conserved intronic response element mediates direct p53-dependent transcriptional activation of both the human and murine bax genes. *Oncogene* **21,** 990–999.
91. Fortin, A., Cregan, S. P., MacLaurin, J. G., Kushwaha, N., Hickman, E. S., Thompson, C. S., et al. (2001) APAF1 is a key transcriptional target for p53 in the regulation of neuronal cell death. *J. Cell Biol.* **155,** 207–216.
92. McCurrach, M. E., Connor, T. M., Knudson, C. M., Korsmeyer, S. J., and Lowe, S. W. (1997) bax-deficiency promotes drug resistance and oncogenic transformation by attenuating p53-dependent apoptosis. *Proc. Natl. Acad. Sci. USA* **94,** 2345–2349.
93. Yin, C., Knudson, C. M., Korsmeyer, S. J., and Van Dyke, T. (1997) Bax suppresses tumorigenesis and stimulates apoptosis in vivo. *Nature* **385,** 637–640.
94. Bennett, M., Macdonald, K., Chan, S. W., Luzio, J. P., Simari, R., and Weissberg, P. (1998) Cell surface trafficking of Fas: a rapid mechanism of p53-mediated apoptosis. *Science* **282,** 290–293.
95. Marchenko, N. D., Zaika, A., and Moll, U. M. (2000) Death signal-induced localization of p53 protein to mitochondria. A potential role in apoptotic signaling. *J. Biol. Chem.* **275,** 16202–16212.
96. Ding, H. F., Lin, Y. L., McGill, G., Juo, P., Zhu, H., Blenis, J., Yuan, J., and Fisher, D. E. (2000) Essential role for caspase-8 in transcription-independent apoptosis triggered by p53. *J. Biol. Chem.* **275,** 38905–38911.
97. Olivetti, G., Quaini, F., Sala, R., Lagrasta, C., Corradi, D., Bonacina, E., et al. (1996) Acute myocardial infarction in

humans is associated with activation of programmed myocyte cell death in the surviving portion of the heart. *J. Mol. Cell Cardiol.* **28,** 2005–2016.

98. Saraste, A., Pulkki, K., Kallajoki, M., Henriksen, K., Parvinen, M., and Voipio-Pulkki, L. M. (1997) Apoptosis in human acute myocardial infarction. *Circulation* **95,** 320–323.

99. Nicotera, P., Leist, M., and Ferrando-May, E. (1998) Intracellular ATP, a switch in the decision between apoptosis and necrosis. *Toxicol. Lett.* **102–103,** 139–142.

III

Approaches to the Study of Apoptosis

16
Analysis of Apoptosis
Basic Principles and Procedures

Li Bai, Jinzhao Wang, Xiao-Ming Yin, and Zheng Dong

INTRODUCTION

Apoptosis is a distinct form of cell death. Originally defined by cellular morphology, apoptosis can now be characterized at molecular, biochemical, and cellular levels. Detection of apoptosis has become more important, not only because of scientific interests but also because of the significance in clinical practice. For example, because apoptosis has been implicated in the development of a variety of devastating diseases such as cancer, a therapeutic approach using apoptosis-inducing drugs is expected. To evaluate the effectiveness of the treatment, one may have to assess the apoptotic response following the treatment. In typical apoptosis, a set of cell structure and biochemical characteristics has been well-defined. In combination, these provide the basis for apoptosis detection in a given setting. The methodology for analyzing these characteristics is as diverse as the research subjects. Several books devoted to the methodology of apoptosis analysis have been published recently *(1–3)*. Readers are advised to review the detailed experimental protocols in these books. This chapter aims to provide an overview of the basic approaches used in analyzing apoptosis, the principles, and the basic methodology, in order to provide a quick reference guide that readers can use to decide what method is available for their own studies. We start with the determination of cell viability and the morphology of dying cells. We then discuss the approaches available to examine apoptotic changes on the cell membrane, in both the cytosol and nucleus.

DETERMINATION OF CELL VIABILITY

Introduction

Regardless of how and when it takes place, apoptosis indicates the death of a cell. Cells undergoing apoptosis end up with compromised metabolism, function, and capacity for proliferation; that is, loss of viability. Therefore, assessment of cell viability should be considered to be one of the primary criteria for apoptosis. Commonly used approaches to measure cell viability can be classified into three categories. The first group of methods relies on the fact that living cells maintain the integrity of their plasma membranes. The second group of methods is based on the ability of a cell to sustain specific metabolic functions. The third group of methods measures the cellular capacity to divide and proliferate.

From: *Essentials of Apoptosis: A Guide for Basic and Clinical Research*
Edited by: X.-M. Yin and Z. Dong © Humana Press Inc., Totowa, NJ

General Procedures

Measurement of Plasma Membrane Integrity by Vital Dye Exclusion Assay

Vital dyes can be generally classified into two types. The first type is not permeable to plasma membranes, and therefore do not enter intact cells. As a result, they do not stain viable cells, but can enter cells with compromised plasma membranes and bind to internal structures or molecules, labeling these cells with color or fluorescence. Good examples in this class include trypan blue, propidium iodide (PI), and ethidium homodimer. The second class of vital dyes is permeable to plasma membranes and usually, after entering the cell, the dyes are modified and trapped inside in the presence of intact plasma membranes. Therefore, these dyes, unlike those in the first class, stain viable cells. Good examples of these dyes include fluorescein diacetate and Calcein-AM *(4)*.

VITAL DYE EXCLUSION ASSAY PROCEDURE

Adherent Cells
1. Cells are rinsed twice with phosphate-buffered saline (PBS) and incubated with PBS containing a vital dye for 5 min.
2. Cells are subsequently rinsed with PBS to remove extracellular dye prior to examination by microscopy.

Cell Suspension
1. Cells are collected by centrifugation, and rinsed with PBS.
2. The cells are subsequently resuspended into PBS containing vital dye.
3. After 5 min of staining, extracellular dye is removed by centrifugation.
4. Cells are resuspended in PBS for microscopic examination.

Measurement of Cytosolic Leakage

This method is based on the consideration that viable cells with intact plasma membranes are able to preserve their cellular contents, particularly macromolecules such as proteins. Once the plasma membrane is broken, cytosol is leaked into the extracellular space. A classic and still commonly used indicator for cytosolic leakage is the release of intracellular enzymes such as lactate dehydrogenase (LDH). LDH is a ubiquitous cytosolic protein of 136 kilodalton with enzymatic activities to catalyze the reaction: pyruvate + NADH = lactate + NAD$^+$. NADH exhibits fluorescence at an excitation wavelength of 360 nm with emission at 450 nm. The velocity of decreases in Ex360 nm/Em450 nm fluorescence in the reaction indicates the conversion of NADH to the nonfluorescent NAD, and therefore LDH activity. Breakdown of the plasma membranes leads to the release of LDH into the incubation medium. The amount of LDH present in the extracellular space can therefore be used as an index of cell leakage or death *(5)*.

CYTOSOLIC LEAKAGE MEASUREMENT PROCEDURE
1. Incubation medium is subjected to brief centrifugation to remove cellular debris.
2. Resultant supernatant is added to enzymatic reaction buffer containing pyruvate and NADH.
3. NADH fluorescence at Ex360 nm/Em 450 nm is monitored during the reaction. Velocity of decreases in NADH fluorescence indicates the conversion of NADH to nonfluorescent NAD$^+$ and therefore LDH activity.
4. Parallel dishes of cells are lysed with 0.1% Triton X-100 to determine total LDH activity.
5. LDH activity obtained from cell incubation medium is divided by the total LDH activity to calculate the percentage of LDH release.

Clonogenic Activity

The ability of cells to divide and form colonies is called clonogenic activity. For proliferative cells, clonogenic activity is a functional measurement of cell viability. Cellular clonogenic activity depends not only on the integrity of plasma membranes but also on a complex set of cellular functions involving energy production, macromolecule synthesis, and mitotic division.

Adherent Cells

1. Cells are collected and separated into individual cells by trypsinization and pipetting.
2. Cells are plated at a very low density (200–1,000 cells/100 mm tissue-culture dish).
3. Cultures are maintained for a period of ~5 cell divisions.
4. Colonies can be counted directly or after MTT (0.5 mg/mL in PBS) staining, under a microscope.

Cell Suspension

1. Cells are collected by centrifugation and cell clumps are broken-up by pipetting.
2. Individual cells are then plated in 0.1–0.2% agarose at low densities.
3. Cultures are maintained and colonies counted as described for adherent cells.

Comments

(a) Although intact plasma membrane is a prerequisite for a cell to live, it does not necessarily indicate that the cell is viable. In the case of apoptosis, usually the integrity of plasma membranes is not compromised until the late stage of the so-called "secondary necrosis." Thus, the measurement of plasma membrane integrity is not very useful in apoptosis detection; rather, they are utilized in examination of cell injury accompanied by membrane damage.

(b) The dose and exposure time of vital dyes should be titrated for specific types of cells. Too much exposure may lead to cytotoxicity and nonspecific permeabilization.

(c) Vital dye staining can be used in conjunction with flow cytometry. This provides quantitative assessment of cell viability.

(d) For enzymatic measurement of cytosolic leakage, one must be sure that the treatment does not interfere with the enzyme directly. Proper controls should be considered.

(e) Obviously, clonogenic activity assay cannot be utilized to assess viability of nondividing cells; for example, primary cultures of neurons usually do not proliferate when grown in vitro.

MORPHOLOGICAL EXAMINATION

Introduction

Apoptosis was originally defined by a sequence of morphologic features *(6)*. Despite recent progress in apoptosis research at the biochemical and molecular levels, morphological changes are still considered the "gold standard" for apoptosis.

By utilizing light microscopy, it can be seen that apoptosis starts with the condensation of nuclear chromatin and shrinkage of the cell. Chromatin after condensation becomes segregated against the nuclear membrane. At this stage, the nucleus appears to be shrunken and condensed as well. After the initial phase, the cell detaches from the neighboring tissues and enters the stage of fragmentation. Cellular fragmentation is characterized by the formation of apoptotic bodies, or blebbing. In the apoptotic bodies, fragments of nucleus and cellular organelles including mitochondria are usually found. In vivo, the apoptotic bodies are rapidly absorbed through phagocytosis by macrophages or neighboring cells. In in vitro experimental models where phagocytosis is not available, the apoptotic cells will eventually undergo degradation by a process similar to necrosis, releasing the cellular contents into the incubation medium. This degradation is also called "secondary necrosis" *(6)*.

Electron microscopy is used to examine ultrastructural features of apoptosis. Typical ultrastructural alterations include the condensation of cytoplasm, fragmentation of nucleus, aggregation of dense masses of chromatin beneath the nuclear membrane, and protrusion of the cell surface or formation of apoptotic bodies.

General Procedures

Standard procedures for light microscopy and transmission electron microscopy are recommended. A light microscope with phase contrast is helpful in documenting the stereotypic morphological changes during apoptosis.

Comments

(a) Morphological features are considered to be the most reliable criteria in defining apoptosis.

(b) Light microscopy provides a quick, convenient, and on-site method for monitoring the progression of apoptosis. It can be semi-quantitative, if the images of representative fields of cells are captured. Electron microscopy provides the definitive morphological evidence of apoptosis; however, it does not provide quantitative data and thus cannot be used to compare apoptosis between experimental conditions.

(c) Microscopic quantitation is subject to individual bias. Development of apoptotic morphology is highly heterogeneous among the cells. A consensus must be reached between investigators as to what is considered apoptotic.

ALTERATIONS IN THE PLASMA MEMBRANE

Introduction

A noticeable change in the plasma membrane of apoptotic cells is the redistribution of phospholipids. In normal cells, phospholipids are present asymmetrically across the plasma membrane. Whereas phosphatidylcholine and sphingomyelin are located mainly in the external leaflet, most of the phosphatidylethanolamine and all of the phosphatidylserine (PS) are restricted to the inner leaflet of the plasma membrane. Early in apoptosis, the lipid asymmetry is lost, resulting in the appearance of PS on the cell surface *(7)*. Measurement of the membrane lipid redistribution therefore provides another indication of apoptosis. The commonly used approach to detect membrane lipid redistribution monitors the appearance of PS on the cell surface. When PS is exposed on the cell surface, it can bind to Annexin V in a Ca^{2+}-dependent way. Annexin V conjugated with fluorophores or biotin is available from commercial sources. One of the popular conjugates is FITC-Annexin V, which binds PS on apoptotic cell surface and labels the cells with fluorescein.

General Procedures

A basic method has been described *(8)*, and can be modified in various experimental models.

Adherent Cells

1. Rinse the cells with PBS.
2. Add FITC-Annexin V prepared in the binding buffer (2.5 mM $CaCl_2$, 150 mM NaCl, 10 mM HEPES, pH 7.4).
3. Incubate cells for 2–5 min at room temperature.
4. Rinse cells thoroughly with binding buffer to remove free FITC-Annexin V for microscopic examination.

Cell Suspension

1. Collect cells by centrifugation.
2. Resuspend cells in FITC-Annexin V in binding buffer.
3. After 2–5 min of staining, wash off the unbound FITC-Annexin V with binding buffer.
4. Examine cells under a fluorescence microscope.

Comments

(a) Binding of Annexin V can be used as a marker only in cells with intact plasma membranes. If the integrity of plasma membranes is compromised, Annexin V enters cells and becomes associated with PS from inside, labeling the cells regardless the mode of cell death. Therefore, to identify apoptosis, another staining, such as with PI, which probes plasma membrane integrity, should be included.

(b) Binding of Annexin to PS requires Ca^{2+}. Thus, care must be taken not to significantly reduce Ca^{2+} concentration in the binding buffer.

(c) Detection of PS exposure by Annexin V binding can be analyzed with flow cytometry to quantify apoptosis.

CHANGES IN THE CYTOSOL

Introduction

A number of apoptotic changes occur in the cytosol. Perhaps the most important events are the activation of caspases and the cleavage of multiple cellular proteins by the activated caspases. Other cellular changes involve the regulatory proteins such as the Bcl-2 family proteins. The background information of caspases and the Bcl-2 family proteins are given in Chapters 1 and 2, respectively. Following is a general discussion of the approaches that can be taken to determine these changes.

Caspases

Activation of caspases is arguably the biochemical hallmark of apoptosis. There are several ways to detect caspase activation. Caspases are a family of cysteine proteases that cleave their substrates after aspartate residues. Existing as zymogens (also called procaspases) within un-stimulated cells, caspases are activated through the cleavage of the zymogen to form the large and small subunits, which in turn form a heterotetramer complex. Thus caspase activation can be determined by Western blots using anti-caspase antibodies to examine whether the zymogen is converted into cleaved active forms. Alternatively, the enzymatic activities of caspases can be measured using synthetic fluorogenic or chromatogenic substrates. Finally, caspase activation status can also be determined by examining the cleavage of endogenous substrates, such as poly (ADP-ribose) polymerase (PARP) (*see* Chapter 1).

General Procedures

WESTERN BLOT FOR DETECTING CASPASE CLEAVAGE AND CASPASE SUBSTRATE CLEAVAGE

Cells undergoing apoptosis are harvested and washed in cold PBS. The cytosol can be prepared by a number of different methods. However, for detecting caspase substrate cleavage, a most effective way is to lyse the cell in a buffer containing 4 *M* urea, 10% glycerol, 2% sodium dodecyl sulfate (SDS), 0.003% bromophenol and 5% 2-ME (added immediately before analysis) *(9)*. Proteins are separated by a standard sodium dodecyl sulfate polyacrylamide gel electrophoresis (SDS-PAGE) followed by Western blot with specific antibodies against either the large or the small subunit of the caspase of interest. Commonly examined caspases are caspases-3, -7, -8, and -9. Alternatively, antibodies against caspase substrates, such as PARP, which is cleaved from 116 kDa to 85 kDa during apoptosis, can be used for the Western blot. There is a selectivity of certain substrates for some caspases, based on the recognition site (*see* Chapter 1).

MEASUREMENT OF CASPASE ACTIVITIES USING SYNTHETIC SUBSTRATES

The caspase recognition site is composed of a tetrapeptide sequence that differs from one type of caspase to another. For example, caspases-3, -6, and -7 will mainly recognize the motif of Asp-Glu-Val-Asp or DEVD, whereas caspases-8, -9, and -10 will recognize the sequence of Ile-Glu-Thr-Asp or IETD. Caspases-1, -4, and -5 will recognize the tetrapeptide sequence of WEHD or YVAD. These tetrapeptide can be synthesized and conjugated to a report group, which can be cleaved by the corresponding caspases. The activity of the measured caspases is in proportion to the amount of the released report group. The most commonly used report groups include p-nitroanilide (pNA, colorimetric detection by absorbance at 405–410 nm), 7-amino-4-methylcoumarin (AMC, fluorometric detection at excitation/emission wavelength of 380 and 460 nm, respectively) and 7-amino-4-trifluoro-methylcoumarin (AFC, fluorometric detection at 405 and 500 nm, respectively, or colorimetric detection at 380 nm).

To measure the caspase activities, cells are first lysed in a caspase assay buffer (20 m*M* PIPES, 100 mM NaCl, 10 mM dithiothretol (DTT), 1 m*M* EDTA, 0.1% CHAPS, 10% sucrose, pH 7.2) *(10)*. About 20–50 µg of proteins are used in each assay, mixed with 20–50 µ*M* of the substrates in a volume of about 200 mL. The reaction is then monitored by fluorometric or colorimetric detection at the specific wavelengths. The values reflect the relative caspase activity, which can be converted to specific activity with proper standards. Commercial kits for caspase activity measurement are now widely available.

MEASUREMENT OF CASPASE ACTIVATION BY IMMUNOSTAINING OR WITH FLUORESCENT SUBSTRATES IN TISSUES AND CELLS

To detect the activation of caspase in a specific anatomic location, one can perform immunohistochemical or immunofluorescence staining on formalin-fixed paraffin sections with antibodies specifically against the activated caspases, which usually target to the subunits, but not the zymogen. Antibodies against activated caspases-3, -7, and -9 are now commercially available (e.g., from Cell Signaling Technology, MA and BD PharMingen, CA). The procedure is no different from the general immunostaining. The primary antibody is the anti-activate caspase and the secondary antibody can be either conjugated to horseradish peroxidase (HRP) or a fluorophore for either histochemical detection or fluorescent detection. It is possible to doubly stain the section with a cell-specific antibody to determine the type of cells that are undergoing apoptosis and therefore contain the activated caspases. A good example of such an application can be found in ref. *(11)*.

To detect caspase activation in cultured cells, other than Western blot, one can also take advantage of several recently developed cell-permeable fluorescent substrates. One such substrate with the trade name of PhiPhiLux (Oncogene Research Products, CA) is a peptide of 18 amino acids, with caspase recognition motifs in the center, and two fluorophores covalently attached near the termini. The two fluorophores quench each other in the native molecule because of intramolecular interactions until caspase hydrolysis breaks the peptide linkage. Thus the substrates fluoresce in response to caspase activation *(12)*. Because these products are cell-permeable, they can label the cells undergoing apoptosis, which can be quickly identified and quantified by fluorescence microscopy or flow cytometry. It is possible to doubly stain the cell to obtain additional information about the nature of the positive cells.

Comments

Caspase activation is determined mainly based on the cleavage of the zymogen and by measuring the proteolytic activities. By combining proper identification methods, one can not only detect caspase activation qualitatively and quantitatively, but also in tissue- or cell-specific ways. However, it is important to note that the synthetic substrates used for measuring various caspase activities are not absolutely specific for a particular type of caspase. Thus DEVD-AFC may detect caspase activities that would be better described as DEVDase activities, which could include those of caspases-3,-7, or -6.

Bcl-2 Family Proteins

The status of the Bcl-2 family proteins can be examined from two aspects: expression level and activation status. For the former, a Western blot or Northern blot could be performed. For the latter, the approach will depend on the individual members. Commonly used assays can determine Bad phosphorylation; Bid cleavage; translocation of Bax, Bid, Bad, or Bim from the cytosol to the mitochondria; and the formation of Bax or Bak oligomers.

General Procedures

All the assays can be based on Western blot using antibodies specific to individual members. Preparation of cell fractions will vary depending on the nature of the experiments. If subcellular localization is not relevant, cells can be lysed in an isotonic buffer with a nonionic detergent (0.5–1%). Cells need to be thoroughly solubilized because many Bcl-2 family proteins are membrane-bound. If subcellular localization is to be examined, then cellular components are subfractionated to obtain the mitochondria and the cytosol fractions (see below).

These fractions are then run on a 12.5% SDS-PAGE followed by Western blot with specific antibodies to individual members to determine whether there is a change in the expression level, a cleavage product (such as p15 of Bid), or a phosphorylation event. Commercial kits are now available to examine the phosphorylation of Bad using specific anti-phosphorylated Bad antibodies (e.g., Upstate

Biotechnology, MA and Cell Signaling Technology, MA). Translocation from the cytosol to the mitochondria is indicated by the appearance of the molecule, such as Bid or Bax, in the mitochondrial fraction following apoptosis stimulation, which would be present only in the cytosol in healthy cells.

To determine whether Bax or Bak forms oligomers that are important to their functions, cross-linking agents are used *(13–15)*. For Bax oligomerization, both a membrane-permeable agent, disuccinimidyl suberate (DSS), and a membrane-impermeable agent, Bis (sulfosuccinimidyl) suberate (BS^3) (Pierce Chemicals), have been used successfully *(13,14)*. For Bak oligomerizatin, Bismaleimidohexane (BMH) (Pierce Chemicals) has been used successfully *(15)*. For the detection of Bax oligomerization *(13)*, the mitochondria fraction (0.5 mg of protein) is suspended in an isotonic buffer (200 m*M* mannitol, 70 m*M* sucrose, 1 m*M* EGTA, 10 m*M* HEPES, pH 7.5), and BS^3 (in 5 m*M* sodium citrate buffer, pH 5.0) or DSS (in dimethyl sulfoxide [DMSO]) is added to a final concentration of 10 m*M*. After 30 min of incubation at room temperature, the crosslinker is quenched by 1 *M* Tris-HCl, pH 7.5, to a final concentration of 20 m*M*. The membranes are then lysed in radioimmunol precipitation assay (RIPA) buffer and cleared by centrifugation at 12,000*g* before analyzed by SDS-PAGE and Western blot with an anti-Bax antibody. For Bak oligomerization, 10 m*M* BMH in DMSO is added to the mitochondria pellet for 30 min at room temperature. The mitochondria are pelleted again and resuspended in protein sample buffer for SDS-PAGE and Western blot with an anti-Bak antibody *(15)*. In both cases, oligomerization can be initiated by an activating molecule, such as Bid, and can be visualized as a high molecular-weight species.

Comments

The location of Bax can be confusing, because in many cultured cells, it is already present in the mitochondria. However, in primary cells, it is likely present only in the cytosol *(16,17)*. It is possible that a suboptimal culture condition may cause Bax translocation even in the absence of an apparent apoptotic signal. Thus, a better way to determine whether or not Bax is activated in these cells following death stimulation is to examine whether or not Bax oligomerization occurs.

MITOCHONDRIAL CHANGES

Introduction

Mitochondria are perhaps the most important organelle involved in apoptosis initiation and regulation. Significant changes can be observed in mitochondrial physiology and morphology during apoptosis. In general, the apoptotic changes can be divided into two categories: release of apoptotic proteins and alteration in mitochondrial functions. The released apoptotic proteins include cytochrome *c*, Smac/Diablo, HtrA2/Omi, endonuclease G, and apoptosis-inducing factor (AIF), in addition to some nonapoptotic proteins (*see* Chapter 6). The redistribution of these proteins can be determined by Western blot, enzyme-linked immunosorbant assay (ELISA), or immunofluorescence staining. One major test for mitochondria functions related to apoptosis is the determination of the transmembrane potentials. Other assays have been developed to determine mitochondria permeability transition, mitochondrial generation of free radicals, mitochondria calcium content, and mitochondria pH changes (for reviews *see* ref. *18–20*).

Mitochondria Release of Cytochrome c

Cytochrome *c* release is perhaps the most common parameter measured to determine whether or not the mitochondria pathway is activated. It is usually determined by Western blot or ELISA using the cytosol, or by immunofluorescence staining on intact cells. In addition, to test whether an agent is capable of inducing cytochrome *c*, this agent can be incubated with isolated mitochondria; the distribution of cytochrome *c* across the mitochondria can then be determined by Western blot or ELISA.

General Procedures

SUBCELLULAR FRACTIONATION

Cells are harvested, washed with cold PBS and then disrupted in the presence of a cocktail of protease inhibitors *(21)*. Cell disruption is conducted with a Dounce homogenizer, a polytron homogenizer, or by repeated passing through a fine needle. Isotonic buffers are used to avoid breakage of mitochondria during homogenization. A commonly used buffer contains 200 mM mannitol, 70 mM sucrose, 1 mM EGTA, 10 mM HEPES, pH 7.5 *(13)*. Alternatively, a KCl-based buffer could be as simple as 150 mM KCl plus 5 mM Tris-HCl, pH 7.4 *(22)*. Experimentation may be required to determine the best way to disrupt a maximal amount of cells without damaging the mitochondria membranes, which could be cell type-dependent and empirically determined. The lysates are then centrifuged at 600g to remove the undisrupted intact cells. The supernatant is then further centrifuged at 12,000g to pellet the heavy membranes, which is mainly comprised of mitochondria. This preparation of mitochondria should be sufficient for the general use described in this chapter. The mitochondria can then be resuspended in an isotonic buffer with energizing agents (250 mM sucrose, 10 mM HEPES, 1 mM ATP, 5 mM sodium succinate, 0.08 mM ADP, 2 mM K_2HPO_4, pH 7.5) *(23)*.

WESTERN BLOT ANALYSIS

Western blot is conducted with both the supernatant and the pellet fractions with a suitable anti-cytochrome *c* antibody. A control for the cytosol proteins can be β-actin; and for the mitochondria fraction, cytochrome *c* oxidase subunit IV. The appearance of cytochrome *c* (~12 kDa) in the supernatant fraction and/or a reduction of cytochrome *c* content in the pellet fraction indicate cytochrome *c* release from mitochondria.

ELISA

The supernatant is prepared as described earlier and subjected to ELISA analysis for the detection of cytochrome *c* using one of the commercial kits (e.g., Oncogene Research Products, CA or R&D Systems, MN). The advantage of this assay is that the released proteins can be easily quantified for comparisons among samples.

IMMUNOSTAINING

If the cell number is too low for subcellular fractionation, cytochrome *c* release may be examined by immunofluorescence staining. Cells undergoing apoptosis are washed and fixed in 4% paraformaldehyde solution. The primary anti-cytochrome *c* antibody applied should be able to recognize the native conformation of the molecule, such as the one from BD PharMingen (clone 6H2.B4, cat. no. 556432). After the application of the secondary antibody conjugated with a fluorophore, cells can be observed under a fluorescence microscope. Normal cells would have a punctuated staining pattern, consistent with the mitochondria distribution of the molecule. In apoptotic cells, the staining may either assume a diffusive cytoplamic distribution, or become very faint or disappear. To further differentiate the mitochondria and cytosol locations, cells can be co-stained with anti-Hsp60 antibodies, or with a mitochondria-specific dye, such as MitoTracker, before fixation.

Comments

These methods can be selected based on different situations. If there are plenty of cells, and it is easy to subfractionate the cells, then Western blot or ELISA can be used for qualitative or quantitative analysis. Otherwise, immunostaining can be used where fewer numbers of cells are available. However, it is important to determine the percentage of cells with a particular cytochrome *c* staining pattern for a direct comparison among different samples.

Mitochondria Transmembrane Potentials

The mitochondria transmembrane potentials ($\Delta\Psi$m) have been used by many investigators as a functional parameter of the mitochondria during apoptosis. The potential is about 180 mV (negative inside). The potentials can be determined with lipophilic cations, which accumulate inside the mitochondria in a potential-dependent way. Commonly used fluorescent lipophilic cations include Rhodamine-123 (Rh123), 3,3'-dihexiloxocarbocyanine iodide ($DiOC_6[3]$), tetramethyrhodamine methyl ester (TMRM), and 5,5',6,6'tetrachloro-1,1',3,3'-tetraethylbenzimidazol-carbocyanine iodide (JC-1), which can be analyzed by flow cytometry or fluorescent microscopy. Nonfluorescent probe are also used, such as tetraphenyl phosphonium ion (TPP^+), which can be analyzed with a specific electrode *(24)*.

General Procedures

The procedures are fairly straightforward and can be applied to both cultured cells and isolated mitochondria, although only the formal is discussed here *(25,26)*. The probes can be directly added into the cell culture at their optimal concentration (10 µg/mL JC-1, 0.5 µM TMRM, or 20 nM $DiOC_6[3]$) and incubated for 15 min at 37°C in the dark. Cells are then harvested and subjected to flow-cytometry analysis. Alternatively, cells can be observed directly under a fluorescent microscope. For probes such as TMRM, the fluorescence intensity is directly coupled with the potentials; the higher the potentials, the greater the fluorescence intensity. On the other hand, JC-1 shifts reversibly from monomeric form to aggregated form upon membrane polarization. The emission wavelength also changes from 530 nm to 590 nm, accordingly, when excited at 488 nm. Thus, cells with a lower $\Delta\Psi$m would be detected in the green channel, and cells with a higher $\Delta\Psi$m would be detected in the orange/red channel. The quotient between green and red fluorescence provides an estimate of the mitochondria potentials of the population, which are independent of mitochondria mass. It is still possible to measure the percentage of positive cells in each channel and the mean fluorescent intensity *(27)*.

It would be ideal to set up control cells that are perfectly healthy with polarized mitochondria and control cells with depolarized mitochondria. The latter could be obtained by adding to the culture either the K^+ ionophore; valinomycin (100 nM or more); or the mitochondrial uncoupler, carbonyl cyanide p-(trifluoromethoxy) phenylhydrazone (FCCP, 250 nM) or carnonyl cyanide m-chlorophenyl hydrazone (CCCP, 10 µM). The chemicals can be added 15 min before the addition of the potential probes and will completely dissipate the potentials, thus providing a good negative control for the assay.

Comments

There are many discussions about the use of different probes for the measurement of $\Delta\Psi$m, concerning specificity, sensitivity, and potential interference with mitochondria function *(25,26,28)*. Rho123 is a classic probe. However, for cells containing mitochondria with different maturation states, as in a continuously growing cell line, Rh123 binding to the mitochondria may not be consistent to the potentials, but affected by the available binding sites *(26)*. It also may not be sensitive enough to a smaller change in potentials *(29)*. $DiOC_6(3)$ is another commonly used probe. However, its incorporation can be affected by the plasma-membrane potentials. It also may be redistributed into endoplasmic reticulum (ER) in cells with depolarized mitochondria. Thus cells may maintain the fluorescence intensity even when mitochondria are depolarized *(29)*. Adjusting the concentration of $DiOC_6(3)$ to a low concentration (20 nM) may reduce some of these concerns *(25)*. Chloromethyltetramethyl-rosamine (CMTMRos) or MitoTracker, which had also been used for detecting membrane potentials, is now found to bind to mitochondria irrespective of the magnitude of potentials *(30)*. Furthermore, it can induce permeability transition once it enters into the mitochondria *(30)*; thus, it is no longer recommended for measurement of $\Delta\Psi$m. Instead, both TMRM and JC-1 are considered to be far more suitable for this purpose, because they are specific to mitochondria, sensitive to potential changes, and do not interfere with mitochondria functions.

CHANGES IN THE NUCLEUS

Introduction

During apoptosis, nucleus of the cell undergoes dramatic morphological changes, including chromatin condensation, peripheral margination, nuclear shrinkage, and subsequent fragmentation *(6)*. The changes are apparently driven by biochemical processes that become active during apoptosis. Key proteins responsible for maintaining nuclear structure and chromatin integrity are proteolysed during apoptosis, releasing specific fragments. In addition, DNA breaks, including internucleosomal cleavage, are generated, and have been recognized as a biochemical hallmark of apoptosis *(31)*.

Nuclear Condensation and Fragmentation

Nuclear condensation and fragmentation in apoptotic cells can be visualized by light and electron microscopy, as noted earlier. An alternative approach is to stain nucleus and its fragments with fluorescent dyes that bind DNA. Two commonly used dyes are Hoechst (bisbenzimide) and DAPI (4',6'-diamidino-2-phenyindole, dilactate). Upon staining, apoptotic nucleus exhibits much stronger fluorescence than the nucleus of normal control cells. Intense staining of apoptotic nucleus may result from the increased permeability of the dyes and higher binding to DNA owing to altered chromatin configuration.

General Procedures

ADHERENT CELLS
1. Cells are rinsed with PBS.
2. Cells are incubated with PBS containing 0.1–1 µg/mL Hoechst or DAPI for approx 2 min.
3. Cells are subsequently rinsed with PBS to remove extracellular dye prior to examination by fluorescence microscopy.

CELL SUSPENSION
1. Cells are collected by centrifugation.
2. Cells are rinsed with PBS.
3. Cells are resuspended into PBS containing Hoechst or DAPI.
4. After 2 min of incubation, extracellular dye is removed by centrifugation and cells are resuspended for fluorescence microscopy.

Comments

Nuclear staining methods are easy to perform and do not take much time. They can confirm apoptotic nuclear morphology and also can be used for semi-quantitative purposes. Under certain conditions, the nucleus of necrotic cells also displays intense Hoechst or DAPI staining. This is owing to increased permeability of the dyes to the necrotic cells with broken plasma membranes. Therefore, integrity of plasma membranes should be examined simultaneously, for example, by PI staining.

DNA Content Staining by Propidium Iodide

An impressive degenerative process during apoptosis is the degradation of nuclear DNA, resulting in decreases in DNA content in the cell or hypoploidy. This provides the basis for sorting and quantification of apoptosis by measuring cellular DNA content with flow cytometry.

General Procedures

A good protocol can be found in ref. *(32)*.
1. Collect and resuspend cells in ice-cold PBS.
2. Fix cells by dropwise addition of 2 vol. of cold methanol and vortex gently.

3. Stain the cells with PBS containing 10–20 μg/mL PI, 0.1% Triton X-100, 2 mM EDTA, and 2 U/mL DNase-free RNase for 30 min.
4. Analyze cells by flow cytometry.

Comments

PI staining using this method is conducted in the presence of Triton X-100, which permeabilizes the plasma membrane. Triton X-100 incubation ensures the same PI exposure to DNA in different cells, regardless of the original plasma-membrane integrity. This is different from PI staining detection of plasma-membrane integrity, where no detergent is included, as mentioned previously.

Not all hypoploid populations represent apoptotic cells. Cellular fragments and debris can be counted into the hypoploid population, resulting in an overestimation of apoptotic cells. To avoid this problem, it is necessary to utilize rigid gating and threshold setting in flow cytometric analysis to exclude cellular fragments or debris. In addition, necrosis could also result in nonspecific DNA degradation, which can also lead to hypoploidity recognized by PI staining.

DNA Fragmentation

Internucleosomal DNA Cleavage

One of the first biochemical hallmarks identified for apoptosis is the cleavage of DNA at internucleosomal sites. This specific cleavage leads to the formation of nucleosomal fragments of 180–200 bp lengths, which upon electrophoresis resolve into a "DNA ladder" *(31)*. Specific endonucleases including DNA fragmentation factor (DFF) or caspase-activated DNase (CAD) are activated during apoptosis, and are responsible mainly for this type of DNA breakdown *(33)*. Internucleusomal DNA cleavage, although a late event in apoptosis, has been used frequently as an indicator of apoptosis.

GENERAL PROCEDURES

Cells are lysed with a hypotonic buffer containing 0.5% Triton X-100, 20 mM EDTA, 10 mM Tris-HCl, pH 7.4. The lysates are centrifuged at 14,000g for 20 min. The supernatant is collected and is subjected to proteinase K and RNase digestion, followed by phenol-chloroform extraction. DNA fragments in the aqueous phase are precipitated with ethanol and subjected to 1.5% agarose gel electrophoresis. After electrophoresis, DNA in the gel is stained with ethidium bromide and visualized under UV transmission *(34)*.

COMMENTS

Internucleosomal DNA cleavage is not always associated with apoptosis. Similar pattern of DNA breakdown has been shown in cells with necrotic morphology, although necrosis usually lead to nonspecific DNA degradation, which more often exibits as a smear in the gel *(34,35)*.

DNA degradation appears to be a late event in apoptosis. In certain apoptotic models, typical apoptosis develops in the absence of internucleosomal DNA cleavage *(36)*.

DNA fragmentation is a qualitative rather than quantitative measurement of apoptosis. The amount of fragmented DNA is not proportional to the frequency of apoptosis. For example, the same apoptotic cells release more DNA fragments as apoptosis develops into later stage *(37)*.

Detection of DNA Breaks by Histochemical Methods (TUNEL)

Based on DNA breakdown in apoptotic cells, two histochemical techniques for apoptosis detection have been developed. The methods are commonly used to examine apoptosis in vivo or in tissue sections. The first method, terminal deoxynucleotidyl transferase (TdT) mediated dUTP-biotin nick end labeling (TUNEL), utilizes the enzyme TdT to incorporate biotinylated dUTP onto the 3' hydroxyl ends of fragmented DNA. The second method, *in situ* end labeling (ISEL), utilizes the enzyme DNA polymerase I or its Klenow fragment to fill in recessed 3' ends of DNA fragments with biotinylated dUTP. For both methods, incorporated biotinylated dUTP can be revealed by fluorescence microscopy after reaction with fluorescein isothiocyanate (FITC)-conjugated avidin. Alternatively, it can be detected with avidin-conjugated horseradish peroxidase (HRP) and the HRP substrate 3,3'-diaminobenzidine tetrahydrochloride (DAB).

GENERAL PROCEDURES

Tissue sections or cells are first fixed with 10% formalin or 4% paraformaldehyde in PBS, followed by a partial digestion with proteinase K. This step is to reduce fixation-induced protein-DNA cross-linkage and increase the accessibility of DNA to TdT and dUTP-biotin. The endogenous peroxidase is inactivated by incubation of the tissues in 3% hydrogen peroxide. Tissue sections are then incubated with a mixture of the TdT enzyme and dUTP-biotin for 1 h at 37°C to label the 3' ends of the broken DNA. After wash, FITC-conjugated avidin can be directly applied for detection. Alternatively, avidin-HRP conjugate can be added, followed by the HRP substrate, DAB.

COMMENTS

The general procedure for ISEL is similar to that of TUNEL assay; instead of TdT, DNA polymerase I or Klenow fragment is used to incorporate biotinlytaed-dUTP.

For a technical control, parallel tissue sections should be processed without enzyme exposure. For example, in TUNEL assay, a control without TdT incubation should be included.

These methods detect DNA breakdown without discriminating the mode of cell death. Positive staining has been shown in cells with either apoptotic or necrotic DNA damage. The staining of necrosis is usually more diffusive than in the latter, which often exhibits a nuclear pattern of staining *(38)*.

REFERENCES

1. Reed, J. C., Abelson, J. N., and Simon, M. I. (2000) Apoptosis, in *Methods in Enzymology* vol. 322 (Reed, J. C., Abelson, J. N., and Simon, M. I., eds.), Academic Press, San Diego, CA.
2. Schwartz, L., Ashwell, J., Wilson, L., and Matsudairaand, P. (2001) Apoptosis, in *Methods Cell Biology* vol. 66 (Schwartz, L., Ashwell, J., Wilson, L., and Matsudairaand, P.), Academic Press, San Diego, CA.
3. LeBlanc, A. C. (2002) Neuromethods, in *Apoptosis: Techniques and Protocols*, 2nd ed., vol. 37 (LeBlanc, A. C., ed.), Humana Press, Totowa, NJ.
4. Dong, Z., Venkatachalam, M. A., Weinberg, J. M., Saikumar, P., and Patel, Y. (2001) Protection of ATP-depleted cells by impermeant strychnine derivatives: implications for glycine cytoprotection. *Am. J. Pathol.* **158**, 1021–1028.
5. Dong, Z., Patel, Y., Saikumar, P., Weinberg J. M., and Venkatachalam, M. A. (1998) Development of porous defects in plasma membranes of adenosine triphosphate-depleted madin-darby canine kidney cells and its inhibition by glycine. *Lab. Invest.* **78**, 657–668.
6. Kerr, J. F., Wyllie, A. H., and Currie, A. R. (1972) Apoptosis: a basic biological phenomenon with wide-ranging implications in tissue kinetics. *Br. J. Cancer* **26**, 239–257.
7. Savill, J. and Fadok, V. (2000) Corpse clearance defines the meaning of cell death. *Nature* **407**, 784–788.
8. Williamson, P., Eijnde van den, S., and Schlegel, R. A. (2001) Phosphatidylserine exposure and phagocytosis of apoptotic cells. *Methods Cell Biol.* **66**, 339–364.
9. Roy, S. and Nicholson, D.W. (2000) Criteria for identifying authentic caspase substrates during apoptosis. *Methods Enzymol.* **322**, 110–125.
10. Stennicke, H. R. and Salvesen, G. S. (2000) Caspase assays. *Methods Enzymol.* **322**, 91–100.
11. Cao, G., Pei, W., Lan, J., Stetler, Y. R., Nagayama, A., Luo, T., et al. (2001) Caspase-activated DNase/DNA fragmentation factor 40 mediates apoptotic DNA fragmentation in transient cerebral ischemia and in neuronal cultures. *J. Neurosci.* **21**, 4678–4690.
12. Komoriya, A., Packard, B. Z., Brown, M. J., Wu, M-L., and Henkart, P. A. (2000) Assessment of caspase activities in intact apoptotic thymocytes using cell-permeable fluorogenic caspase substrates. *J. Exp. Med.* **191**, 1819–1828.
13. Gross, A., Jockel, J., Wei, M. C., and Korsmeyer, S. J. (1998) Enforced dimerization of Bax results in its translocation, mitochondrial dysfunction and apoptosis. *EMBO J.* **17**, 3878–3885.
14. Eskes, R., Desagher, S., Antonsson B., and Martinou, J. C. (2000) Bid induces the oligomerization and insertion of Bax into the outer mitochondrial membrane. *Mol. Cell. Biol.* **20**, 929–935.
15. Wei, M. C., Lindsten, V. T., Mootha, S., Weiler, K. A., Gross, A., Ashiya, M., Thompson, C. B., and Korsmeyer, S. J. (2000) tBid, a membrane-targeted death ligand, oligomerizes Bak to release cytochrome *c. Genes Dev.* **14**, 2060–2071.
16. Desagher, S., Osen-Sand, A., Nichols, A., Eskes, R., Montessuit, S., Lauper, S., et al. (1999) Bid-induced conformational change of Bax is responsible for mitochondrial cytochrome *c* release during apoptosis. *J. Cell. Biol.* **144**, 891–901.

17. Zhao, Y., Li, S., Childs, E. E., Kuharsky D. K., and Yin, X. M. (2001) Activation of pro-death bcl-2 family proteins and mitochondria apoptosis pathway in TNFα-induced liver injury. *J. Biol. Chem.* **276**, 27432–27440.

18. Lemasters, J. J., Qian, T., Elmore, S. P., Trost, L. C., Nishimura, Y., Herman, B., et al. (1998) Confocal microscopy of the mitochondrial permeability transition in necrotic cell killing, apoptosis and autophagy. *Biofactors* **8**, 283–285.

19. Reynolds, I. J. (1999) Mitochondrial membrane potential and the permeability transition in excitotoxicity. *Ann. NY Acad. Sci.* **893**, 33–41.

20. Matsuyama, S. and Reed, J. C. (2000) Mitochondria-dependent apoptosis and cellular pH regulation. *Cell Death Differ.* **7**, 1155–1165.

21. Hsu, Y. T., Wolter, K. G., and Youle, R. J. (1997) Cytosol-to-membrane redistribution of Bax and Bcl-x(L) during apoptosis. *Proc. Natl. Acad. Sci. USA* **94**, 3668–3672.

22. Ott, M., Robertson, J. D., Gogvadze, V., Zhivotovsky B., and Orrenius, S. (2002) Cytochrome *c* release from mitochondria proceeds by a two-step process. *PNAS* **99**, 1259–1263.

23. Gross, A., Yin, X. M., Wang, K., Wei, M. C., Jockel, J., Milliman, C., et al. (1999) Caspase cleaved Bid targets mitochondria and is required for cytochrome *c* release, while Bcl-x$_L$ prevents this release but not tumor necrosis factor-r1/fas death. *J. Biol. Chem.* **274**, 1156–1163.

24. Kamo, N., Muratsugu, M., Hongoh, R., and Kobatake, Y. (1979) Membrane potential of mitochondria measured with an electrode sensitive to tetraphenyl phosphonium and relationship between proton electrochemical potential and phosphorylation potential in steady state. *J. Membr. Biol.* **49**, 105–1021.

25. Zamzami, N., Metivier D., and Kroemer, G. (2000) Quantitation of mitochondrial transmembrane potential in cells and in isolated mitochondria. *Methods Enzymol.* **322**, 208–213.

26. Cossarizza, A. and Salvioli, S. (2001) Analysis of mitochondria during cell death. *Methods Cell Biol.* **63**, 467–486.

27. Cossarizza, A., Baccarani-Contri, M., Kalashnikova G., and Franceschi, C. (1993) A new method for the cytofluorimetric analysis of mitochondrial membrane potential using the j-aggregate forming lipophilic cation 5,5',6,6'-tetrachloro-1,1',3,3'-tetraethylbenzimidazolcarbocyanine iodide (JC-1). *Biochem. Biophys. Res. Commun.* **197**, 40–45.

28. Bernardi, P., Scorrano, L., Colonna, R. V., Petronilli, V., and Di Lisa, F. (1999) Mitochondria and cell death. Mechanistic aspects and methodological issues (published erratum appears in *Eur. J Biochem.* 1999;**265(2)**, 847). *Eur. J. Biochem.* **264**, 687–701.

29. Salvioli, S., Ardizzoni, A., Franceschi C., and Cossarizza, A. (1997) JC-1, but not DiOC6(3) or rhodamine 123, is a reliable fluorescent probe to assess delta psi changes in intact cells: implications for studies on mitochondrial functionality during apoptosis. *FEBS Lett.* **411**, 77–82.

30. Scorrano, L., Petronilli, V., Colonna, R., Di Lisa F., and Bernardi, P. (1999) Chloromethyltetramethylrosamine (mitotracker orange) induces the mitochondrial permeability transition and inhibits respiratory complex I. Implications for the mechanism of cytochrome *c* release. *J. Biol. Chem.* **274**, 24657–24663.

31. Wyllie, A. H. (1980) Glucocorticoid-induced thymocyte apoptosis is associated with endogenous endonuclease activation. *Nature* **284**, 555–556.

32. Loo, D. T. and Rillema, J. R. (1998) Measurement of cell death. *Methods Cell Biol.* **57**, 251–264.

33. Wyllie, A. (1998) Apoptosis. An endonuclease at last. *Nature* **391**, 20–21.

34. Dong, Z., Saikumar, P., Weinberg, J. M., and Venkatachalam, M. A. (1997) Internucleosomal DNA cleavage triggered by plasma membrane damage during necrotic cell death. Involvement of serine but not cysteine proteases. *Am. J. Pathol.* **151**, 1205–1213.

35. Collins, R. J., Harmon, B. V., Gobe, G. C., and Kerr, J. F. (1992) Internucleosomal DNA cleavage should not be the sole criterion for identifying apoptosis. *Int. J. Radiat. Biol.* **61**, 451–453.

36. Oberhammer, F., Wilson, J. W., Dive, C., Morris, I. D., Hickman, J. A., Wakeling, A. E., et al. (1993) Apoptotic death in epithelial cells: cleavage of DNA to 300 and/or 50 kb fragments prior to or in the absence of internucleosomal fragmentation. *EMBO J.* **12**, 3679–3684.

37. Darzynkiewicz, Z., Li X., and Bedner, E. (2001) Use of flow and laser-scanning cytometry in analysis of cell death. *Methods Cell Biol.* **66**, 69–109.

38. Li, Y., Sharov, V. G., Jiang, N., Zaloga, C., Sabbah, H., and Chopp, M. (1995) Ultrastructural and light microscopic evidence of apoptosis after middle cerebral artery occlusion in the rat. *Am. J. Pathol.* **146**, 1045–1051.

Index